Genetics and Biotechnology of
Lactic Acid Bacteria

Genetics and Biotechnology of Lactic Acid Bacteria

Edited by

MICHAEL J. GASSON
AFRC Institute of Food Research
Norwich
UK

and

WILLEM M. DE VOS
Department of Biophysical Chemistry
Netherlands Institute for Dairy Research (NIZO)
and
Department of Microbiology
Wageningen Agricultural University
The Netherlands

BLACKIE ACADEMIC & PROFESSIONAL
An Imprint of Chapman & Hall
London · Glasgow · New York · Tokyo · Melbourne · Madras

Published by
Blackie Academic & Professional, an imprint of Chapman & Hall,
Wester Cleddens Road, Bishopbriggs, Glasgow G64 2NZ

Chapman & Hall, 2–6 Boundary Row, London SE1 8HN, UK

Blackie Academic & Professional, Wester Cleddens Road, Bishopbriggs, Glasgow G64 2NZ, UK

Chapman & Hall Inc., One Penn Plaza, 41st Floor, New York NY10019, USA

Chapman & Hall Japan, Thomson Publishing Japan, Hirakawacho Nemoto Building, 6F, 1-7-11 Hirakawa-cho, Chiyoda-ku, Tokyo 102, Japan

DA Book (Aust.) Pty Ltd., 648 Whitehorse Road, Mitcham 3132, Victoria, Australia

Chapman & Hall India, R. Seshadri, 32 Second Main Road, CIT East, Madras 600 035, India

First edition 1994

© 1994 Chapman & Hall

Typeset in 10/12 pt Times New Roman by ICON Graphic Services, Exeter
Printed in Great Britain by St Edmundsbury Press, Bury St Edmunds, Suffolk

ISBN 0 7514 0098 X

A catalogue record for this book is available from the British Library.

Library of Congress Catalog Card Number: 93–73770

Printed on acid-free text paper, manufactured in accordance with ANSI/NISO Z39.48-1992 (Permanence of Paper).

Preface

Over the past decade a dramatic expansion of research interest in lactic acid bacteria has taken place both in academia and industry. This reflects the industrial importance of these bacteria for a wide variety of fermentation processes, and also the interesting genetic and biochemical properties that their more detailed analysis has revealed. We have seen the field grow from a time when there was a mere handful of laboratories interested in the genetics and molecular biology of lactic acid bacteria to the present situation, when such research is prominent in all of the molecular scientific journals. Despite the importance of lactic acid bacteria and the availability of review articles, a book devoted to their genetics and biochemistry is timely.

We have attempted to cover both the genetic technology and the genetics of the major industrially important traits. An emphasis on lactococci will be apparent, reflecting the fact that whilst rapid progress in the analysis of all species of lactic acid bacteria is being made, the mesophyllic lactococci are the best characterized. The first two chapters address general aspects of the genetics and molecular biology of lactic acid bacteria. The emphasis of chapter 1 is on gene transfer processes, including conjugation, transduction and transformation and it also covers the growing range of transposition phenomena that have been recognized in lactic acid bacteria. The exploitation of generalized recombination processes, especially for chromosomal integration and gene replacement techniques, is included and the recent success in constructing genetic and physical maps of the chromosome is covered. Chapter 2 focuses on the more molecular aspects of lactic acid bacterial genetics. It describes the development of gene cloning vector systems based on both plasmids and chromosomal integration. All aspects of gene expression are described in detail, including transcription, translation, regulation and secretion. The most current available vectors for use in lactic acid bacteria are also described.

Chapters 3 to 5 deal with three aspects of the lactic acid bacteria that are prominent in their current and potential industrial applications. Chapter 3 describes the proteolytic systems of lactic acid bacteria and includes both the proteinase and peptidase enzymes and their genes, as well as amino-acid and peptide transport systems. Bacteriophages represent one of the most enduring and intractable problems faced by industrial processes involving lactic acid bacteria. A comprehensive description of bacteriophages is included in chapter 4, which also describes the very interesting and varied mechanisms of bacterio-phage defence that have evolved in lactic acid bacteria. The latter have already been exploited in the dairy industry and their continuing in-depth analysis should lead to improved strategies to minimize the bacteriophage problem.

Chapter 5 describes the range of antimicrobial proteins that is produced by lactic acid bacteria. Sophisticated analysis of the biosynthesis of both small peptides and lantibiotics such as nisin is described. These various molecules provide the opportunity to use a range of natural antimicrobial agents in food preservation and protection from food poisoning organisms, and they represent a potential application of lactic acid bacteria beyond their established role in food fermentation. Because of the emphasis on lactococci throughout the book, the final chapter is specifically concerned with the important species of *Lactobacillus, Leuconostoc* and *Streptococcus thermophilus,* and ensures that good coverage of all lactic acid bacteria is achieved.

All authors are established and active researchers in the field of lactic acid bacterial genetics. We hope that, as a result, this book represents an in-depth critical and comprehensive review of what is currently being achieved in the genetics of lactic acid bacteria. The book should appeal to anyone currently active in lactic acid research, but will also serve as an introduction to the molecular genetics of lactic acid bacteria for advanced students of food science and as an introduction to lactic species for geneticists and molecular biologists working outside the field.

We wish to acknowledge the support that lactic acid bacteria research in Europe has received from the EC. Most contributors to the book have benefited from support for collaborative research through the BEP, BAP and BRIDGE programmes of the EC, which has greatly facilitated both transnational interaction and scientific progress.

MJG
WMdeV

Contents

3   Bacteriophages and bacteriophage resistance   106

T. R. KLAENHAMMER and G. F. FITZGERALD

4   The proteolytic system of lactic acid bacteria   169

J. KOK and W. M. DE VOS

5   Bacteriocins of lactic acid bacteria   **211**
H. M. DODD and M. J. GASSON

6   Genetic engineering of lactobacilli, leuconostocs and *Streptococcus thermophilus*   **252**
A. MERCENIER, P.H. POUWELS and B. M. CHASSY

Contributors

Bruce M. Chassy	Department of Food Science, University of Illinois, ABL103, 1302 West Pennsylvania, IL 61801, USA
Willem M. de Vos	Department of Biophysical Chemistry, Netherlands Institute for Dairy Research, PO Box 20, NL–6710 BA, Ede, The Netherlands and also Department of Microbiology, Wageningen Agricultural University, The Netherlands
Helen M. Dodd	AFRC Institute of Food Research, Norwich Research Park, Colney, Norwich NR4 7UA, UK
Gerald F. Fitzgerald	Department of Food Microbiology, University College, Cork, Ireland
Mike J. Gasson	AFRC Institute of Food Research, Norwich Research Park, Colney, Norwich NR4 7UA, UK
Todd R. Klaenhammer	Department of Food Science, Southeast Dairy Foods Research Center, North Carolina State University, North Carolina, USA
Jan Kok	Department of Genetics, State University of Groninger, Kerklaan 30, 9751, NN Haren, The Netherlands
Annick Mercenier	Transgene SA, 11 rue Molsheim, 67000 Strasbourg, France
Peter H. Pouwels	TNO Medical Biological Laboratory, P O Box 5815, 2280 HV Rijswijk, The Netherlands
Guus F. M. Simons	KeyGene, P O Box 216, 6700 A E Wageningen, The Netherlands

1 Gene transfer systems and transposition

M.J. GASSON and G.F. FITZGERALD

1.1 Introduction

The lactic acid bacteria have evolved and been selected for exploitation in a variety of food fermentation processes. The strains currently available have been subject to gene introduction and genetic reorganization by a variety of natural mechanisms. The study of these mechanisms of gene transfer and DNA rearrangement has led to the development of genetic techniques that can be exploited for strain improvement. These systems complement the genetic engineering methods that are described in chapter 2. This chapter covers the characterization and exploitation of conjugation, transduction and transformation in lactic acid bacteria and the exploitation of generalized recombination and chromosomal integration. The lactic acid bacteria harbour a large number of transposable genetic elements and these undoubtedly contribute to the genetic instability that appears characteristic of many strains. The characterization of the known IS elements and the nisin transposon are described as well as the exploitation of heterologous transposons for genetic analysis. The recent development of pulsed field gel electrophoresis for analysis of whole chromosomes together with some elegant genetic strategies has led to progress in physical and genetic mapping of lactococcal genomes which is also reviewed.

1.2 Conjugation

1.2.1 Heterologous conjugation systems

The first example of conjugation in lactic acid bacteria was the introduction and transfer of the broad host range drug resistance plasmid pAMβ1 which was originally identified in *Enterococcus faecalis* (Gasson and Davies, 1980a; Gibson *et al.*, 1979). Subsequently this and related plasmids such as pIP501 have been used to effect conjugal bridges within and between a wide variety of lactic acid bacterial species (Vescovo *et al.*, 1983; West and Warner, 1985; Langella and Chopin, 1987; Pucci *et al.* 1987; Shrago *et al.*, 1986; Gonzalez and Kunka, 1983). The main exploitation of these heterologous plasmids is to mobilize homologous traits, especially where the latter are plasmid encoded. Examples of this in lactococci include pAMβ1 mobilization of plasmids encoding genes for bacteriophage resistance, proteinase production, lactose catabolism and polysaccharide production. De Vos and Davies (1984) used pAMβ1 to mobilize the bacteriophage resistance plasmid pSK112 which prevents bacteriophage adsorption to the cell surface. A plasmid responsible for polysaccharide production and a mucoid phenotype in *L. lactis* subsp. *cremoris* was transferred to a plasmid free derivative of *L. lactis* subsp. *lactis* 712 in a similar way by Smart and Gasson (unpublished data). The mobilization of proteinase plasmids pCI301 and pCI203 from their original multiple plasmid carrying host strains to a plasmid free strain has been analysed in detail revealing something of the mechanism involved. Transfer of these latter plasmids depended on co-transfer of pAMβ1 and proteinase plasmids recovered from the transconjugants proved to be cointegrates (Hayes *et al.*, 1990a). Subsequent analysis of pCI301 revealed that a specific recombination site was involved in the generation of pCI301::pAMβ1 fusion plasmids. This site could be sub-cloned onto an ordinarily non-mobilizable *L. lactis-* *E. coli* shuttle vector, thereby facilitating the conjugal transfer of this vector by pAMβ1 mobilization. This process involved the creation of site and orientation specific cointegrates of the two plasmids and their perfect resolution into the component plasmids was also observed (Hayes *et al.*, 1990b).

Another example of pAMβ1 and pIP501 mobilization is provided by the introduction of the tetracycline resistance plasmid pMV158 into lactococci using *Enterococcus faecalis* donors (Van der Lelie *et al.*, 1990). It was also demonstrated that transfer of pMV158 between *L. lactis* subsp. *lactis* strains IL1403 and MG1363 took place in the absence of pAMβ1, perhaps depending on an uncharacterized chromosomally located sex factor present in strain IL1403. These authors demonstrated that pMV158 transfer depended on its production of a protein Pre (plasmid recombination enzyme) which in other plasmids has been implicated in the formation and resolution of plasmid cointe-grates (Gennaro *et al.*, 1987). In other lactic acid bacteria, Ahn *et al.* (1992) reported the use of pAMβ1 to mobilize a naturally occurring chloramphenicol resistance plasmid from *Lactobacillus plantarum* into *Carnobacterium* species.

One application of plasmid mobilization is the introduction of a vector plasmid to a strain that is recalcitrant to transformation. In such an approach a readily transformable intermediate strain is exploited for primary cloning or clone introduction by transformation. This is followed by conjugal mobilization into the strain of interest. The system was first demonstrated in *Enterococcus faecalis* by Smith and Clewell (1984) who exploited pVA797, a derivative of the broad host range plasmid pIP501 that remains conjugative and carries a gene for chloramphenicol resistance. It was used in conjunction with the cloning vector pVA838 which combines the *E. coli* vector pACYC184 with an erythromycin resistance gene from pAMβ1 and in common with pVA797 has a replicon derived from *Streptococcus ferus*. The two plasmids can replicate in a wide range of Gram-positive host strains but because they have the same replicon they are incompatible. By introducing pVA838 into a host already carrying pVA797 and selecting for resistance to both chloramphenicol and erythromycin, cointegration of the two plasmids takes place and conjugal transfer of pVA797 thus results in high efficiency mobilization of pVA838. Resolution of the two individual plasmids together with segregation of pVA797 takes place in the progeny once selection for chloramphenicol resistance is released. This strategy was successfully exploited to capitalize on the efficient transformation system of *Streptococcus sanguis*. Introduction of pVA838 following cloning experiments was combined with conjugal delivery into strains of *E. faecalis* which at the time were difficult to transform. This same approach was used by Romero *et al.* (1987) to effect the mobilization of various gene cloning vectors including pVA838 and pSA3 to a range of different lactococci as well as strains of *S. thermophilus*. In these experiments protoplasts of *L. lactis* strain MG1363 were used for primary transformation.

An extension to this approach is provided by the recent discovery that conjugation can be effected between such taxonomically distant microorganisms as Gram-negative and Gram-positive bacteria including lactic acid bacteria. Trieu-Cout *et al.* (1987) developed a vector delivery system in which *E. coli* transformation preceded conjugal transfer into a variety of different Gram-positive species including *L. lactis*. This system made use of vector pAT187 which combined the conjugal transfer origin of the P group plasmid RK2, replication regions of both pAMβ1 and pBR322 and a kanamycin resistance marker that was expressed both in Gram-negative and Gram-positive hosts. This vector was efficiently mobilized from *E. coli* by a 'trans-helper' plasmid pRK212-1, which was incompatible with pAT187. Delivery of the vector from *E. coli* into *L. lactis* ILI419 was achieved at a frequency of 3×10^{-7} transconjugants per donor. Whilst these various systems illustrate the value of conjugal bridges in innovative gene cloning strategies, it is fair to say that in practice their use has been somewhat overtaken by the ease and wide applicability of direct electro-transformation.

1.2.2 Homologous conjugation systems

Indigenous conjugation systems in lactococci are very common, perhaps reflecting the abundance of plasmid DNA. The first example to be discovered was conjugal transfer of lactose plasmids in *L. lactis* but a variety of other traits have been shown to be transmissible either by self transfer or by mobilization. Examples include bacteriocin production, proteinase production, bacteriophage resistance and the genetically linked sucrose fermentation and nisin production phenotypes. Transfer of bacteriophage resistance genes is of applied value in the construction of bacteriophage insensitive starter cultures and this is discussed in detail, in chapter 3. Genes for sucrose fermentation and nisin biosynthesis are encoded by a conjugative transposon, rather than a plasmid and this is described in detail later in section 1.6.3. There are relatively few examples of homologous conjugation in lactic acid bacteria other than the lactococci; transfer of the lactose plasmid of *L. casei* was the first to be reported (Chassy and Rokaw, 1981).

Detailed study of conjugation is restricted to that associated with the lactose plasmid of related *L. lactis* subsp. *lactis* strains ML3, 712 and C2. This system is of special interest because these hosts are well characterized, the transfer system can be almost maximally efficient (i.e. achieve nearly 100% gene transfer) and conjugation involves an interesting cell aggregation phenomenon. The characterization of this system is well documented but because of its complexity and general relevance it warrants a detailed description. The original mating experiments (Gasson and Davies, 1980b) involved transfer of the lactose plasmid pLP712 between genetically marked strains derived from *L. lactis* subsp. *lactis* 712 and transfer frequencies were found to be very low, 2×10^{-7} progeny per

Figure 1.1 The cell aggregation phenotype encoded by cointegrates of the lactose plasmid and sex factor in *L. lactis* subsp. *lactis* ML3/C2/712 strains that are capable of very high frequency conjugation.

recipient being typical. Interest might have faded here except for the chance observation that some of the rare progeny colonies had an unusual morphology. Rather than the soft texture typical of lactococcal colonies these were quite hard, retaining their integrity when touched with a loop. Broth cultures derived from these variant colonies exhibited a dramatically different clumping morphology which is illustrated in Figure 1.1. Most significant was the observation that when used as donors in subsequent mating experiments very high conjugation frequencies were observed. Clearly a major change had taken place and an obvious explanation was the derepression of a transfer system by the selection of a mutation in a repressor/operator control circuit. This subsequently proved too simple an explanation.

Whilst the *L. lactis* subsp. *lactis* 712 was being studied in England similar work was undertaken in the USA in the laboratory of McKay using the related strains ML3 and C2. The generation of variant colonies was also observed following conjugation and the physical analysis of lactose plasmids in these progeny revealed complex changes (Walsh and McKay, 1981, 1982). The lactose plasmid in the *L. lactis* subsp. *lactis* ML3 donor strain was 55 kb in size but progeny strains were found to contain a novel lactose plasmid of 105 kb. Further analysis revealed that the original lactose plasmid pSK08 had been enlarged by the integration of foreign DNA which was shown to arise from a previously overlooked plasmid pRS01 which was 48 kb in size. Restriction endonuclease digestion of both component plasmid molecules and a range of cointegrate plasmids led to the identification of the fusion points on pSK08 and pRS01. Most cointegrate plasmids involved one site on pSK08 but a wide variety of pRS01 sites were used. A critical observation was the fact that the sum of individual restriction fragments involved in the creation of fusion fragments was consistently 0.8–1.0 kb less than the observed size of the fusion fragments. This suggested that cointegration was caused by the activity of an IS element located at the fixed fusion point on the lactose plasmid pSK08. A transposition process involving such an IS element would lead to duplication of the element and since they are typically close to 1 kb in size this would account for the increased molecular weight of the cointegrate plasmids. This explanation was enhanced by the finding that cointegration still took place when a mutant deficient in generalized recombination was used, again suggesting the involvement of a transposition mechanism (Anderson and McKay, 1984). The role of an IS element was eventually confirmed when the fusion points on pSK08, pRS01 and a representative co-integrate plasmid pPW2 were sequenced. The latter plasmid was shown to have one copy of a duplicated IS element at each of its pSK08–pRS01 junctions and details of the transposition process were revealed by DNA sequence comparison of these junctions with the donor and target regions of pSK08 and pRS01 respectively. A single copy of the IS element was found at the fusion point on the lactose plasmid pSK08. Whilst most cointegrate plasmids involved the same fusion point on pSK08, occasionally a different site was used. The explanation for this was found to be the presence of two copies of the IS element on pSK08. These elements were 800 bp in size and have been named ISSIS and ISSIT, the former being more frequently used in cointegrate formation (Polzin and Shimizu-Kadota, 1987). Structural details of this insertion sequence and its involvement in other *in vivo* transposition events are described later in section 1.6.1.

The discovery of this IS element allowed a straightforward plasmid to plasmid cointegration process to explain the creation of enlarged lactose plasmids in progeny derived from *L. lactis* subsp. *lactis* matings. However, this explanation proved insufficient to explain the results of an analysis of enlarged lactose plasmids in *L. lactis* subsp. *lactis* 712 (Gasson *et al.*, 1992). Whilst broadly similar observations were made, the structures of enlarged lactose plasmids were variable and inconsistent with their creation from the fusion of two component plasmids in their entirety. In *L. lactis* subsp. *lactis* 712 it also proved difficult to reliably isolate a plasmid equivalent to pRS01. Labile plasmid DNA from the otherwise plasmid-free derivative strain MG1363 was occasionally banded on caesium chloride gradients and this appeared analogous to pRS01, being shown by restriction endonuclease digestion to be a component of enlarged lactose plasmids. One explanation for the difficulty in isolating this plasmid was that the *L. lactis* subsp. *lactis* 712 equivalent of pRS01 was a chromosomally integrated sex factor which occasionally excised, and only in that event was it isolated as plasmid DNA. This was shown to be the case in DNA probe experiments using DNA fragments cloned from a representative cointegrate plasmid. The most compelling data came from pulsed field gel analysis in which it was demonstrated that DNA equivalent to pRS01 was integrated within a 600 kb *Sma*I chromosomal DNA fragment. It was further shown that this was also true of *L. lactis* subsp. *lactis* ML3. This integrated sex factor was found to be missing from the chromosome of some strains, leading to an approximately 50 kb deletion in the 600 kb *Sma*I fragment. Isolates of *L. lactis* subsp. *lactis* 712, C2 and ML3 from different sources varied with respect to the presence of the chromosomally integrated element and its spontaneous loss from a laboratory maintained strain was also demonstrated. As would be expected, the absence of this element prevented the isolation of progeny with the cell aggregation phenotype.

More complex models for enlarged lactose plasmid formation could thus be postulated in which chromosomal excision of a sex factor was involved and this provided for size variation in independently isolated enlarged lactose plasmids, as was observed in *L. lactis* subsp. *lactis* 712. These latter observations were made using a mini-lactose plasmid pMG820 in an otherwise plasmid free strain. Plasmid pMG820 was derived in two steps, by transductional shortening followed by deletion of the proteinase region from the 55 kb lactose plasmid pLP712 (Maeda and Gasson, 1986). The resultant mini-lactose plasmid was only 23 kb in size and it retained a single copy of the IS element IS*SIT*. Conjugal transfer led to the isolation of enlarged lactose plasmids which had high transfer efficiency and exhibited the cell aggregation phenotype. Whilst independently isolated enlarged lactose plasmids had related foreign DNA inserted at the site of IS*SIT*, the insert DNA varied markedly. Restriction endonuclease digestion was used to analyse these plasmids and an interesting observation was the existence of more closely related families of plasmids in which the insert DNA appeared to have one fixed end and one variable end, suggestive of a one-ended transposition phenomenon. Other than variation in insert DNA, enlarged plasmids derived from pMG820 had properties analogous to those derived from pSK08 in that the IS element IS*SIT* was shown to be duplicated at the junctions between pMG820 and the insert DNA. The latter was shown to be homologous to the sex factor that existed in the chromosome. As described above this element could excise leading to sex factor cured strains as well as its isolation as rather labile CCC plasmid DNA. Related models were developed to explain these somewhat complex observations and these are shown in Figure 1.2 which also summarizes the known properties of the sex factor. Table 1.1 contains examples of the conjugation properties for a variety of sex factors and lactose plasmid combinations.

Table 1.1 Conjugation properties for different conformations of lactose plasmid and sex factor DNA

DONOR			PROGENY	
Lactose genes	Sex factor	Phenotype	Frequency	Phenotype
Plasmid	Chromosomally integrated	Normal Clu⁻	2×10^{-7}	Mix Clu⁺ and Clu⁻
Plasmid	Cured	Normal Clu⁻	5×10^{-9}	Clu⁻ only
Plasmid cointegrate	Plasmid cointegrate	Clumping Clu⁺	5×10^{-2}	Clu⁺ only
Chromosomally integrated	Chromosomally integrated Clu⁻	Normal	2×10^{-6}	Clu⁻ only
Chromosomally integrated	Cured	Normal Clu⁻	2×10^{-9}	Clu⁻ only

Typical transfer frequency data expressed as progeny per donor are shown for various combinations of sex factor and lactose plasmid in the *L. lactis* C2/ML3/712 strain complex. Plasmid cointegrates described in *L. lactis* subsp. *lactis* ML3 have a range of properties, including the example cited. In this case distinct phenotypes include no aggregation and intermediate transfer frequency and depend on the site of cointegration on the sex factor (Anderson and McKay, 1984).

The association of cell aggregation with efficient conjugation is not unique to lactococci and was first described for haemolysin plasmid transfer in *Enterococcus faecalis*. The involvement of a 'sex pheromone' in the induction of cell aggregation is a general feature of these systems and analysis of the cell aggregation process has revealed the involvement of an inducible surface protein that binds with another cell surface component. A 137 kDa surface protein has been characterised in *E. faecalis* and evidence for the involvement of lipoteichoic acid as its receptor has been obtained (Ehrenfeld *et al.*, 1986; Galli *et al.*, 1990). In the lactococcal system there is no evidence for the involvement of sex pheromones and as yet detailed analysis of the mechanism of cell aggregation has not been reported, although recent work has identified the gene for an aggregation protein (Godon and Gasson, unpublished). Another interesting aspect is the failure of some host strains to exhibit cell aggregation even though they harbour an enlarged Clu⁺ lactose plasmid that could be demonstrated to cause the clumping phenotype when it was present in other strains such as those derived from *L. lactis* subsp. *lactis* 712. Study of a large number of lactococcal strains revealed that they fell into two groups, Agg⁺ strains including *L. lactis* 712 clumped when carrying a Clu⁺ lactose plasmid whereas Agg⁻ strains such as *L. lactis* IL1403 showed no phenotypic change (Van der Lelie *et al.*, 1991).

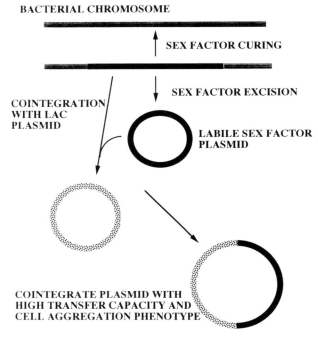

Figure 1.2(a) The experimentally established properties of the sex factor present in the *L. lactis* C2/ML3/712 strain complex are summarized. Cointegrate formation between an autonomous sex-factor plasmid (pRS01) and the lactose plasmid (pSK08) is the basis of the Anderson and McKay (1984) model which includes two iso-IS*S1* elements on pSK08 and random integration sites on the sex factor plasmid pRS01.

Transfer frequencies remained high in matings between strains from the two different groups and between individual Agg+ strains. Significantly, transfer between individual Agg− strains was extremely inefficient. It was further observed that high transfer frequency was always associated with the capacity of the mating mixture to form visible clumps, even when neither strain exhibited constitutive cell aggregation. This work established the role of cell aggregation in effecting efficient cell to cell contact prior to DNA transfer. It further suggests that cell aggregation is a two component system, probably involving a cell surface protein Clu that is induced by cointegrate plasmid formation and an uncharacterized component that is present in Agg+ strains but absent or masked in Agg− strains. The mechanism by which cointegrate formation leads to the expression of cell aggregation has not been elucidated and this will require detailed analysis of the genes involved in conjugal transfer. However one can speculate that an IS*S1*/IS*S1T* located promoter, which is known to exist (see section 1.6.1), could switch on the expression of relevant genes following the transposition events involved in cointegrate formation.

Outside of the lactococci, the role of *Lactobacillus* cell aggregation as a stimulus to conjugal transfer has also been studied. Reniero *et al.* (1992)

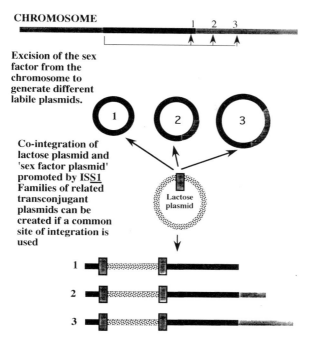

CHROMOSOME

Excision of the sex factor from the chromosome to generate different labile plasmids.

Co-integration of lactose plasmid and 'sex factor plasmid' promoted by ISS1 Families of related transconjugant plasmids can be created if a common site of integration is used

Lactose plasmid

Figure 1.2 (b) For *L. lactis* subsp. *lactis* 712 a more complex model is required to explain the observed structures of cointegrate plasmids. Whilst the basis of this model is as shown in (**a**) provision is needed for variable excision of the sex factor from the chromosome so as to generate sex factor plasmids of different sizes. As described by Gasson *et al.*, (1992) there is also the possibility that the lactose plasmid integrates into the chromosome within or adjacent to the sex factor prior to excision of plasmid cointegrates. Although experimental evidence for this alternative is not available, the structure of cointegrates would be analogous to those formed by cointegration between the lactose plasmid and the sex factor plasmid.

reported the involvement of a secreted 34 kDa protein in the autoaggregation of *Lactobacillus plantarum* 4B2. Cell culture supernatants containing this protein promoted cell aggregation in a variety of other lactic acid bacteria and the protein appeared to act as a molecular bridge which bound to kojibiose substitutions on lipoteichoic or teichoic acids. These authors demonstrated that high efficiency transfer of the heterologous plasmid pAMβ1 was facilitated by use of the aggregation protein to effect cell clumping in mating mixtures.

As has been observed for co-transfer of plasmids with pAMβ1, conjugation involving homologous plasmids sometimes leads to the formation of cointegrates and more complex rearrangement of plasmid DNA. Transfer of a 131 kb bacteriocin plasmid and a 64 kb lactose plasmid from L. *lactis* subsp. *lactis* WM4 led to the appearance of a novel 117 kb recombinant plasmid in the progeny. This plasmid was shown by restriction endonuclease analysis to be composed of DNA originating from both the bateriocin plasmid and the lactose plasmid present in the donor strain (Scherwitz *et al.*, 1983; Harmon and McKay, 1987). During conjugation experiments involving *L. lactis* subsp. *lactis* ME2,

100 kb cointegrates were formed between a 72 kb lactose plasmid and a 28 kb plasmid encoding genes for a restriction and modification activity (Higgins *et al.*, 1988). Jarvis (1988) reported the formation of a 106 kb cointegrate plasmid carrying both lactose and heat sensitive bacteriophage resistance genes and this was created from individual plasmids of 60 and 68 kb. Similarly Baumgartner *et al.* (1986) reported cointegrate plasmids following the transfer of lactose plasmid pCI726 and bacteriophage resistance plasmid pCI750 from *L. lactis* subsp. *cremoris* UC653 to strain MG1363. Steele and McKay (1989) observed a more complex interaction between a lactose plasmid pKB32 and a cryptic plasmid pJS88 in *L. lactis* subsp. *lactis* 11007. Following their co-transfer complete resolution of these two plasmids was observed as were a series of deleted cointegrate plasmids. An instance of chromosomal integration of one of the deleted cointegrate plasmids was observed but complete pKB32::pJS88 cointegrates were never found. The properties of the complex interaction between the component plasmids suggested that a transposition process involving a transposable element of the Tn*3* family could be involved (Steele *et al.*, 1989).

1.3 Transduction

1.3.1 Transduction in Lactococcus *spp.*

Transduction is a mechanism of gene exchange in which bacterial genetic information is erroneously packaged into bacteriophage heads and is passively transferred to a suitable phage sensitive host. Among the lactic acid bacteria this type of gene transfer has been frequently demonstrated in lactococci and less commonly in *S. thermophilus* and *Lactobacillus*. While transduction is less significant than other gene transfer systems for the development of genetics in lactic acid bacteria, nevertheless it has proved a valuable analytical tool. It was first demonstrated in lactococci as early as 1962, just a decade after it was described in *Enterobacteriaceae* (Zinder and Lederberg, 1952) when transfer of streptomycin resistance (Allen *et al.*, 1963) and tryptophan independence (Sandine *et al.*, 1962) by virulent phages was reported. However, most of the subsequent reports describing transductional transfer amongst lactic acid bacteria involved the use of temperate phage induced from a donor strain. A number of reviews describe the development of transduction as a gene transfer system especially in lactococci (Fitzgerald and Gasson, 1988; Gasson, 1990; Davidson *et al.*, 1990), and here instances where transduction has provided new insight to the genetic makeup of lactic acid bacteria are highlighted.

In the 1970s when lactococcal genetics was a developing research area transduction proved valuable in establishing the plasmid encoded nature of lactose metabolism (Lac) and proteinase activity (Prt). In a series of pioneering papers in the early 1970s McKay's laboratory demonstrated the transduction of Lac by UV-induced lysogenic phage from wild-type *L. lactis* subsp. *lactis* C2 to a

Lac⁻ derivative (McKay *et al.*, 1973) as well as the intermittent co-transduction of the Prt marker with Lac (McKay and Baldwin, 1974; Klaenhammer and McKay, 1976). The detailed analysis of Lac⁺ Prt⁻ and Lac⁺ Prt⁺ transductants and the high frequency with which isolates defective in these phenotypes could be isolated, provided preliminary evidence that both traits were plasmid encoded, conclusions which were supported by Molskness *et al.* (1974) who also demonstrated transduction of Lac in *L. lactis* subsp. *lactis* C2. McKay *et al.* (1976) demonstrated the presence of plasmid molecules of between 20–22 MDa in Lac⁺ Prt⁺ transductants providing the first physical evidence that Lac was encoded by plasmid DNA. This molecule was lost in Lac⁻ isolates of these hosts. Curing experiments with the original C2 host had indicated that Lac was encoded by a plasmid of approximately 30 MDa, and McKay *et al.* (1976) suggested that the smaller plasmid observed in transductants arose by a process called transductional shortening whereby the limited capacity of the transducing phage head dictated that only deleted derivatives of the Lac plasmid (in this case molecules of between 20–22 MDa) were of the appropriate size to be packaged. This was supported by the fact that Klaenhammer and McKay (1976) had previously shown that two defective transducing phage could be induced from *L. lactis* C2 which had slightly different head sizes capable of accommodating 22.6 and 23.8 MDa of phage DNA.

A further feature of the transduction system observed with *L. lactis* subsp. *lactis* C2 was the enhanced frequency of transfer of plasmid encoded markers in second generation transduction experiments, i.e. when Lac⁺ transductants were used as donors, transfer of Lac occurred at a frequency approximately 100-fold higher than originally observed. An explanation for this high frequency transduction (HFT) phenomenon was later provided by Gasson and co-workers who also demonstrated transductional shortening and HFT in the related *L. lactis* subsp. *lactis* strain 712. Analysis of transductionally shortened plasmids isolated from either Lac⁺ Prt⁺ or Lac⁺ Prt⁻ transductants indicated that both were in fact deleted derivatives of a large plasmid pLP712 which carried genes for both Lac and Prt in strain 712 (Gasson, 1990). Transducing phage particles thus contained no phage DNA and the HFT phenomenon could not be explained by genetic linkage of phage and Lac plasmid-derived DNAs as had been previously speculated (Klaenhammer and McKay 1976). Davies and Gasson (1981) produced evidence that HFT was dependent on the presence in the donor strain, of transductionally shortened plasmids which, in the case of pLP712, were generated by a relatively high frequency of spontaneous deletion events. Mapping studies of plasmids isolated from Lac⁺ Prt⁺ and Lac⁺ Prt⁻ transductants confirmed this hypothesis (Gasson, 1983; 1990; Gasson and Davies, 1984). The majority of Lac⁺ Prt⁺ transductionally shortened plasmids contained a single, relatively large deletion, which left *lac*, and *prt* genes as well as the replication region of pLP712 intact. The high frequency with which such plasmids were isolated suggested a preferential deletion event which is consistent with the known pattern of pLP712 instability (Gasson, 1983; Maeda and Gasson, 1986).

In the case of the Lac$^+$ Prt$^-$ transductants, a single event resulting in deletion of *prt* genes was never observed and this was explained by the size reduction being insufficient for the plasmid to be packaged in the transducing phage head. Rather, multiple deletions were detected which did produce plasmid derivatives with a size suitable for packaging. Thus, on the basis of experimental evidence, a plausible model to explain the HFT phenomenon was proposed whereby prophage induced from wild-type strain 712 could erroneously package preformed spontaneously deleted derivatives of pLP712, which were of a size roughly equivalent to that of the phage genome (Fitzgerald and Gasson, 1988; Gasson, 1990). When transductants which only contained this deleted plasmid were used as donors, a significantly higher number of bacteriophage heads were packaged with lactose plasmid DNA and thus phage preparations induced from transductants transduced Lac at an elevated frequency.

The characterization of transductants derived from the C2/712 strains revealed other interesting features. The analysis of transductionally shortened plasmids in derivatives of *L. lactis* subsp. *lactis* 712 led to the localization of the *lac* and *prt* genes on pLP712 (Gasson, 1983; Gasson *et al.*, 1987). It also resulted in the recognition of the inherent instability of this plasmid which has more recently been explained by the activity of IS elements.

Yet another significant development was the stabilization of the plasmid encoded Lac and Prt traits in derivatives of *L. lactis* subsp. *lactis* C2 by their chromosomal integration (McKay and Baldwin, 1978). The fact that infrequently Lac$^+$ and Lac$^+$ Prt$^+$ transductants retained these markers following extensive measures to cure them suggested that in some instances the transduced DNA had integrated into the recipient chromosome. This was supported by observation of the effect of UV irradiation on transduction frequencies (Arber, 1960). Transduction of the putative chromosomal *lac* genes was stimulated by UV irradiation while transfer of plasmid *lac* traits was depressed following this treatment. Stabilization of plasmid located traits has also been reported in *L. lactis* subsp. *lactis* strain 712 (Davies and Gasson, 1981) and ML3 (Snook *et al.*, 1981) and whilst the molecular basis for these events has not been determined one stabilized transductant derived from the 712 strain has been characterized in detail (Swindell *et al.*, 1993). Such events would be favoured particularly if transductional shortening resulted in deletion of the replication region of the transduced plasmids thereby making them non-viable as extrachromosomal molecules after transduction. It has been established that IS elements occur commonly on lactococcal plasmids and transpositional recombination between plasmid and chromosomal DNA has been shown to take place during the high-frequency conjugal transfer of pLP712 in *L. lactis* (as described in section 1.2.2).

One practical observation stemming from the stabilization of *lac* and *prt* genes in *L. lactis* subsp. *lactis* C2 was that the transductant exhibited a significantly lower level of proteinase activity compared to the parental strain. When the isolate was used as a starter in the manufacture of Cheddar cheese, it yielded a product which was considered to be less bitter even though starter cell numbers in the cheese were similar to those in control cheese made with the parental C2 strain (Kempler *et al.*, 1979).

In addition to plasmid encoded traits, chromosomal markers have also been transferred by transduction. McKay *et al.* (1973) transduced maltose and mannose markers from a *L. lactis* subsp. *lactis* C2 donor to isogenic derivatives deficient in these traits. Interestingly, while the initial transfer occurred at a somewhat higher frequency than Lac transfer, the HFT phenomenon was not observed. Although the mechanism by which chromosomal markers are transduced has not been determined, it is likely to result from the erroneous packaging of fragmented DNA segments which integrate into the equivalent location in the recipient chromosome by homologous recombination. Anderson and McKay (1983) used transductional evidence to confirm the authenticity of Rec⁻ derivatives of *L. lactis* subsp. *lactis* ML3. These mutants could be transduced by a lactose plasmid but not by a chromosomal marker. The failure to demonstrate transfer in the latter case was due to inactivation of the homologous recombination system.

While transduction in lactococci has been applied and analysed most extensively in the related *L. lactis* subsp. *lactis* strains C2, 712 and ML3, phage mediated transfer has also been demonstrated in *L. lactis* subsp. *lactis* C20 (McKay *et al.*, 1980), *L. lactis* subsp. *cremoris* strains C3 (Snook *et al.*, 1981) and the H2/BK5 system (Lakshmidevi *et al.*, 1988). Interspecies transduction of the Lac marker from strain C3 to *L. lactis* subsp. *lactis* C2 and ML3 derivatives was observed at a very low frequency and three classes of transductants were identified on the basis of their plasmid content – those containing no detectable plasmids, those harbouring a 23 MDa molecule and those containing a 23 and a 30 MDa plasmid (Snook *et al.*, 1981). The stability of the Lac marker in some transductants in which no plasmid DNA could be detected suggested that the transduced genetic material had integrated in the recipient chromosome. The mechanism by which transductants acquired two plasmids of 23 and 30 MDa was not clear.

The ability of the well studied temperate phage BK5-T from *L. lactis* subsp. *cremoris* BK5 to transduce both plasmid and chromosomal markers has also been demonstrated and a number of interesting features unique to this system have been highlighted (Davidson *et al.*, 1990). When this phage was propagated vegetatively on its lytic host *L. lactis* subsp. *cremoris* H2, Lakshmidevi *et al.* (1988) detected the presence of high molecular weight DNA in preparations obtained from the lysate which did not appear to be derived from the BK5-T phage genome and which were in fact found to be derived from the H2 chromosome. The phage preparation was capable of transducing chromosomal markers at a relatively high frequency (approximately 10^{-4} per pfu) whereas plasmid DNA was transduced at a much lower frequency (approximately 10^{-7} per pfu). A likely explanation for the elevated frequency of transfer of the chromosomal markers was provided by Lakshmidevi *et al.* (1988) who detected a number of regions on the H2 chromosome which hybridized to a specific restriction fragment of BK5-T which harboured the phage *pac* site. Thus, it was proposed that the host genome contained a number of *pac*-like regions which would facilitate the incorporation of chromosomal DNA into phage heads which, in the case of BK5-T, are packaged by a head-full mechanism (Lakshmidevi *et al.*, 1988), thereby generating transducing particles.

Transduction of other plasmids has also been demonstrated between lactococcal hosts. Davies and Gasson (1981) have transferred the MLS resistance pAMβ1 plasmid between *Lactococcus lactis* subsp. *lactis* strains at a relatively low frequency. HFT could not be demonstrated, probably because the plasmid was already of a size suitable for packaging. The efficient transduction of the small lactococcal cloning vector pCK1 suggests that a minimum size limit for transduction may not exist (Fitzgerald and Gasson, 1988). This plasmid, contains a pSH71 replicon (Anderson and Gasson, 1985) which replicates by generating single stranded intermediates (Gruss and Ehrlich, 1989), and the resultant concatamers could be packaged into phage heads. Interestingly transduction of the theta replicating plasmid vector pIL253 has also been demonstrated (Gasson and Underwood, unpublished data).

1.3.2 Transduction in Streptococcus thermophilus

There has been only a single report describing transduction in *S. thermophilus*. Mercenier *et al.* (1988) used virulent phage to transfer cloning vectors between isogenic and non-isogenic strains and they were able to demonstrate transduction at frequencies ranging between 10^{-1} and 10^{-6} per pfu. Transduction efficiencies increased with decreasing multiplicity of infection (MOI) but these appeared to be unaffected by plasmid size or the selective marker located on the vector. Although the very high frequencies of transfer (up to 10^{-1} per pfu in some instances) could be due to short regions of homology between the vector and phage genome facilitating efficient packaging of vector DNA into the transducing phage head, no homology was detected by hybridization analysis. Transfer efficiencies were, however, significantly affected by the presence or absence of restriction systems in the recipient strains. The ease with which cloning vectors could be introduced into commercial strains of *S. thermophilus* led Mercenier and Lemoine (1989) to suggest that transduction could be a powerful technique for the genetic modification of these bacteria.

1.3.3 Transduction in Lactobacillus spp.

Although lytic and lysogenic bacteriophage for *Lactobacillus* species have been described quite frequently in the literature and the genetic analysis of some *Lactobacillus* phages is quite advanced (Sechaud *et al.*, 1988; Lahbib-Mansias *et al.*, 1988; Trautwetter *et al.*, 1986) there have been very few reports of transduction in this genus. Generalized transduction of auxotrophic markers by induced phage was described in *Lactobacillus salivarius* by Tohyama *et al.* (1971) while Raya *et al.*, (1989) reported the transfer of vector plasmids in *Lactobacillus acidophilus* (now reclassified *Lactobacillus gasseri*; Raya and Klaenhammer, 1992) by phage which were either induced from, or grown vegetatively on, this host. It was notable that while transduction of vectors encoding chloramphenicol resistance (Cm^r) was detected at low frequencies ranging from approximately 10^{-8} to 10^{-10} transductants per pfu, transfer of erythromycin resistance (Em^r) vectors was not detected at all probably due to the poor expression of this marker in *Lactobacillus* hosts (Luchansky *et al.*, 1989; 1988). In contrast to the observations made with *S. thermophilus*, the highest transduction frequencies were observed with small plasmids. Interestingly, transduction of the Cm^r vector pGK12 occurred at similar low frequencies irrespective of whether the phage lysate was obtained by induction or lytic growth on the *Lactobacillus* host (Raya *et al.*, 1989).

In a subsequent study, Raya and Klaenhammer (1992) were able to increase the transduction frequency of plasmid pGK12 in *Lactobacillus gasseri* ADH by 10^2- to 10^5-fold by inserting restriction fragments of bacteriophage φ*adh* into the vector. The frequency achieved generally correlated with the degree of homology existing between the phage and plasmid DNAs. However, unlike transduction systems in some other bacteria, the presence of the phage *cos* sites in the transduced plasmid did not appear to be solely responsible for the enhanced transfer frequency. Raya and Klaenhammer (1992) produced evidence which indicated that multimers of the recombinant pGK12 plasmids were integrated into the φ*adh* genome and that it was this cointegrate which was packaged in the phage particles. However, following their introduction into recipient cells by transduction, these recombinant molecules resolved yielding monomeric plasmid DNA. pGK12-derived plasmids replicate by a rolling circle mechanism, resulting in the production of single-stranded plasmids which generate tandem multimers. This probably facilitated the formation of high-molecular-weight and highly recombinogenic forms of the vector which, together with the homology between phage and plasmid DNAs, was responsible for the high frequency of transduction.

1.4 Transformation

1.4.1 Transformation systems

The ability to introduce free DNA in its native form or following *in vitro* manipulation into a recipient cell is an essential element in the application of genetic technology to any bacterial system. There are four intrinsically different ways by which bacterial cells can be transformed. Natural competence is a physiological state which can be induced in some strains of bacteria (e.g. *S. pneumoniae*, *S. mutans*, *S. sanguis*, *Haemophilus influenza*) during which they have the capacity to actively take up and process DNA. Artificial competence can be achieved with some, usually Gram-negative, bacteria whereby treatment with chemicals often combined with a heat shock, results in uptake of DNA. Osmotically protected protoplasts of bacteria, in which the peptidoglycan barrier is removed, may allow passage of DNA through the cytoplasmic membrane. Electrotransformation involves subjecting cells to a high voltage electric discharge, which makes cells permeable to DNA. These different types of bacterial transformation systems have been described in detail in an excellent review by Mercenier and Chassy (1988).

In lactic acid bacteria, transformation by either natural or artificially induced competence has not been reported although Moller-Madsen and Jensen did describe transformation of whole cells of *L. lactis* subsp. *lactis* as early as 1962. The subsequent failure to demonstrate natural competence in any strain of lactic acid bacteria despite extensive screening efforts casts some doubt on the validity of this work.

Although significant progress in the development of genetic systems in the lactococci was achieved during the 1970s, the absence of a transformation process for these and other lactic acid bacteria was a major barrier to further advances. Thus, the success of Kondo and McKay (1982a) in introducing Lac plasmid DNA into *L. lactis* subsp. *lactis* protoplasts represented a major milestone in lactic acid bacterial genetics and meant that the engineering of strains to generate altered or enhanced traits by recombinant DNA technology was a realistic and achievable goal. Subsequent to that first report, protoplast transformation has been optimized and applied to other lactic acid bacteria, particularly species of *Lactobacillus*. However, the availability of electrotrans-formation which is a relatively simple and more widely applicable technique has revolutionized gene technology in lactic acid bacteria and has ensured that genetic manipulation and the construction of recombinant strains is not confined to the few isolates for which protoplast transformation had been developed.

1.4.2 Protoplast transformation

Protoplast transformation as applied to *Lactococcus* strains and subsequently to other lactic acid bacteria is based on the original protocol developed for transformation of *Bacillus* by Chang and Cohen (1979). This involves the controlled removal of the cell wall using a muralytic enzyme (usually lysozyme or mutanolysin), the stabilization of protoplasts in an osmotically protective medium, the incubation of the protoplasts with the transforming DNA in the presence of polyethylene glycol (PEG) and the regeneration of transformed protoplasts. Experience has indicated that the basic procedure needs to be modified for different hosts and even then it frequently fails to yield transformants for many strains. Furthermore, for those strains which can be transformed successfully, the frequency appears to be critically dependent on a vast number of experimental parameters.

The effective production of protoplasted cells depends on the phase of growth in which they are harvested and generally early to mid log phase cells have proved most suitable (Simon *et al.*, 1986; Kondo and McKay, 1984). The effect of growing cells in the presence of agents which weaken the cell wall (e.g. glycine, DL-threonine) has not been studied. It appears that the composition of the growth medium does not significantly affect the ultimate efficiency of protoplast transformation with lactococci typically being cultivated in M17 medium (Terzaghi and Sandine, 1975), and MRS or LCM media often being used for lactobacilli (de Mann *et al.*, 1960; Eftymiou and Hansen, 1962).

Lysozyme (Kok *et al.*, 1985; von Wright *et al.*, 1985; Simon *et al.*, 1986), mutanolysin (Kondo and McKay, 1982a; 1984) or a combination of both (Lin and Savage, 1986; Morelli *et al.*, 1987; Mercenier *et al.*, 1988; McCarthy *et al.*, 1988) have been used to remove cell walls. While mutanolysin has been shown to be more effective than lysozyme in digesting the cell wall of *L. lactis* (Kondo and McKay, 1982b) in general higher transformation frequencies have been achieved with lysozyme-generated *L. lactis* protoplasts (Simon *et al.*, 1986). However, this may be strain specific as Woskow and Kondo (1987) have reported that transformation of protoplasts produced following lysozyme-treatment was inefficient with their *L. lactis* strain. Mercenier *et al.* (1988) found that regeneration efficiencies for lysozyme treated *S. thermophilus* cells were significantly higher than those obtained with mutanolysin-generated protoplasts. Over-treatment with muralytic enzymes is counter-productive as this adversely affects regeneration of the cell wall (Simon *et al.*, 1986; Mercenier and Chassy, 1988).

Typically, the composition of protoplast buffers is quite complex and includes an osmotic stabilizer (e.g. sucrose, raffinose, malate or succinate at concentrations ranging between 0.5 and 0.2 M), bovine serum albumin, and $MgCl_2$ or $CaCl_2$, (usually between 0.001 and 0.025 M). The pH of protoplast buffer is usually between 6.5 to 7.0. Simon *et al.* (1986) found that inclusion of M17 medium in the protoplast buffer resulted in increased transformation efficiency of *L. lactis* while Von Wright *et al.* (1985) showed that replacing $MgCl_2$ with $CaCl_2$ increased frequencies more than 10-fold.

PEG is an essential element in the protoplast transformation process. Generally PEG with molecular weights between 1000 and 6000 at concentrations ranging between 20 to 40% has been used for transformation of most species of lactic acid bacteria with treatment times ranging between 2 to 20 minutes. Kondo and McKay (1984) found that longer times of incubation, up to 20 minutes, in PEG yielded higher transformation efficiencies with *L. lactis* protoplasts.

The efficiency of protoplast transformation appears to be inversely proportional to the size of the transforming DNA. Simon *et al.* (1986) generated 7.8×10^4 transformants of *L. lactis* subsp. *lactis* per µg of pAMβ1 (26.5 kb) DNA whereas 6.9×10^6 transformants per µg were obtained with pIL204 (5.5 kb) DNA. Transformation efficiency is also linearly proportional to the concentration of DNA used (Kondo and McKay, 1984; Simon *et al.*, 1986) and the efficiency is reduced severely (between 10- and 100-fold) when linear DNA or ligation mixtures are used (De Vos, 1987; Kondo and McKay, 1984).

Efficient regeneration of transformed protoplasts is perhaps the most critical step in the process. Typically, protoplasts are plated on conventional complex growth media supplemented with osmotic stabilizers and a selective agent (usually at the lowest concentration which will distinguish transformed cells from the non-transformed recipients). In the case of lactococci, pour plating in agar overlays as opposed to spread plating on agar surfaces had a dramatically beneficial effect on protoplast regeneration frequencies and therefore transformation efficiencies (Kondo and McKay, 1984).

While the frequencies reported by Kondo and McKay (1982a) for the protoplast transformation of *L. lactis* subsp. *lactis* LM0230 (a plasmid cured derivative of the widely used laboratory strain C2) were very low (8.5 transformants per μg of DNA) the basic protocol was soon improved and frequencies of $4 \times 10^4/\mu g$ DNA for this strain (Kondo and McKay, 1984) and 5×10^6 for subsp. *lactis* IL1403 (Simon *et al.*, 1986) were achieved. These frequencies were sufficiently high to allow direct cloning of genes into *L. lactis* hosts. Kondo and McKay (1984) described the cloning of *lac* genes into strain LM0230 using their optimized protocol and Loureiro dos Santos and Chopin (1987) cloned a large chromosomal fragment harbouring an inserted Tn*916* element into host IL1403.

Significant variations to the basic protocol described above include the use of liposomes, composed of cardiolipin and phosphatidylcholine, to entrap the transforming DNA, which results in significant increases in transformation frequencies in *L. lactis* (Van der Vossen *et al.*, 1988) and *S. thermophilus* (Mercenier *et al.*, 1988). Woskow and Kondo (1987) have shown that the addition of one of a range of proteolytic enzymes immediately after exposure of cells of *L. lactis* to mutanolysin increased transformation frequencies by almost 100-fold.

Simon *et al.* (1985) developed a co-transformation protocol in which cryptic plasmids could be passively transformed into *L. lactis* subsp. *lactis* IL1403 during selection of an Em[r] indicator plasmid, pHV1301. A total of 41.5% of the pHV1301-containing isolates also harboured one or other of the four cryptic lactococcal plasmids used in the experiments.

Sanders and Nicholson (1987) described a protocol for the transformation of non-protoplasted cells of *L. lactis* with plasmid DNA achieving frequencies as high as 5×10^5 transformants per μg of DNA with an optimized procedure. However, realistic frequencies were achieved only when PEG and an osmotic stabilizer were used suggesting that the protocol still rendered the cells osmotically fragile, to some degree at least. The parameters which affected the efficiency of transformation appeared to be similar to those previously identified for true protoplast transformation, i.e. PEG concentration and PEG buffer composition (30% in a succinate-based buffer applied at 0°C gave optimal results), size and nature of transforming DNA, and a reduction in frequency when ligated DNA was used. A heat shock (42°C for 5 min) raised transformant recovery two-fold.

The development and optimization of protoplast transformation for lactic acid bacteria was an effective response to an obvious requirement for the genetic analysis and manipulation of these hosts. However, the procedure is time-consuming, very sensitive to experimental variation and operator changes and is extremely strain specific. With lactococci, high frequencies of trans-formation were achieved with relatively few strains of *L. lactis* subsp. *lactis* while frequencies obtained for a range of *Lactobacillus* species including *Lactobacillus acidophilus* (Lin and Savage, 1986; Morelli *et al.*, 1987; McCarthy *et al.*, 1988), *Lactobacillus reuteri* (Morelli *et al.*, 1987) and *Lactobacillus plantarum* (Badii *et al.*, 1989) were generally very low. Transformation of the industrially significant *L. lactis* subsp. *cremoris* (Woskow and Kondo, 1987; van der Lelie *et al.*, 1988) and other *Lactobacillus* species occurred at very low frequencies if at all, while there are no reports of protoplast transformation of *Leuconostoc* or *Pediococcus* strains. Thus, there was still a requirement for a broadly applicable, more reproducible transformation method for lactic acid bacteria and this has largely been fulfilled by the development of electrotransformation technology.

1.4.3 Electrotransformation

The application of high voltage electric field pulses to prokaryotic or eukaryotic cells results in permeabilization of the cell membrane allowing the transient passage of macromolecules across the membrane structure. This has been exploited to achieve membrane fusions, and to either remove or introduce DNA from or into cells. Electroporation was originally applied to transform eukaryotic cells (Neumann *et al.*, 1982) but its obvious potential for transfor-mation of bacterial cells was quickly recognized. Shivarova *et al.* introduced DNA into *Bacillus cereus* protoplasts by this method in 1983 and Harlander (1987) was the first to demonstrate electroporation of whole cells when she generated transformants of *L. lactis* at frequencies of approximately 10^3 per μg of plasmid DNA.

Species of most bacterial genera have now been transformed by electropo-ration with Gram-negative hosts generally yielding higher frequencies than Gram-positive bacteria. This difference may be attributed to the thick, more dense cell wall of Gram-positive cells. The principles of electroporation, the development of electroporation technology and the parameters affecting electrotransformation efficiency in bacteria have been reviewed by Chassy *et al.* (1988).

All genera of the group 'lactic acid bacteria' have been transformed by electroporation. The basic protocol involves preparation of cells which are usually harvested in mid-log phase from media which may or may not contain supplements designed to weaken the cell wall (e.g. glycine, DL-threonine).

Following a series of washing steps (often in distilled water), the cells are resuspended in buffer with or without sucrose (up to 0.5 M). Chilled cells and the appropriate concentration of DNA are then placed in a customized cuvette which incorporates positive and negative electrodes and an electric field pulse is delivered. After application of the pulse the preparation may be held, often on ice, for a period (5 to 15 minutes), diluted in ice cold medium (again containing sucrose) and plated on the appropriate selective medium.

It is difficult to identify which parameters in the generalized electroporation protocol will effect transformation. Typically, these need to be determined for the particular host being transformed. However, the specific voltage applied does have a major impact on the efficiency of the process. For lactococci field strengths of 5.0 kV/cm (van der Lelie et al., 1988), 6.25 kV/cm (Aukrust and Nes, 1988; Powell et al., 1988), 7.5 kV/cm (Dornan and Collins, 1990), 10 kV/cm (Holo and Nes, 1989) and 12.5 kV/cm (Ward and Jarvis, 1991) have been reported as being effective. Generally higher field strengths yielded greater numbers of transformants per μg of DNA particularly when older cultures were used (McIntyre and Harlander, 1989a, b). Chassy and Flickinger (1987) were the first to report the successful electroporation of a *Lactobacillus* species and they obtained transformants at efficiencies of between approximately 10^3 and 10^5/μg of DNA for a range of *Lactobacillus casei* strains. While a field strength of 5.0 kV/cm was found to be optimal, no transformants were obtained when 6.25 kV/cm was used. Zink et al. (1991) examined field strengths between 2 and 8 kV/cm and found that highest transformation frequencies (1.3×10^3/μg DNA) for *Lactobacillus* sp. were obtained when a voltage of 6.25 kV/cm was used. David et al. (1989) found that the efficiency of transformation of *Leuconostoc paramesenteroides* increased almost exponentially with increasing field strength up to the highest voltage applied (6.25 kV/cm) when 4×10^3 transformants per μg of DNA were obtained, while Wyckoff et al. (1991) obtained 2×10^6 transformants per μg of DNA for *Leuconostoc cremoris* using 8 kV/cm. Kim et al. (1992) transformed *Pediococcus acidilactici*, albeit at low frequency, using 12.5 kV/cm. For *S. thermophilus*, Somkuti and Steinberg (1988) found that the number of transformants obtained increased with voltage, reaching a plateau (approximately 4×10^2/μg of DNA) at 4 kV/cm. Luchansky et al. (1988) reported that more transformants (3.8×10^4 per μg DNA) were obtained with higher voltages (up to 6.25 kV/cm) compared to lower voltages for *Lactobacillus acidophilus* and using an optimized protocol they were able to demonstrate transformation of a number of species of *Lactobacillus* including *Lactobacillus casei* (1.7×10^2/μg), *Lactobacillus fermentum* (9.7×10^5/ug), *Lactobacillus plantarum* (8.5×10^3/ug), *Lactobacillus reuteri*, (7.0×10^3) in addition to *Leuconostoc* (1.1×10^4/μg) and *Pediococcus* (5.4×10^1/μg) species. While the high voltages used will have a major killing effect which is generally proportional to the severity of the treatment, the composition of the electroporation buffer and the specific species or strain of lactic acid bacteria will also determine the degree of lethality. The capacitance used is generally

25 μF which is the highest setting available in the most commonly used commercial electroporator.

The composition of the electroporation buffer can also influence transformation frequency. As well as sucrose, $MgCl_2$ or $CaCl_2$ are often included together with buffer salts. Holo and Nes (1989) have reported that transformants of *L. lactis* can be obtained at efficiencies of greater than 10^7 per μg of DNA using a suspending medium containing just 0.5 M sucrose and 10% glycerol in water. McIntyre and Harlander (1989a) also found that a suspending medium of deionized distilled water yielded the highest number of *L. lactis* transformants particularly at high field strengths. The substitution of either maltose or raffinose for sucrose in the electroporation buffer resulted in a 10-fold decrease in the transformation frequency of *Leuconostoc mesenteroides* subsp. *cremoris* (Wyckoff *et al.*, 1991) but Somkuti and Steinberg (1988) have shown that substitution of raffinose with sucrose did not significantly affect the number of *S. thermophilus* transformants.

Based on these observations it is clear that the ionic strength of the medium in which cells destined for electroporation are suspended will affect the efficiency of transformation. Luchansky *et al.* (1988) reported that when either phosphate- or Hepes-based buffers were used at concentrations greater than 1 × (both also contained sucrose and $MgCl_2$ and were adjusted to pH values of 7.4 and 7.3, respectively) a higher number of *Lactobacillus acidophilus* transformants was obtained. In the case of *S. thermophilus* replacement of K^+ with Na^+ in the buffer salts resulted in a 10-fold reduction in transformation frequency of the strains examined (Somkuti and Steinberg, 1988).

There is evidence to suggest that the composition of the growth medium is also significant. McIntyre and Harlander (1989b) have found that growth of *L. lactis* in a defined, rather than a complex, medium resulted in a significant improvement in transformation frequency and this was enhanced still further when DL-threonine was included. Holo and Nes (1989) reported that the presence of sucrose and glycine in the growth medium (the former to protect osmotically fragile cells generated as a consequence of the effect of glycine on the cell wall) had a dramatically positive effect on the transformation of *L. lactis* subsp. *cremoris*. Van der Lelie *et al.* (1988) were unable to generate transformants of *L. lactis* subsp. *cremoris* without the inclusion of 40 mM DL-threonine in the growth medium. Dornan and Collins (1990) also used DL-threonine in the growth medium but they indicated that extensive washing of the *L. lactis* cells prior to electroporation was a significant factor in obtaining reproducible transformation frequencies. The presence of glycine (1.2%) also yielded a significant improvement in the transformation efficiency of *Lactobacillus helveticus* subsp. *jugurti* (Hashiba *et al.*, 1990). When Aukrust and Nes (1988) examined the effect of glycine concentration on the transformation of *Lactobacillus plantarum*, they obtained transformants only when 1% was included in the growth medium. Cosby *et al.* (1989) did obtain relatively low numbers of transformants of *Lactobacillus plantarum* in the absence of glycine. However, it is possible that this low frequency was attributable to the temperature sensitive replicon of the transforming plasmid DNA, pTV1ts, (Youngman, 1987).

Unlike protoplast transformation, generally the phase of growth at which cells are harvested prior to electroporation does not appear to be critical. Typically cells are harvested in mid-log phase although McIntyre and Harlander (1989a, b) were able to obtain significantly higher numbers of transformants of *L. lactis* when cells were harvested in stationary phase. In the case of *Lactobacillus helveticus* subsp. *jugurti*, few transformants were obtained when cells were harvested at either early exponential or stationary phase while mid- to late-exponential phase cells yielded the best transformation efficiencies (3.6 × 10^2 per μg of DNA; Hashiba *et al.*, 1990).

Pretreatment of cells with lysozyme may also improve transformation frequency. Power *et al.* (1988) obtained a 300- to 1000-fold increase when *L. lactis* subsp. *lactis* LM0230 was incubated with 2 kU/ml of lysozyme for 20 minutes at 37°C in electroporation buffer. Electroporated cells were plated on selective media containing 0.5 M sucrose. Kim *et al.* (1992) also treated *Pediococcus acidilactici* cells with lysozyme prior to electroporation but in this case, transformants were plated on media to which sucrose was not added. Practically all electroporation protocols describe the chilling of cells before the application of the electric field pulse followed by incubation of the treated cells on ice for a short period before dilution and plating on selective media.

As with protoplast transformation, the frequency of electroporation of most bacteria generally decreases with increasing size of the transforming DNA although the generally higher efficiency of electrotransformation usually makes these differences less critical. However, Powell *et al.* (1988) and McIntyre and Harlander (1989a) reported no relationship between plasmid size and transformation frequency of *L. lactis* subsp. *lactis* LM0230. As a general rule, the total number of transformants recovered increases with increasing concentration of added DNA but the efficiency of electroporation is usually highest at lower DNA concentrations (Powell *et al.*, 1988; McIntyre and Harlander, 1989a, b; Luchansky *et al.*, 1988; Kim *et al.*, 1992). While ligated DNA electrotransforms less efficiently than covalently closed circular molecules, in many bacterial systems this reduction is tolerable since the overall efficiency of the process is good.

A significant advantage of electroporation is that cell preparations can be stored frozen and transformed when required at a later time (McIntyre and Harlander, 1989b; Holo and Nes, 1989; Wyckoff *et al.*, 1991). Holo and Nes (1989; 1991) have found that fresh and thawed cells of *L. lactis* subsp. *cremoris* which had been frozen at −85°C in a sucrose/glycerol solution, reproducibly yielded transformants at similar frequencies while Wyckoff *et al.* (1991) observed a 10-fold decrease in transformation frequency when *Leuconostoc mesenteroides* subsp. *cremoris* was stored in electroporation buffer for 90 days at −70°C.

Other factors which affect electroporation efficiency are generally relevant to all bacterial transformation systems. The purity of the transforming DNA preparation is of major significance in protoplast transformation (Sanders and

Nicholson, 1987) but appears to be less critical for electroporation (Holo and Nes, 1989; Ward and Jarvis, 1991). The ability of the transforming DNA to be maintained and expressed in the recipient is essential while the existing plasmid complement of the recipient is also likely to affect its transformability. Van der Lelie *et al.* (1988) have found that incompatibility between native and trans-forming plasmids has a major bearing on the frequency of transformation of a range of *L. lactis* subsp. *cremoris* hosts with the highest efficiencies being obtained with plasmid-free strains using either electro or protoplast transfor-mation protocols. The negative effect appeared to be lessened when the copy number of the resident incompatible plasmid was low. Incompatibility between vector pTRK13 and a resident plasmid in *Lactobacillus acidophilus* was also considered to be responsible for the poor recovery of transformants (Luchansky *et al.*, 1988). However, while Posno *et al.* (1991) also observed incompatibility between resident plasmids in *Lactobacillus pentosus* and incoming vectors, this did not appear to be responsible for the relatively low transformation frequencies. Rather, the authors concluded that the lower efficiencies obtained with this host (10^2–10^3/μg) compared with other *Lactobacillus* species (e.g. 10^6–10^7/μg for *Lactobacillus casei* and *Lactobacillus plantarum*) were most likely due to the use of non-optimized procedure. It has also been observed that a higher recovery of transformants is achieved when the transforming DNA can be selected for chloramphenicol resistance rather than erythromycin resistance in *Leuconostoc* and *Lactobacillus* hosts (Wyckoff *et al.*, 1991; Luchansky *et al.*, 1988) although this did not appear to be the case with *Lactobacillus plantarum* (Cosby *et al.*, 1989).

The restriction/modification status of both the recipient and the host from which the transforming DNA is isolated would be expected to influence the efficiency of the transformation process. Luchanksy *et al.* (1988) have shown that marginally higher numbers of transformants of *Lactobacillus acidophilus* were obtained when vector DNA from this rather than another *Lactobacillus acidophilus* or *E. coli* host was used. There was a 30-fold increase in the number of *Lactobacillus* transformants obtained when homologous rather than heterol-ogous DNA was used (Zink *et al.*, 1991). Hashiba *et al.* (1990) attributed the two-log reduction in the efficiency of transformation of *Lactobacillus helveticus* subsp. *jugurti* to the presence of a restriction/modification system in these hosts. Langella and Chopin (1989) have specifically examined the effect of restriction/ modification systems on transformation efficiency of *L. lactis* subsp. *lactis*. They found that in two strains, each harbouring a restriction/modification system with different specificities, the efficiency of protoplast transformation with unmodified DNA was reduced by more than three orders of magnitude while, surprisingly, no difference was obtained when the same hosts were electroporated with modified and unmodified DNAs. It was suggested that this may be due to a transient inactivation of the restriction endonuclease due to damage inflicted on the cell by the electric pulse. Posno *et al.* (1991) have shown that the efficiency of transformation of *Lactobacillus casei* was similar

(10^6–10^7/μg of DNA) irrespective of the host from which the DNA was isolated (i.e. *E. coli* or *Lactobacillus casei*). In addition, vectors containing *E. coli* DNA transformed as well as plasmids which contained DNA exclusively derived from Gram-positive bacteria. David *et al.* (1989) have also reported that DNA isolated from *Leuconostoc paramesenteroides*, *L. lactis* subsp. *lactis* or *E. coli* was capable of transforming a *Leuconostoc paramesenteroides* recipient at similar efficiencies but in this case the restriction/modification status of the hosts was not described.

An interesting application of electroporation technology has been the electro-transfer of plasmids between *L. lactis* and *E. coli* (Ward and Jarvis, 1991). This was achieved by electroporating the donor strain (i.e. the *Lactococcus* or *E. coli* depending on the desired direction of the transfer) using optimal conditions, harvesting the supernatant which was then mixed with the recipient host and subjected to an electroporation treatment. Following incubation in expression broth appropriate for the recipient (i.e. *L. lactis* or *E. coli*), the electroporated cells were plated on selective media. Transfer of the shuttle vector pFX3 between *E. coli* and *L. lactis* occurred at a frequency of approximately 1×10^3 transformants per electroporation while the rate of transfer in the opposite direction was approximately 2×10^2. The advantage of this approach is that plasmid DNA does not have to be isolated and purified in preparation for trans-formation experiments.

Clearly, the development of electrotransformation has removed a major bottleneck in the full application of recombinant DNA technology to the lactic acid bacteria. Its broad applicability, reproducibility and the relative simplicity of the technique has meant that it has largely superseded protoplast transfor-mation as the method of choice for introducing free DNA into these hosts.

1.4.4 Transfection

Transfection is a variation of transformation in which free bacteriophage DNA, rather than plasmid or chromosomal DNA, is introduced into recipient cells. The major difference between the two methods is that expression of the transfected DNA results in the production of plaques on an appropriate indicator strain rather than viable colonies. In lactic acid bacteria transfection has generally been employed to develop optimal conditions for regular transformation, particularly protoplast transformation. It is of particular value in this regard since outgrowth of viable cells is not required after a successful transfection event, a step which has often proved problematic in protoplast transformation. Furthermore, a medium containing a selective agent (usually an antibiotic) which might inhibit the growth of weakened or injured transformed cells is not required. In addition, the ability of the host to express the transfected DNA is usually known in advance since this can be evaluated by routine plaque assay. In transformation this will only be the case where some other way of introducing the DNA into the host exists or the DNA is originally isolated from an isogenic strain.

Geis (1982) was the first to report the successful transfection of a *Lactococcus* strain when he introduced phage P008 DNA into lysozyme-generated protoplasts of *L. lactis* subsp. *lactis* var *diacetylactis* in the presence of PEG (mol. wt. 6000). The highest transfection efficiency was obtained with low DNA and protoplast concentrations and at PEG concentrations of 30% or more. Woskow and Kondo (1987) transfected *L. lactis* subsp. *lactis* protoplasts, obtained by mutanolysin and chymotrypsin treatment, at frequencies between 10^4 to 10^5 per µg while Sanders and Nicholson (1987) also demonstrated transfection of whole (i.e. non-protoplasted) cells of the same host at a somewhat lower frequency. In both of these cases the transfection experiments were done to demonstrate DNA uptake as a prelude to developing a plasmid transformation system. Powell *et al.* (1988) also transfected two unrelated *L. lactis* subsp. *lactis* hosts with homologous phage DNA and found a 10-fold difference in the frequencies obtained which they concluded could be due to either host or phage factors or a combination of both.

Chassy and Flickinger (1987) successfully used transfection to demonstrate that large molecules could be introduced into *Lactobacillus casei* by electroporation. Protoplasts of *Lactobacillus plantarum* obtained by mutanolysin/lysozyme digestion were transfected at efficiencies between 25 and 250 pfu per µg of DNA (i.e. 2.2×10^{-5} to 4.7×10^{-4} pfu per recovered protoplast) under conditions in which no plasmid transformants could be obtained (Cosby *et al.*, 1988). There have also been a number of reports citing the use of liposome-encapsulated phage DNA to transform various *Lactobacillus* species. Shimizu-Kadota and Kudo (1984) obtained a low frequency of transfection of *Lactobacillus casei* spheroplasts using phage J1 DNA enclosed in liposomes prepared with lecithin. Later, Chassy (1987) described an optimized liposome mediated transfection system for *Lactobacillus casei* protoplasts and he indicated that the time of treatment with mutanolysin was the most critical factor affecting transfection frequencies. The optimized protocol, which included a simplified method for forming liposomes, yielded transfectants at a frequency greater than 10^6 per µg of phage DNA. Boizet *et al.* (1988) also transfected *Lactobacillus delbrueckii* protoplasts with phage DNA at a high frequency of 5×10^7 pfu per µg of DNA but this was achieved only after incubation of transfected cells for 24 h prior to plating in an overlay agar containing the indicator host. However, they did show that bursting of cells and subsequent infection of other cells was not responsible for the high number of plaques obtained.

The development of optimized conditions for protoplast transformation of *S. thermophilus* was assisted by transfection assays which identified specific factors relating to muralytic enzyme treatment and regeneration conditions, required for efficient and reproducible transformation. These assays were optimized to yield up to to 6×10^5 pfu per µg of transfecting phage DNA.

Transfection of *Pediococcus* and dairy *Leuconostoc* species has not been described probably because phages for these hosts are less readily available.

1.5 Protoplast fusion

An alternative strategy for exchanging genetic material between two cell types involves the fusion of protoplasts in the presence of a fusant such as PEG, after which recombination and exchange between both DNA complements can occur. Since it is a relatively simple and unsophisticated technique and is neither strain- nor species-specific, it has been considered as a viable approach to achieving gene exchange in lactic acid bacteria, particularly before the advent of electropo- ration, when other gene transfer methods were only applicable to a few strains.

Gasson (1980) first demonstrated protoplast fusion in lactococci. Using lysozyme-generated protoplasts derived from isogenic *L. lactis* subsp. *lactis* strains, recombination of chromosomal markers (*mal, str*) was observed at frequencies between 10^{-4} and 10^{-5} but only when PEG was used. While transfer of the Lac and pAMβ1 plasmids was also detected even in the absence of PEG, the transfer frequencies were significantly higher in the presence of the fusant. Okamoto *et al.* (1983, 1985) also successfully employed protoplast fusion between *L. lactis* subsp. *lactis* strains and between *L. lactis* subsp. *cremoris* and *L. lactis* subsp. *lactis* to generate a range of chromosomal recombinants.

The formation and regeneration of protoplasts or spheroplasts of a number of species of *Lactobacillus* has been reported including *Lactobacillus* species *Lactobacillus reuteri*, *Lactobacillus gasseri* (Connell *et al.*, 1988, Vescovo *et al.*, 1984) and *Lactobacillus casei* (Lee-Wickner and Chassy, 1984). Watanabe *et al.* (1987) have also produced protoplasts of *Lactobacillus casei* using phage OL-1 lysin. Protoplast fusion between isogenic derivatives of *Lactobacillus fermentum* has been reported by Iwata *et al.* (1986) and fusants harbouring tetra- cycline- and erythromycin-resistance plasmids derived from each of the parental strains were isolated. Kanatani *et al.* (1990) have described protoplast fusion between isogenic *Lactobacillus plantarum* derivatives during which recombi- nation between a range of chromosomal markers (including those for isoleucine, leucine and phenylalanine auxotrophy) occurred at frequencies ranging from 4.7 to 6.3×10^{-7} of the plated protoplasts.

Intergeneric protoplast fusion has also been demonstrated amongst lactic acid bacteria. Cocconcelli *et al.* (1986) have described intergeneric protoplast fusion between *L. lactis* and *Lactobacillus reuteri* in which transfer of pAMβ1 occurred at a frequency of approximately 10^{-5}, while recombinants of the *Lactobacillus* strain capable of fermenting trehalose were recovered at a frequency of 8.3×10^{-5}. Transfer of vector plasmid pGKV21, which is capable of replicating in both *Lactococcus* and *Bacillus* species, between protoplast fused cells of these hosts was observed at frequencies of approximately 10^{-5} per recipient which was similar to the frequency of transfer between *L. lactis* strains (van der Vossen *et al.*, 1988). Therefore, it appears that the genetic relatedness of strains is not a significant factor in influencing the efficiency of plasmid transfer between fused protoplasts. Van der Lelie *et al.* (1988) have also demonstrated transfer of plasmids between *Bacillus* and *L. lactis* subsp.

cremoris strains by protoplast fusion at frequencies ranging from 10^{-5} to 10^{-7} per recipient. The ability of some recipients to acquire plasmids at higher frequencies was attributed to their superior regeneration capacities. While protoplasts of *Leuconostoc mesenteroides* have been produced and regenerated at relatively good frequencies there was considerable strain variation (Otts and Day, 1987) and protoplast fusion or protoplast transformation of *Leuconostoc* has not been described.

1.6 Transposition

1.6.1 Insertion sequences

A range of insertion sequences has been discovered in lactic acid bacteria (Table 1.2) often as a result of analysis of *in vivo* DNA rearrangement or spontaneous mutation. The first IS element to be described was isolated from a virulent bacteriophage that was shown to share extensive DNA homology with a prophage that was integrated in the host strain. The difference between the virulent and temperate bacteriophages was shown to be caused by the presence of a novel IS element in the virulent phage genome. This element IS*L1* may have caused the inactivation of an operator/repressor control circuit or it may have introduced a new promoter which expressed bacteriophage genes. In any event the copy of IS*L1* prevented the maintenance of lysogeny by the variant bacteriophage thereby creating its virulent phenotype. Hybridization experiments confirmed that IS*L1* was also present in both chromosomal and plasmid DNA of *Lactobacillus casei* but it was not widely distributed, being detected in only 3 of 19 strains that were examined (Shimizu-Kadota *et al.*, 1985, 1988).

The first lactococcal IS element to be identified was IS*S1S* which is associated with lactose plasmid cointegrate formation following conjugal transfer (Polzin and McKay, 1991). The molecular rearrangement promoted by IS*S1S* creates a cell aggregation phenotype and facilitates high frequency conjugal transfer of the lactose plasmid. The isolation of IS*S1S* and the related element IS*S1T* is described above in section 1.2.2. In addition iso-IS*1* elements, IS*S1W* and IS*S1N* have been found adjacent to the *prt*M gene in proteinase plasmids of *L. lactis* subsp. *cremoris* WgL and SK11, respectively (Haandrikman *et al.*, 1990). The former element was flanked on one side by part of a second IS element suggesting that the proteinase plasmid pWV05 may have evolved as a deletion derivative of another plasmid, which originally carried at least two copies of the element. Another iso-IS*1* element named IS*946* was isolated from the bacteriophage resistance plasmid pTR2030 following mobilization of vector plasmids pGK12 and pSA3 in conjugation experiments. IS*946* promoted cointegrate formation between pTR2030 and either pGK12 or pSA3 and resolution of the small vector plasmid following conjugation generated derivatives carrying a copy of the IS element (Romero and Klaenhammer, 1990). Yet another iso-IS*1* was discovered adjacent to the bacte-

Table 1.2 IS elements in lactic acid bacteria

IS element	Size (bp)	Inverted repeats (bp)	Target duplication (bp)	Possible transposase (a.a.'O)	Homology	Known distribution	Known iso-elements
ISL1	1256	40	3	274	IS2/IS3	*Lactobacillus*	H
ISS1S	808	18	8	226	IS26	*Lactococcus*	ISS1T; ISS1N; ISS1W IS946
IS904	1241	39	4	253	IS2/IS3	*Lactococcus*	IS1076
IS981	1222	40	5	280	IS2/IS3	*Lactococcus*	
IS905	1313	28	8	392	IS256	*Lactococcus*	
IS1165	1553	39	3 or 8	412		*Leuconostoc* *Pediococcus* *Lactobacillus*	

Summary of the properties of the IS elements described in the lactic acid bacteria.

riophage resistance gene *abi416* isolated from *L. lactis* subsp. lactis IL*416*. This IS element was shown to provide the promoter for expression of the bacterio-phage resistance gene (Cluzel *et al.*, 1991).

During analysis of nisin genes a new IS element, IS*904* was found to be located near to one end of the conjugative transposon Tn*5301* that also encoded genes for nisin biosynthesis and sucrose fermentation. This IS element appeared to play no role in the transposition or conjugal transfer of Tn*5301* (Dodd *et al.*, 1990). A closely related insertion sequence, IS*1076* has been characterized by Huang *et al.* (1991). Following transformation of *L. lactis* subsp. *lactis* LM0230 with a recombinant plasmid that carried bacteriophage resistance genes, a derivative was isolated in which the gene encoding an abortive infection system was inactivated. DNA sequence analysis revealed that this gene was disrupted by the integration of a novel insertion sequence IS*981* (Polzin and McKay, 1991). During the construction of a lactococcal expression system for the production of protein engineered nisin Dodd *et al.* (1992) selected a strain for resistance to nisin in an attempt to switch on the nisin biosynthetic operon. This was achieved and further analysis showed that the promoter activity was provided by the integration adjacent to the nisin biosynthetic genes of a novel IS element, IS*905*. Sequence analysis of this element revealed strong homology with a known staphylococcal IS element IS*246* which encodes an outward reading promoter that is responsible for expression of the widespread *acc*A-*aph*D aminoglycoside drug resistance determinant in this genus (Dodd *et al.*, 1993).

An IS element has also recently been characterized in *Leuconostoc mesen-teroides* subsp. *cremoris*. This element IS*1165* was isolated following the investigation of random *Leuconostoc* clones as DNA probes. One clone was found to have homology with multiple fragments of digested *Leuconostoc* chromosomal DNA in Southern blots. This proved to be due to the presence of multiple copies of IS*1165* in the genome. Analysis of further strains of this species revealed this IS element to be widely distributed and IS*1165* or closely related elements were detected in *Leuconostoc lactis*, *Leuconostoc oenos*, *Pediococcus* sp., *Lactobacillus helveticus* and *Lactobacillus casei* but not in *Lactococcus* (Johansen and Kibenich, 1992).

The presence of multiple copies of IS elements within the genome has also been demonstrated for IS*904*, IS*905*, IS*981* and iso-IS*S1* elements although in the latter case fewer chromosomal copies are normally present. The value of IS element probes as a means of strain identification has also been established (Gasson, 1990; Schäfer *et al.*, 1991). The characteristic features of these various IS elements such as their overall size, inverted repeats, target site duplications and transposase genes are summarized in Table 1.2.

In addition to the *in vivo* processes that are controlled by IS element activity, insertion sequences have also been exploited as genetic tools, primarily to effect integration into the lactococcal chromosome. This was first demonstrated using the iso-IS*S1* element IS*946*, which was cloned on a suicide vector that could not replicate in lactococci. Transformation of this construct into lactococcal strains, including one deficient in generalized recombination, led to random insertion of the plasmid into the chromosome and in one case a resident plasmid (Romero and Klaenhammer, 1992). Similar observations were made using IS*981* (Polzin and McKay, 1992). An iso-IS*S1* element was also exploited to promote random chromosomal integration of a vector designed to aid physical mapping of the genome (Le Bourgeois *et al.*, 1992a) and the use of this latter plasmid is described in section 1.8.

Romero and Klaenhammer (1991) described the construction of artificial composite transposons based on two inversely repeated copies of IS*946* flanking a chloramphenicol resistance gene (Tn-CmA) and an erythromycin resistance gene encoded by vector pSA3 (Tn-EmA). An assay for *in vivo* transposition was based on mobilization of the antibiotic resistance markers by the sex factor plasmid pRS01. Rec-independent transposition of artificial transposon Tn-EmA onto pRS01 was shown to depend on the presence of one functional copy of IS*946* and the terminal inverted repeats of both elements. Sequence analysis of the junctions between pRS01 DNA and Tn-EmA revealed the presence of new 8 bp direct repeats thus proving that a genuine transposition event had taken place.

1.6.2 Heterologous transposons

Amongst the genetic tools developed for Gram-positive bacteria, transposons of the Tn*916*/Tn*919* family and Tn*917* have been exploited in the lactic acid bacteria. Transposons Tn*916* and Tn*919* are similar conjugative elements that were originally isolated in *Enterococcus faecalis* DS16 (Franke and Clewell, 1981) and *Streptococcus sanguis* FC1 (Fitzgerald and Clewell, 1985), respectively. It is well established that these elements promote their own transfer to a wide variety of different Gram-positive strains and in many cases they integrate randomly into the host strain's genome. Hill *et al.* (1985) first demonstrated that Tn*919* could be transferred into lactic acid bacteria from *E. faecalis*. Frequencies of transfer varied between 4×10^{-8} and 5×10^{-5} per recipient and the transposon was successfully introduced into various lactococci, *Leuconostoc cremoris* and *Lactobacillus plantarum*. The main value of transposons in genetic analysis is their potential to integrate at random causing mutations by insertional inactivation and thereby physically marking genes of interest. Disappointingly, analysis of independently isolated Tn*919* insertions into the chromosome of the genetically well characterized strain MG1363 suggested that a common site was used and that this represented a hot-spot for Tn*919* integration. Whilst this at first appeared to prevent the use of Tn*919* for transposon mutagenesis in this host, more recent work in which Tn*916* was delivered from *E. faecalis* or *B. subtilis* suggests that more random integration in strain MG1363 can in fact take place (Bringel *et al.*, 1991, 1992). Another limitation to the successful exploitation of transposon mutagenesis is the need for efficient conjugal transfer. The frequency of transposition events is relatively low, and it is essential that their delivery into a target strain takes place at a significantly higher frequency. The difference between this delivery frequency and the frequency of transposition dictates the number of transposon insertion events that can actually be recovered. Hill *et al.* (1987) used the high frequency conjugation system that has been developed in *L. lactis* subsp. *lactis* 712 to increase the efficiency of transposon delivery in *L. lactis*. A donor strain that carried both Tn*919* and a cointegrate lactose plasmid that was capable of very high frequency conjugation was used and the transposon was delivered more efficiently to other *L. lactis* subsp. *lactis* strains. In the case of *L. lactis* subsp. *lactis* biovar *diacetylactis* 18–16, random integration of Tn*919* was observed and the generation of mutations was achieved by insertional inactivation of genes for maltose catabolism and citrate utilization (Hill *et al.*, 1991). In a separate study Renault and Heslot (1987) demonstrated the introduction of Tn*916* from *Enterococcus faecalis* into *L. lactis* and isolated transposon mutants in genes for the malolactic fermentation.

Transposon Tn*916* has been the subject of detailed molecular analysis and there is a very good understanding of its mode of transposition. Recent work using *L. lactis* subsp *lactis* MG1363 has generated interesting data that are relevant to the exploitation of this transposon. Bringel *et al.* (1991) showed that both Tn*916* and Tn*919* retain the same chromosomal location following their transfer between strains of the *L. lactis* C2/712 family. Furthermore this transfer still took place when a mutant transposon Tn*916-int*1 was used. This element has a defect in the integrase gene and the fact that its transfer still occurred suggested that in strains such as MG1363, Tn*916* is unable to excise causing it to retain its original chromosomal location. Under these circumstances the transposon is transferred passively as a component of a distinct conjugation process involving mobilization of the whole lactococcal chromosome. In MG1363 this appears to be promoted by the chromosomally integrated sex factor that is also involved in cointegrate formation with the lactose plasmid (Bringel *et al.*, 1992; Gasson *et al.*, 1992). In matings using *L. lactis* as donor, no transfer into *B. subtilis* or *E. faecalis* was detected and this was concluded to reflect the narrow host range of the lactococcal chromosome transfer system. Analysis of several derivatives of MG1363 carrying Tn*916* at separate chromosomal locations has recently revealed different frequencies of conjugal transfer with the highest frequencies in strains with Tn*916* and the sex factor physically close to each other.

Tn*917* is another transposon that has been widely exploited as a genetic tool in Gram-positive bacteria. Youngman and his colleagues have developed sophisticated derivatives of Tn*917* that combine its potential for gene disruption with the expression of various reporter genes under the control of host promoters that flank the site of transposon integration (Youngman, 1987). These elegant genetic systems have been used to study sporulation genes in *Bacillus* (Youngman *et al.*, 1985; Camilli *et al.*, 1990) and equally they have great potential for use in the lactic acid bacteria. Direct transformation of *L. lactis* and

Lactobacillus plantarum with a Tn917 carrying plasmid that appeared unable to replicate in these strains, led to the transposon's integration at different chromosomal sites (Sanders and Nicholson, 1987; Aukrust and Ness, 1988). More recently Israelson and Hansen (1993) have shown that the replication of the pE194ts replicon that is used in Tn917 delivery vectors can function in *L. lactis*. This depended on the antibiotic resistance gene that was used for selection, chloramphenicol resistance allowing plasmid maintenance whereas erythromycin resistance led to plasmid loss and selection for chromosomal integration. This observation was exploited to force transposition in *L. lactis* and relatively random integration was observed as well as the expression of a promoterless *E. coli* β-galactosidase gene that is carried by the Tn917 variant LTV1. Although the replicon used in this work was temperature sensitive in *Bacillus* the maximum growth temperature for *L. lactis* was insufficient to inhibit plasmid maintenance. The exploitation of antibiotic selection as described above circumvented this problem but it was necessary to use replica plating of primary transformants as a method to prevent the isolation of a single dominant transposition derivative.

The isolation of a temperature sensitive mutation in the replication region of a lactococcal vector is a major advance that should facilitate the use of transposons as well as chromosomal integration promoted by homologous DNA. The pG+host series of vectors was developed by Maguin *et al.* (1992) using the classic broad host range lactococcal vector pGK12. The temperature sensitive derivative has a maximum growth temperature of 35°C, above which plasmid replication is prevented. The major advantage of such a temperature sensitive vector is that even poorly transformable strains can receive the plasmid and once introduced a large population of cells carrying DNA destined for transposition or recombinational integration is available thereby facilitating good frequencies of the sought after event. The use of this vector to promote chromosomal integration of homologous DNA is described in section 1.7.1.

1.6.3 Nisin transposons

The ability of certain strains of *Lactococcus lactis* to ferment sucrose and produce the antimicrobial peptide nisin is a linked genetic trait that can be cured and transferred by conjugation (Gasson, 1984; Gonzales and Kunka, 1985; Steele and McKay, 1986; Tsai and Sandine, 1987). Despite this and the extensive analysis of plasmid complements, physical characterization of the genetic element that encodes nisin and sucrose genes was only achieved after the genes for nisin biosynthesis were cloned and used as DNA probes. Dodd *et al.* (1990) produced the first data suggesting that these genes were chromosomally encoded and not carried by a plasmid as had been speculated in the literature for over a decade. The most significant observation was that a chromosomal restriction fragment from a plasmid free recipient strain was disrupted after the conjugal transfer of sucrose and nisin genes and this led to the suggestion that a conjugative transposon was involved. Furthermore, the presence of a copy of the insertion sequence IS904 on the introduced DNA close to its junction with chromosomal DNA indicated that a composite transposon might exist. This was subsequently disproved when the novel conjugative transposon Tn5301 was discovered and characterized (Horn *et al.*, 1991). Pulsed-field gel electrophoresis of total chromosomal DNA that had been digested with restriction endonuclease *Sma*I proved that the recipient chromosome was enlarged by 70 kb following conjugal transfer of sucrose and nisin genes. Analysis of 10 independent transconjugants that resulted from matings with eight different nisin producing donor strains showed that there were just two preferred target sites for the orientation-specific integration of the transposon. In two cases an identical macrogenomic rearrangement appeared to accompany introduction of the transposon. Sequence analysis of one chromosomal target before and after transposition revealed duplication of a six base pair sequence 5'-TTTTTG-3' that flanked the introduced transposon. The ends of the introduced transposon did not exhibit inverted repeats, a feature that is charac-teristic of some other conjugative transposons. These features of the nisin transposon Tn5301 originating in *L. lactis* subsp. *lactis* NCFB 894 were confirmed by Rauch and De Vos (1992) who independently characterized Tn5276 that is present in *L. lactis* subsp. *lactis* NIZO R5. Similarly, Thompson *et al.* (1991) provided data on chromosomally located transposons Tn5306 and Tn5307 in *L. lactis* subsp. *lactis* strains K1 and ATCC 11454 respectively and Gireesh *et al.* (1992) described a 68 kb conjugative transposon in *L. lactis* subsp. *lactis* DL11.

The analysis of Tn5276 revealed that recently isolated transconjugants often harboured multiple copies of the transposon. There was, however, clear evidence for an orientation specific hot spot for integration. In this study the chromosomal sequences flanking Tn5276 were determined in the donor strain *L. lactis* subsp. *lactis* NIZO R5 as well as in a transconjugant derived from strain MG1614. A comparison of this target region for Tn5276 in the distinct strains showed that 19 of 25 bases were identical adding to the evidence for sequence specific integration of the transposon. As with Tn5301, a 5'-TTTTTG-3' hexanucleotide sequence was present as a direct repeat flanking both junctions of Tn5276 in the donor strain and a transconjugant. These studies do not however establish the mechanism by which this repeat sequence is created. In many transposition processes the transposition mechanism causes the duplication of a pre-existing target sequence which subsequently flanks the newly inserted transposon as a direct repeat and the size of this sequence is characteristic for each transposable genetic element. Less commonly the repeated sequence is physically introduced as part of the incoming transposon and the transposition process involves an integration/excision system. This mechanism is characteristic of the Gram-positive transposons Tn554, Tn916 and Tn1545 (Murphy, 1989). Recent studies of Tn5276 revealed the presence of open reading frames showing homology with known *int* and *xis* genes and the formation of a putative circular transposition intermediate which carried a single copy of the 5'-TTTTTG-3' sequence (Rauch and De Vos, unpublished data). These observations strongly suggest that the nisin transposon transposes by this latter mechanism and that one of the directly repeated 5'-TTTTG-3' sequences is present on and physically introduced with the element.

The large 70 kb size of the transposon reflects the presence of a complex operon for the biosynthesis of nisin, as well as the genes necessary for sucrose fermentation, conjugation and transposition. It has also been shown that nisin transposons encode genes for a bacteriophage insensitivity phenomenon (Gonzalez and Kunka, 1985) and for production of the enzyme N5-(carboxyethyl) ornithine synthase (Donkershoot and Thompson, 1990).

1.7 Generalized recombination

1.7.1 Integration processes

The transfer of chromosomal genes by *in vivo* processes generally involves their integration into a recipient strain chromosome by recombination between homologous regions of DNA. Whilst plasmid vectors are widely used for genetic engineering in the lactic acid bacteria, the integration of a cloned gene into the bacterial chromosome is of special interest due to the resultant enhanced stability and of course, it is also one approach to developing a food compatible technology. The exploitation of integration during gene cloning is described in chapter 2 and here the study of integration and recombination processes will be highlighted. The experimental investigation of integration usually involves a plasmid carrying a cloned region of DNA homologous to a particular target within the recipient strain chromosome and this plasmid will be chosen or designed so that it cannot replicate in the recipient strain. In the lactic acid bacteria this has been achieved by the use of inherent replicon specificity (Leenhouts *et al.*, 1989), the use of a temperature sensitive mutation in a replication gene (Maguin *et al.*, 1992) or by the development of an engineered replicon which only functions in the presence of a trans-active replication protein that is absent in the recipient (Leenhouts *et al.*, 1991c).

In the most straightforward approach a conventional *E. coli* plasmid vector which does not replicate in lactic acid bacteria is used. The introduction of such a vector will fail to yield transformants unless integration into the recipient genome takes place. If a single cross-over event occurs within a region of homology between the plasmid and the recipient chromosome the entire plasmid integrates and is flanked by copies of the homologous DNA. This Campbell-like integration event, illustrated in Figure 1.3, was first demonstrated in *L. lactis* by Leenhouts *et al.* (1989) using plasmids composed of a pBR322 replicon, a chloramphenicol resistance gene known to be selectable in lactococci and a random 1.3 kb DNA fragment cloned from the lactococcal chromosome.

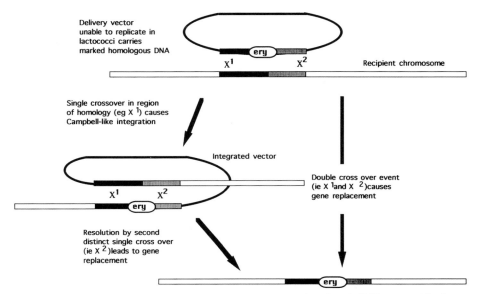

Figure 1.3 Integration into the bacterial chromosome by Campbell-like recombination, replacement recombination and the resolution of Campbell-like integration events into gene replacements are shown.

Although low frequency integration of the vector was detected in the absence of this homologous fragment of chromosomal DNA, the latter stimulated integration up to 100-fold. Southern hybridization was used to confirm that Campbell integration had occurred and to demonstrate that after growth in the presence of chloramphenicol amplification of the chromosomally integrated plasmid took place up to a level of about 15 copies per cell. More recently it has been realised that the chloramphenicol marker used in these experiments is poorly expressed in lactococci and this causes a selection for the amplification of multiple copies of the chloramphenicol resistance gene. Interestingly, Leenhouts *et al.* (1989) observed that the amplified copies of the integrated vector were gradually lost once selection for chloramphenicol resistance was released. Leenhouts *et al.* (1990) investigated the stability of a number of different chromosomally integrated plasmids using an erythromycin resistance gene as the selective marker. The good expression of resistance by a single copy of this gene facilitated the stable maintenance of single copies of some integrated plasmids that were based on the replicons of pSL101 and pTB19. Stable multicopy integration of pBR322 derived vectors was also observed, but this was probably due to the integration of preformed oligomeric plasmid molecules rather than amplification following single copy integration. Integration of a plasmid based on the replicon of pUB110 led to unstable constructs and this was concluded to be due to the residual replicative activity of this staphylococcal replicon in *L. lactis*.

Chopin *et al.* (1989) demonstrated Campbell integration of the staphylo-coccal erythromycin resistance plasmid pE194 into a lysogenic strain of *L. lactis* when fragments of prophage DNA were cloned into this plasmid's unique *Xba*I site. Amplification of the integrated plasmid was also demonstrated, but this required the use of the alternative antibiotic clindamycin which was chosen because it failed to induce high level expression of the erythromycin resistance gene. As with the use of chloramphenicol, this selection forced increased copy numbers of the integrated drug resistance gene by *in situ* amplification. The Southern hybridization analysis of several integrated plasmids revealed that whilst most had arisen by Campbell-like recombination involving a single cross over event, one example of replacement recombination had also been selected. In this case a region of the prophage was deleted and replaced with the plasmid vector. In the experiments of Chopin *et al.* (1989) a ligation mix, rather than intact CCC plasmid DNA, was transformed directly into *L. lactis* providing the opportunity for a linear molecule to be generated with two distinct non-tandem regions of homology. Double cross-over events involving these separate sites would then integrate the vector between these two regions of homology and simultaneously delete the pre-existing recipient chromosomal DNA between them. This replacement recombination event is illustrated in Figure 1.4.

Leenhouts *et al.* (1991b) took the application of replacement recombination a stage further by exploiting the *pep*XP gene that encodes X prolyl dipeptidyl aminopeptidase. The activity of this enzyme in colonies of *L. lactis* is readily detectable in a colorimetric plate assay facilitating the differentiation of strains with and without an active gene. The *pep*XP gene was cloned as a 5.3 kb chromosomal fragment onto an *E. coli* pUC18 derived vector and the cloned gene was inactivated by insertion of an erythromycin resistance gene. When this vector was transformed into a *L. lactis* recipient most transformants arose by

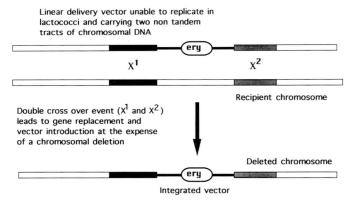

Figure 1.4   The gene replacement event described by Chopin *et al.*, (1989) is illustrated in which direct transformation of a ligation mixture leads to the formation of a linear recombination molecule that promotes a gene replacement event leading to simultaneous deletion of chromosomal DNA.

Campbell-like integration of the whole vector giving rise to erythromycin resistant, Pep XP⁺ colonies. However, in 2% of transformants an erythromycin resistant, Pep XP⁻ phenotype was found and Southern analysis confirmed that these arose by replacement recombination leading to integration of the erythromycin resistance gene within a single copy of the chromosomal *Pep*XP gene. As well as the detection of the double cross-over event following transformation, Leenhouts *et al.* (1991b) used the colorimetric plate assay to demonstrate the resolution of Campbell-like integration events into replacement recombination events by a second distinct single cross-over. The frequency of this resolution was determined by observing the conversion of *Pep*XP⁺ colonies into *Pep*XP⁻ colonies and the recombination events between 1.6 kb nontandem repeats within the chromosome varied between $<2.8 \times 10^{-6}$ and 8.5×10^{-6}. Replacement recombination and the resolution of Campbell-like integration events are illustrated in Figure 1.5.

These systems have already been of considerable value for the chromosomal integration and stabilization of biotechnologically important traits including the lactococcal cell wall bound proteinase (Leenhouts *et al.*, 1991a) the bacteriophage insensitivity gene *abi* (Casey *et al*, 1992) and amylase and endoglucanase genes in silage strains of *Lactobacillus plantarum* (Scheirlinck *et al.*, 1989). A distinct application is the directed inactivation of a chromosomally located gene

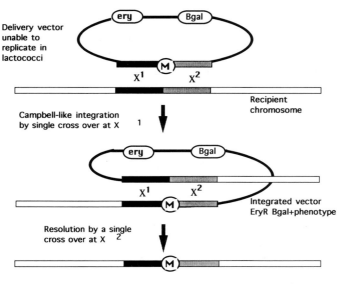

Figure 1.5 The use of an integration vector to introduce a variant gene is illustrated. The selection for a Campbell-like integration event followed by resolution of a gene replacement event that can be detected by loss of the *E. coli* β-galactosidase gene facilitates the introduction of a replacement gene with no other change to the genome.

as demonstrated for *pep*XP. This has been exploited to inactivate the *nis*A gene during the development of a lactococcal expression system for engineered nisins (Dodd *et al.*, 1992) as is described in chapter 5.

The examples of chromosome integration cited so far have relied on the use of a heterologous plasmid vector which is unable to replicate in the recipient species of lactic acid bacteria. An elegant alternative approach, based on the well characterized lactococcal plasmid pWV01, has also been developed. This relies upon the fact that the replication of this plasmid is controlled by a trans-acting replication protein encoded by the gene *rep*A. It is established that deletion of the *rep*A gene can generate a plasmid which retains an origin of replication and can still replicate in the presence of another intact plasmid which supplies the Rep A protein *in trans* (DeVos, 1987; Leenhouts *et al.*, 1990) Leenhouts *et al.* (1991c) developed a strain of *B. subtilis* which expressed a chromosomally integrated *rep*A gene. This host strain supported the replication of a pWV01 derived vector which lacked *rep*A but retained an origin of replication. Similar derivatives of *L. lactis* and *E. coli* have also been developed. In strains of *L. lactis* lacking the *rep*A gene this vector failed to replicate and it could thus be used to direct chromosomal integration of cloned DNA.

This delivery system has been further developed to facilitate the introduction of genetic changes without the retention of any heterologous DNA (Leenhouts and Venema, unpublished). As illustrated in Figure 1.6 the combination on this vector of a drug resistance gene and the *E. coli* β-galactosidase gene expressed under the control of a lactococcal promoter facilitated a two step strategy for the introduction of chromosomal mutations. First integration of the vector was selected using the drug resistance gene. The derivative colonies resulted from Campbell-like integration of the entire vector and thus carried an integrated β-galactosidase gene which caused blue colonies on media containing X-gal. Integration is directed by the homology of a cloned gene which could carry a mutation and following integration of the vector this gene is present in two copies as nontandem repeats. One gene is the original functional version, the other may carry an introduced mutation. Resolution of the Campbell-like integration strain could be detected by searching for white colonies that are no longer drug resistant. Such colonies arose by a second cross-over event between the nontandem repeats leading to excision and loss of the vector sequences together with the drug resistance gene, the β-galactosidase gene and one copy of the duplicated target gene. Depending on the precise location of this second cross-over event, either the original genotype is recreated or the mutant gene replaces the wild-type original. In the latter case no foreign DNA remains and a mutation generated for example by site-specific mutagenesis in a cloned segment of chromosomal DNA can be reintroduced into the chromosome as the sole change.

Another strategy to force integration of plasmid vector DNA is to use a derivative of the pWV01 replicon in which the *rep*A gene carries a temperature sensitive mutation. Such a vector was developed by Maguin *et al.* (1992) and it

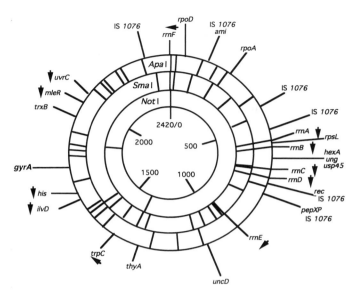

Figure 1.6 The physical and genetic map for *L. lactis* subsp. *lactis* determined by Le Bourgeois *et al.*, 1992b is shown. Numbers represent the largest restriction fragments present in *Not*1, *Sma*1 and *Apa*1 digests of the chromosome. Bars on the pointers indicate the range over which genes are actually located, reflecting the current limits of mapping data. The mapped homologous genes are for biosynthesis of isoleucine and valine (*ilv*), histidine (*his*), tryptophan (*trp*), X-propyl dipeptidyl aminopeptidase (*pepXP*), malolactic enzyme (*mleR*), the highly secreted protein Usp45 (*usp*45) as well as *trxB* and *uvrC*. The heterologous 'housekeeping genes' are for 16s (*rrn*), ribosomal protein (*rpsL*), RNA polymerase (*rpoA*), ATPase (*uncD*), heat shock protein (*groE*), recombination (*rec*), DNA mismatch repair (*hexA* and *ung*), DNA topology (*gyrA*) and oligonucleotide transport (*ami*).

has been exploited in lactic acid bacteria to effect Campbell-like integration and the resolution of replacement recombination events. In this system the vector is introduced by transformation at the permissive temperature and the entire plasmid-carrying cell culture is then grown at a higher temperature that is inhibitory to replication of the vector. Selection for a vector-based drug resistance gene results in the efficient recovery of strains in which the vector has integrated into the chromosome. As described above subsequent excision of the vector can be induced and used to replace a pre-existing gene with an engineered variant or alternatively to add a cloned gene to the genome at a precise site. In this system expression of the plasmid replication function can be used to increase the frequency with which excision occurs, simply by growth of the Campbell-like integration strain at the permissive temperature. The instability of an integrated plasmid caused by rolling-circle replication has also been described by Leenhouts *et al.* (1990b) during a study of pUB110 integration.

As well as directed integration using various plasmid vector systems and cloned genes, several examples of *in vivo* integration of lactococcal plasmids into the chromosome have been reported. Examples include plasmids encoding genes for lactose utilization and proteinase production (Hayes *et al.*, 1990a;

McKay and Baldwin, 1978; Steele *et al.*, 1989) and bacteriocin production (Neve *et al.*, 1984). The integration of plasmid pSKIIL was characterized by Petzel and McKay (1992) who demonstrated that the process depended on generalized recombination and involved a discreet region within a 1.65 kb fragment of the plasmid. The parent lactose plasmid pSKIIL and similar plasmids derived from other *L. lactis* subsp. *cremoris* strains exhibited temperature sensitive replication after they were transformed into heterologous strains of *L. lactis* subsp. *lactis* (Feirtag *et al.*, 1991; Horng *et al.*, 1991) and this property promoted the recovery of strains that were stabilized by plasmid integration following growth at high temperature. Sub-cloning of the pSKIIL region involved in chromosomal integration onto a plasmid vector that did not replicate in lactococci facilitated the chromosomal integration of this vector and Petzel and McKay (1992) suggested that this was another mechanism that could be exploited for the directed integration of cloned genes.

The stabilization of lactose and proteinase genes following transduction of plasmids in *L. lactis* subsp. *lactis* C2 and 712 was described above. In the case of one strain derived from *L. lactis* subsp. *lactis* the extent of the integration region has been mapped and located within the chromosome (Swindell *et al.*, 1993). The site of integration was shown to be preserved following repeated rounds of transduction and conjugation. Whilst this might imply site-specific integration it may equally reflect passive transfer of the integrated DNA and its subsequent integration into a new recipient by generalized recombination between flanking regions of homologous chromosomal DNA. The chromosomally integrated lactose plasmid DNA was shown to be 25 kb in size, considerably less than the head capacity of the transducing bacteriophage ØT712. This leaves space available for the packaging of flanking chromosomal DNA which would facilitate the subsequent integration of lactose genes as if they were a conventional chromosomal marker. Also it was established that conjugal transfer of the integrated lactose genes was dependent on the presence of an unlinked sex factor strongly suggesting that this process depended upon chromosomal transfer analogous to that in an *E. coli* Hfr strain. This possibility is also suggested by the observation of conjugal transfer of integration defective Tn*916* transposons in the same strain, as reported by Bringel *et al.* (1992) and described in section 1.6.2.

1.7.2 Recombination genes

A derivative of *L. lactis* subsp. *lactis* ML3 that was deficient in generalized recombination was isolated by Anderson and McKay (1983) using ethyl methane sulphonate mutagenesis and a screen for methyl methane sulphonate and UV sensitive survivors indicative of a defective SOS repair system. This Rec⁻ strain has been extensively used in studies of *in vivo* DNA rearrangement where it has proved invaluable for the differentiation of generalized recombination from transposition processes. Recombination deficient mutants may also stabilize strains in which homologous DNA is present and they thus have considerable potential for more widespread use. There is a possibility that prophage induction could be prevented by inactivating the recombinogenic part of the multifunctional RecA protein. One approach to the identification of the lactococcal equivalent of this well characterized *E. coli* gene is to attempt to complement an *E. coli rec*A mutant strain with a library of *L. lactis* DNA. Novel *et al.* (1990) succeeded in this by selecting *E. coli* transformants which were resistant to methyl methane sulfonate. The cloned lactococcal gene complemented a variety of other *E. coli rec*A defects and produced a protein that reacted with an antibody to *E. coli* RecA protein, however, no evidence for complementation of the *rec*A recombinase activity was found. Interestingly the cloned gene complemented the SOS repair defect of the *L. lactis* recombination defective mutant isolated by Anderson and McKay (1983) but its effect on generalized recombination was not determined.

Duwat *et al.* (1992) used a different strategy to clone and sequence the lactococcal *rec*A gene. Degenerate oligonucleotide primers were designed from conserved regions of the amino acid sequence of 20 known RecA proteins. The primers were used to amplify an internal fragment of a putative lactococcal *rec*A gene and this was used as a probe to isolate the intact gene from cloned fractionated DNA of *L. lactis* subsp. *lactis* ML3. Sequence analysis of the lactococcal *rec*A gene revealed it to encode a protein of 365 amino acids with a predicted molecular weight of 38 900. This protein reacted with an antibody to the *E. coli* RecA protein. The availability of the *rec*A gene opens new opportunities for the construction of *rec*A phenotypes in lactococci by gene replacement including the possibility of splitting the pleiotropic phenotypes known to be associated with the gene.

1.8 Chromosome mapping

The advent of pulsed field gel electrophoresis (PFGE) has provided the opportunity for physical mapping of bacterial chromosomes in a wide variety of species. This process exploits the capacity of PFGE to separate and size large restriction fragments generated by the cleavage of intact chromosomal DNA with rare cutting restriction endonucleases. Such endonucleases are usually those with recognition sites exclusively made up of GC or AT nucleotides chosen to oppose the bias in the target organism's DNA (i.e. GC sites for AT rich lactococcal DNA). The use of DNA probes in conjunction with PFGE facilitates the assignment of genes to individual restriction fragments which together with the ordering of the restriction fragments, ultimately leads to the generation of a physical and genetic map of the chromosome. In lactic acid bacteria this relatively new technology has mainly been applied to a few strains of *L. lactis*.

The first application of PFGE to lactic acid bacteria was reported by Le Bourgeois *et al.* (1989), who used restriction endonucleases with recognition sites composed exclusively of GC nucleotides to digest the chromosomes of a range of lactococcal strains as well as one strain of *S. thermophilus*. Of the restriction endonucleases used, *Sma*I (GCCGGG) and *Apa*I (GGGCCC) gave distributions of restriction fragments that appeared well suited to genome analysis. Tanskanen *et al.* (1990) also reported *Sma*I digest patterns for 29 different strains of *L. lactis* and highlighted the value of this technique for strain identification. These studies established a genome size for *L. lactis* in the range 2.0–2.7 Mbp.

The first physical map of a lactococcal strain was constructed for *L. lactis* subsp. *lactis* strain DL11, a derivative of the well studied nisin producing strain ATCC11454 (Tulloch *et al.*, 1991). This map was produced by two dimensional PFGE of DNA digested with restriction endonucleases *Not*I and *Sma*I. These enzymes were chosen because respectively they had 6 and 21 well spaced sites on the chromosome of *L. lactis* DL11. Following digestion with one enzyme and PFGE separation, the individual bands were cut from the gel, digested with the second enzyme and again subjected to PFGE. Comparison of the single digests patterns with the double digest patterns generated data that was used to map all of the *Not*I sites and 10 of the *Sma*I sites. To extend the map *Not*I digested DNA was partially digested with *Sma*I and the map was completed by the introduction of a third restriction endonuclease, *Sal*I. Tulloch *et al.* (1991) also used an rRNA gene probe in Southern hybridization experiments with PFGE separated DNA. In this way six regions of the lactococcal chromosome were shown to encode rRNA operons.

The mapping approach used for *L. lactis* DL11 relied on the relatively even distribution of sites for restriction endonucleases *Not*I and *Sma*I. Other lactococcal strains such as MG1363 and IL1503 have been used extensively in genetic analysis and are prime candidates for mapping projects. The distribution of sites in MG1363 is inconvenient with only 3 sites for *Not*I which are close together in one region of the chromosome. Le Bourgeois *et al.* (1992a) have developed an alternative mapping strategy that to some extent overcomes the problem of inconvenient site distribution and also provides a valuable new approach for the precise location of cloned genes within individual PFGE fragments. Their strategy makes use of plasmid vectors which cannot replicate in *L. lactis* and which contain unique cleavage sites for the restriction endonucleases used for map construction, including *Sma*I, *Apa*I and *Not*I. The maintenance of such plasmids in lactococci depends on their integration into lactococcal DNA and this was achieved by provision of an erythromycin resistance marker for selection and either a cloned lactococcal gene or an iso-IS*SI* element. The former directed integration into the homologous gene by the generalized recombination mechanism whereas the latter promoted random integration by transposition. In both cases the site of integration was physically marked by the additional sites for the mapping enzymes *Sma*I, *Apa*I and *Not*I. Analysis of PFGE patterns subsequently facilitated the precise location and orientation of cloned genes within individual chromosomal fragments and the IS-promoted introduction of extra sites provided a novel approach to physical map solution.

Le Bourgeois *et al.* (1992b) have recently published a physical and genetic map of *L. lactis* IL1503 consisting of restriction sites *Not*I, *Sma*I and *Apa*I together with 24 genetic markers. The construction of the physical map involved firstly cloning the *Not*I sites and the use of these cloned fragments as probes against chromosomal DNA previously digested with *Not*I, *Sma*I, and *Apa*I, as well as double digested with *Not*I plus *Sma*I and *Not*I plus *Apa*I. This revealed the relative positions of those *Apa*I and *Sma*I fragments that contained the *Not*I sites. Analysis of 35 recombinant clones in which the iso-IS*SI* containing vector pRL1 was randomly integrated into the chromosome led to the mapping of 65% of the *Sma*I and *Apa*I sites. The remaining unmapped sites were located by a strategy known as indirect end labelling (Canard and Cole, 1989). This allowed restriction fragments to be ordered with respect to a fixed point and involved the use of a vector probe and partial chromosomal digestion of representative recombinant clones each carrying a chromosomally integrated vector. The genetic map was determined using gene probes derived from cloned lactococcal genes and various heterologous 'housekeeping' genes. In some cases chromosomal integration of the vector described above was used to introduce additional sites for the restriction endonucleases *Not*I, *Sma*I and *Apa*I adjacent to a target gene thereby facilitating its precise location and orientation. The current physical and genetic map for *L. lactis* subsp. *lactis* IL403 is reproduced in Figure 1.6 and the precise size of this genome was calculated to be 2420 kb. An equivalent map for *L. lactis* subsp. *lactis* MG1363 is close to completion. Comparison of the available maps for *L. lactis* revealed that whilst there is wide variation in the physical maps the genetic organization appears to be conserved as has previously been found in other species.

1.9 Concluding remarks

Natural gene transfer processes are of significance both in the evolution of lactic acid bacteria and as a tool for the construction of genetically manipulated strains. For some purposes the later approach may have advantages over recombinant DNA methods which require regulatory approval and consumer acceptance. It is well established that lactic acid bacteria exhibit genetic instability which manifests itself in spontaneous loss of industrially relevant plasmids as well as extensive structural instability. The discovery and analysis of transposition processes is beginning to account for at least some of the observed instability. Notable examples are the insertional inactivation of genes by insertion sequences and instances of gene expression by insertion sequence based promoters. Insertion sequences have also been exploited to effect chromosomal integration of DNA particularly in association with novel strategies for chromosome mapping. Such approaches have been vital to the rapid progress towards the construction of physical and genetic maps of *Lactococcus*. As well as transposition significant progress has been made in the integration of DNA by generalized recombination which is a useful means to achieve stabilized clones and one that can have food compatible characteristics. Gene transfer processes are important to gene cloning methodology and the most significant development in recent years is the perfection of electrotransformation as an efficient and generally applicable method for DNA introduction. For the future natural gene transfer will continue to play an important role in the biotechnology of lactic acid bacteria both as non-recombinant approach to strain construction and as a contribution to genetic engineering strategies.

References

Ahn, C., Collins-Thompson, D., Duncan, C and Stiles, M.E. (1992) Mobilization and location of the genetic determinant of chloramphenicol resistance from *Lactobacillus plantarum* ca TC2R. *Plasmid* **27**, 169–176.

Allen, L.K., Sandine, W.E. and Elliker, P.R. (1963) Transduction in *Streptococcus lactis*. *J. Dairy Res.* **30**, 351–357.

Anderson, D.G. and McKay, L.L. (1983) Isolation of a recombination–deficient mutant of *Streptococcus lactis* ML3. *J. Bacteriol.* **155**, 930–932.

Anderson, D.G. and McKay, L.L. (1984) Genetic and physical characterization of recombinant plasmids associated with cell aggregation and high frequency conjugal transfer in *Streptococcus lactis* ML3. *J. Bacteriol.* **156**, 954–962.

Anderson, P.H. and Gasson, M.J. (1985) High copy number plasmid vectors for use in lactic streptococci. *FEMS Microbiol. Lett.* **30**, 193–196.

Arber, W. (1960) Transduction of chromosomal genes and episomes in *Escherichia coli*. *Virology* **11**, 273–288.

Aukrust, T. and Nes, I.F. (1988) Transformation of *Lactobacillus plantarum* with the plasmid pTV1 by electroporation. *FEMS Microbiol. Lett.* **52**, 127–132.

Badii, R., Jones, S. and Warner, P.J. (1989) Spheroplast and electroporation-mediated transformation of *Lactobacillus plantarum*. *Lett. Appl. Microbiol.* **9**, 41–44.

Baumgartner, A., Murphy, M., Daly, C. and Fitzgerald, G.F. (1986) Conjugative co-transfer of lactose and bacteriophage resistance plasmids from *Streptococcus cremoris* UC653. *FEMS Microbiol. Lett.* **35**, 233–237.

Beresford, T.P.J. (1991) A physiological and genetic study of ribosomal RNA synthesis in *Lactococcus lactis* subsp. *lactis*. PhD Thesis, National University of Ireland, University College, Cork.

Boizet, B., Flickinger, J.L. and Chassy, B.M. (1988) Transfection of *Lactobacillus bulgaricus* protoplasts by bacteriophage DNA. *Appl. Environ. Microbiol.* **54**, 3014–3018.

Bringel, F., van Ustine, G.L. and Scott, J.R. (1991) A host factor absent from *Lactococcus lactis* subsp. *lactis* MG1363 is required for conjugative transposition *Molec. Microbiol.* **5**, 2983–2993.

Bringel, F., van Alstine, G.L. and Scott, J.R. (1992) Transfer of Tn*916* between *Lactococcus lactis* subsp. *lactis* strains is nontranspositional: evidence for a chromosomal fertility function in strain MG1363. *J. Bacteriol.* **174**, 5840–5847.

Broadbent, J.R. and Kondo, J.K. (1991) Genetic construction of nisin-producing *Lactococcus lactis* subsp. *cremoris* and analysis of a rapid method for conjugation. *Appl. Environ. Microbiol.* **57**, 517–524.

Camilli, A., Portnoy, D.A. and Youngman, P. (1990) Insertional mutagenesis of *Listeria monocytogenes* with a novel Tn*917* derivative that allows direct cloning of DNA flanking transposon insertions. *J. Bacteriol.* **172**, 3738–3744.

Canard, B. and Cole, S.J. (1989) Genome organisation of the anaerobic pathogen *Clostridium perfringens*. Proc. Natl. Acad. Sci. U.S.A. **86**, 6676–6680.

Casey, J., Daly, C. and Fitzgerald, G.F. (1992) Controlled integration into the *Lactococcus* chromosome of the pCI829-encoded abortive infection gene from *Lactococcus lactis* subsp. *lactis* UC811. *Appl. Environ, Microbiol.* **58**, 3283–3291.

Chang, S. and Cohen, S.N. (1979) High frequency transformation of *Bacillus subtilis* protoplasts by plasmid DNA. *Mol. Gen. Genet.* **168**, 111–115.

Chassy, B.M. and Rokaw. E. (1981) Conjugal transfer of plasmid-associated lactose metabolism in *Lactobacillus casei* subsp. *casei*. in: Molecular Biology, Pathogenesis and Ecology of Bacterial Plasmids, eds. Levy, S., Clowes, R. and Koenig, E., Plenum Press, New York, p 590.

Chassy, B.M. (1987) Prospects for the genetic manipulation of lactobacilli. *FEMS Microbiol. Rev.* **46**, 279–312.

Chassy, B.M. and Flickinger, J.L. (1987) Transformation of *Lactobacillus casei* by electroporation. *FEMS Microbiol. Lett* **44**, 173–177.

Chassy, B.M., Mercenier, A. and Flickinger, J. (1988) Transformation of bacteria by electroporation. *TIBTECH* **6**, 303–309.

Chopin, M-C., Chopin, A., Rouault, A. and Galleron, N. (1989) Insertion and amplification of foreign genes in the *Lactococcus lactis* subsp. *lactis* chromosome. *Appl. Environ. Microbiol.*, **55**, 1769–1774.

Cluzel, P.-J., Chopin, A., Ehrlich, S.D. and Chopin, M.-C. (1991) Phage abortive infection mechanism from *Lactococcus lactis* subsp. *lactis*, expression of which is mediated by an iso-IS*S1* element. *Appl. Environ. Microbiol.* **57**, 3547–3551.

Cocconcelli, P.S., Morelli, L., Vescovo, M. and Bottazzi, V. (1986) Intergeneric protoplast fusion in lactic acid bacteria. *FEMS Microbiol. Lett.* **35**, 211–214.

Connell, H., Lemmon, J. and Tannock, G.W. (1988) Formation and regeneration of protoplasts and spheroplasts of gastrointestinal strains of lactobacilli. *Appl. Environ. Microbiol.* **54**, 1615–1618.

Cosby, M.W., Axelsson, L.T. and Dobrogosz, W.J. (1989) Tn*917* transposition in *Lactobacillus plantarum* using the highly temperature–sensitive plasmid pTV1Ts as a vector. *Plasmid* **22**, 236–243.

Cosby, W.M., Casas, I.A. and Dobrogosz, W.J. (1988) Formation, regeneration, and transfection of *Lactobacillus plantarum* protoplasts. *Appl. Environ. Microbiol.* **54**, 2599–2602.

David, S., Simons, G. and De Vos, W.M. (1989) Plasmid transformation by electroporation of *Leuconostoc paramesenteroides* and its use in molecular cloning. *Appl. Environ. Microbiol.* **55**, 1483–1489.

Davidson, B.E., Powell, I.B. and Hillier, A.J. (1990) Temperate bacteriophages and lysogeny in lactic acid bacteria. *FEMS Microbiol. Rev.* **87**, 79–90.

Davies, F.L. and Gasson, M.J. (1981) Reviews of the progress of dairy science: Genetics of lactic acid bacteria. *J. Dairy Res.* **48**, 363–367.

De Mann, J.C., Rogosa, M. and Sharpe, M.E. (1960) A medium for the cultivation of lactobacilli. *J. Appl. Bacteriol*, **23**, 130–135.

De Vos, W.M. and Davies, F.L. (1984) Plasmid DNA in lactic streptococci: bacteriophage resistance and proteinase plasmids in *Streptococcus cremoris* SK11. In *Third European Congress on Biotechnology*, Verlag Chemie, Weisheim **3** 201–205.

De Vos, W.M. (1987) Gene cloning and expression in lactic streptococci. *FEMS Microbiol. Rev.* **46**, 281–295.

Dodd, H.M., Horn, N. and Gasson, M.J. (1990) Analysis of the genetic determinant for production of the peptide antibiotic nisin. *J. Gen. Microbiol.* **136**, 555–560.

Dodd, H.M., Horn, N., Hao, Z. and Gasson, M.J. (1992) A lactococcal expression system for engineered nisins. *Appl. Environ. Microbiol.* **58**, 3683–3693.

Dodd, H.M., Horn, N. and Gasson M.J. (1993) Characterization of IS*905*, a new multi-copy IS element identified in lactococci. *Molec. Microbiol.* in press.

Donkershoot, J.A. and Thompson, J. (1990) Simultaneous loss of N^5-(carboxyethyl) ornithine synthase, nisin production, and sucrose-fermenting ability by *Lactococcus lactis* K1. *J. Bacteriol.* **172**, 4122–4126.

Dornan, S. and Collins, M.A. (1990) High efficiency electroporation of *Lactococcus lactis* subsp. *lactis* LM0230 with plasmid pGB301. *Lett. Appl. Microbiol.* **11**, 62–64.

Duwat, P., Ehrlich, S.D. and Gruss, A. (1992) Use of degenerate primers for polymerase chain reaction cloning and sequencing of the *Lactococcus lactis* subsp. *lactis recA* gene. *Appl. Environ. Microbiol.* **58**, 2674–2678.

Eftymiou, C. and Hansen, P.A. (1962) An antigenic analysis of *Lactobacillus acidophilus*. *J. Infect. Dis.* **110**, 258–267.

Ehrenfeld, E.E., Kessler, R.E. and Clewell, D.B. (1986) Identification of pheromone-induced surface proteins in *Streptococcus faecalis* and evidence of a role for lipoteichoic acid in formation of mating aggregates. *J. Bacteriol.*, **168**, 6–12.

Feirtag, J.M., Petzel, J.P., Pasalodos, E., Baldwin, K.A. and McKay, L.L. (1991) Thermosensitive plasmid replication, temperature-sensitive host growth, and chromosomal plasmid integration conferred by *Lactococcus lactis* subsp. *cremoris* lactose plasmids in *Lactococcus lactis* subsp. *lactis. Appl. Environ. Microbiol.* **57**, 539–548.

Fitzgerald, G.F. and Clewell D.B. (1985) A conjugative transposon (Tn*919*) in *Streptococcus sanguis. Infect. Immun.*, **47**, 415–420.

Fitzgerald, G.F. and Gasson, M.J. (1988) *In vivo* gene transfer systems and transposons. *Biochimie* **70**, 489–502.

Franke, A. and Clewell, D.B. (1981) Evidence for a chromosome-borne resistance transposon (Tn*916*) in *Streptococcus faecalis* that is capable of 'conjugal' transfer in the absence of a conjugative plasmid. *J. Bacteriol.* **145**, 494–502.

Galli, D., Lottspeich, F., Wirth, R (1990) Sequence analysis of *Enterococcus faecalis* aggregation substance encoded by the sex-pheromone plasmid pAD1. *Mol. Microbiol.* **4**, 895–904.

Gasson, M.J. (1980) Production, regeneration and fusion of protoplasts in lactic streptococci. *FEMS Microbiol. Lett.* **9**, 99–102.

Gasson, M.J. and Davies. F.L. (1980a) Conjugal transfer of the drug resistance plasmid pAMβI in the lactic streptococci. *FEMS Microbiol. Lett.* **7**, 51–53.

Gasson, M.J. and Davies, F.L. (1980b) High frequency conjugation associated with *Streptococcus lactis* donor cell aggregation. *J. Bacteriol.* **143**, 1260–1264.

Gasson, M.J. (1983) Plasmid complements of *Streptococcus lactis* NCDO712 and other lactic streptococci after protoplast–induced curing. *J. Bacteriol.* **154**, 1–9.

Gasson, M.J. (1984) Transfer of sucrose fermenting ability, nisin resistance and nisin production into *Streptococcus lactis* 712. *FEMS Microbiol. Lett.* **21**, 7–10.

Gasson, M.J. and Davies, F.L. (1984) The genetics of dairy lactic acid bacteria. In: *Advances in the Microbiology and Biochemistry of Cheeses and Fermented Milks* (Davies, F.L. and Law, B.A., eds), Elsevier Applied Science Publishers, New York, pp. 99–126.

Gasson, M.J., Hill, S.H.A. and Anderson, P.H. (1987) Molecular genetics of metabolic traits in lactic streptococci. In *Streptococcal Genetics* eds. J. Ferretti and R. Curtiss III. American Society for Microbiology pp. 242–245.

Gasson, M.J. (1990) *In vivo* genetic systems in lactic acid bacteria. *FEMS Microbiol. Rev.* **87**, 43–60.

Gasson, M.J., Maeda, S., Swindell, S. and Dodd, H.M. (1992) Molecular rearrangement of lactose plasmid DNA associated with high frequency and cell aggregation in *Lactococcus lactis* 712. *Mol. Microbiol.* **6**, 3213–3223.

Geis, A. (1982) Transfection of protoplasts of *Streptococcus lactis* subsp. *diacetylactis. FEMS Microbiol. Lett.* **15**, 119–122.

Gennaro, M.L., Kornblum, J. and Novick, R.P. (1987) A site-specific recombination function in *Staphylococcus aureus* plasmids. *J. Bacteriol.* **154**, 1–9.

Gibson, E.M., Chace, N.M., London, S.B. and London. J. (1979) Transfer of plasmid mediated antibiotic resistance from streptococci to lactobacilli. *J. Bacteriol.* **137**, 614–617.

Gireesh, T., Davidson, B. and Hillier, A.J. (1992) Conjugal transfer in *Lactococcus lactis*, of a 68-kilobase-pair chromosomal fragment containing the structural gene for the peptide bacteriocin nisin. *Appl. Environ. Microbiol.* **58**, 1670–1676.

Gonzalez, C.J. and Kunka, B.S. (1983) Plasmid transfer in *Pediococcus* sp.: intergenic and intragenic transfer of pIP501. *Appl. Environ. Microbiol.* **46**, 81–89.

Gonzalez, C.F. and Kunka, B.S. (1985) Transfer of sucrose fermenting ability and nisin production phenotype among lactic streptococci. *Appl. Environ. Microbiol.* **49**, 627–633.

Gruss, A. and Ehrlich, S.D. (1989) The family of highly interrelated single-stranded deoxyribonucleic acid plasmids. *Microbiol. Rev.* **53**, 231–241.

Haandrikman, A.J., van Leeuwen, C., Kok, J., Vos, P., De Vos, W.M. and Venema, G. (1990) Insertion elements on lactococcal proteinase plasmids. *Appl. Environ. Microbiol.* **56**, 1890–1896.

Harlander, S.K. (1987) Transformation of *Streptococcus lactis* by electroporaton. In: *Streptococcal Genetics* (ed. J. Ferretti and R. Curtiss), American Society for Microbiology, Washington, D.C., USA. pp 229–233.

Harmon, K.S. and McKay, L.L. (1987) Restriction enzyme analysis of lactose and bacteriocin plasmids from *Streptococcus lactis* subsp. *diacetylactis* WM4 and cloning of BcLI fragments coding from bacteriocin production. *Appl. Environ. Microbiol.* **53**, 1171–1174.

Hashiba, H., Takiguchi, R., Iskii, S. and Aoyama, K. (1990) Transformation of *Lactobacillus helveticus* subsp. *jugurti* with plasmid pLHR by electroporation. *Agric. Biol. Chem.* **54**, 1537–1541.

Hayes, F., Caplice, E., McSweeney, A., Fitzgerald, G.F. and Daly, C. (1990a) pAMβ1-associated mobilization of proteinase plasmids from *Lactococcus lactis* subsp. *lactis* UC317 and *L. lactis* subsp. *cremoris* UC205. *Appl. Environ. Microbiol.* **56**, 195–201.

Hayes, F., Daly, C. and Fitzgerald, G.F. (1990b) High frequency, site specific recombination between lactococcal and pAMβ1 plasmid DNAs. *J. Bacteriol.* **172**, 3485–3489.

Hayes, J., Law, J., Daly, C. and Fitzgerald, G.F. (1992) Integration and excision of plasmid DNA in *Lactococcus lactis* subsp. *lactis*. *Plasmid* **24**, 81–89.

Higgins, D.L., Sanozky-Dawes, R.B. and Klaenhammer, T.R. (1988) Restriction and modification activities from *Streptococcus lactis* ME2 are encoded by a self–transmissible plasmid pTN20, that forms cointegrates during modification of lactose fermenting ability. *J. Bacteriol.* **170**, 3435–3442.

Hill, C., Daly, C. and Fitzgerald, G.F. (1985) Conjugative transfer of the transposon Tn*919* to lactic acid bacteria. *FEMS Microbiol. Lett.* **30**, 115–119.

Hill, C., Daly, C. and Fitzgerald, G.F. (1987) Development of high-frequency delivery system for transposon Tn*919* in lactic streptococci: random insertion in *Streptococcus lactis* subsp. *diacetylactis* 18-16. *Appl. Environ. Microbiol.* **53**, 74–78.

Hill, C., Daly, C. and Fitzgerald, G.F. (1991) Isolation of chromosomal mutation of *Lactococcus lactis* subsp. *lactis* biovar. *diacetylactis* 18-16 after introduction of Tn*919*. *FEMS Microbiol. Lett.* 135–140.

Holo, H. and Nes, I.F. (1989) High–frequency transformation, by electroporation, of *Lactococcus lactis* subsp. *cremoris* grown with glycine in osmotically stabilized media. *Appl. Environ. Microbiol.* **55**, 3119–3123.

Horn, N., Swindell, S., Dodd, H.M. and Gasson, M.J. (1991) Nisin biosynthesis genes are encoded by a novel conjugative transposon. *Mol. Gen. Genet.* **228**, 129–135.

Horng, J.S., Polzin, K.M. and McKay, L.L. (1991) Replication and temperature-sensitive maintenance functions of lactose plasmid pSK11L from *Lactococcus lactis* subsp. *cremoris*. *J. Bacteriol.* **173**, 7573–7581.

Huang, D.C., Novel, M. and Novel, G. (1991) A transposon-like element on the lactose plasmid of *Lactococcus lactis* subsp. *lactis* 2270. *FEMS Microbiol. Lett.* **77**, 101–106.

Israelson, H. and Hansen, E.B. (1993) Insertion of transposon Tn*917* derivatives into the *Lactococcus lactis* subsp. *lactis* chromosome. *Appl. Environ. Microbiol.* **59**, 21–26.

Iwata, M., Mada, M. and Ishiwa, H. (1986) Protoplast fusion of *Lactobacillus fermentum*. *Appl. Environ. Microbiol.* **52**, 392–393.

Jarvis, A.W. (1988) Conjugal transfer in lactic streptococci of plasmid-encoded insensitivity to prolate and small isometric–headed bacteriophage. *Appl. Environ. Microbiol.* **54**, 777–783.

Johansen, E. and Kibenich, A. (1992) Isolation and characterization of IS1165 an insertion sequence of *Leuconostoc mesenteroids* subsp. *cremoris* and other lactic acid bacteria. *Plasmid* **27**, 200–206.

Kanatani, K., Yoshida, K., Tahara, T., Sakamoto, M. and Oshimura, M. (1990) Intraspecific protoplast fusion of *Lactobacillus plantarum*. *Agric. Biol. Chem.* **54**, 225–227.

Kempler, G.M., Baldwin, K.A., McKay, L.L., Morris, H.A., Halambeck, S. and Thorsen, G. (1979) Use of genetic alterations to improve *Streptococcus lactis* C2 as a potential Cheddar cheese starter. *J. Dairy Sci.* **62**, (Suppl. 1) 42.

Kim, W.J., Ray, B. and Johnson, M.C. (1992) Plasmid transfers by conjugation and electroporation in *Pediococcus acidilactici*. *J. Appl. Bacteriol.* **72**, 201–207.

Klaenhammer, T.R. and McKay, L.L. (1976) Isolation and examination of transducing bacteriophage particles from *Streptococcus lactis* C2. *J. Dairy Sci.* **59**, 396–404.

Kok, J., van Dijl, J.M., van der Vossen, J.M.B.M. and Venema, G. (1985) Cloning and expression of a *Streptococcus cremoris* proteinase in *Bacillus subtilis* and *Streptococcus lactis*. *Appl. Environ. Microbiol.* **50**, 94–101.

Kondo, J.K. and McKay, L.L. (1982a) Transformation of *Streptococcus lactis* protoplasts by plasmid DNA. *Appl. Environ. Microbiol.* **43**, 1213–1215.

Kondo, J.K. and McKay, L.L. (1982b) Mutanolysin for improved lysis and rapid protoplast formation in dairy streptococci. *J. Dairy Sci.* **65**, 1428–1431.

Kondo, J.K. and McKay, L.L. (1984) Plasmid transformation of *Streptococcus lactis* protoplasts: Optimization and use in molecular cloning. *Appl. Environ. Microbiol.* **48**, 252–259.

Lahbib-Mansias, Y., Mata, M. and Ritzenthaler, P. (1988) Molecular taxonomy of *Lactobacillus* phages. *Biochimie* **70**, 429–435.

Lakshmidevi, G., Davidson, B.E. and Hillier, A.J. (1988) Circular permutation of the genome of a temperate bacteriophage from *Streptococcus cremoris* BK5. *Appl. Environ. Microbiol.* **54**, 1039–1045.

Langella, P. and Chopin, A. (1987) Evaluation of conjugative gene transfer systems in lactic acid bacteria. *FEMS Microbiol. Rev.* **46**, 6.

Langella, P. and Chopin, A. (1989) Effect of restriction-modification systems on transfer of foreign DNA into *Lactococcus lactis* subsp. *lactis. FEMS Microbiol. Lett.* **59**, 301–306.

Le Bourgeois, P., Mata, M. and Ritzenthaler, P. (1989) Genome comparison of *Lactococcus* strains by pulsed–field gel electrophoresis. *FEMS Microbiol. Lett.* **59**, 65–70.

Le Bourgeois, P., Lautier, M., Mata, M. and Ritzenthaler, P. (1992a) New tools for the physical and genetic mapping of *Lactococcus* strains. *Gene* **111**, 109–114.

Le Bourgeois, P., Lautier, M., Mata, M. and Ritzenthaler, P. (1992b) Physical and genetic map of the chromosome of *Lactococcus lactis* subsp. *lactis* IL1403. *J. Bacteriol.* **174**, 6752–6762.

Leenhouts, K.J., Kok, J. and Venema, G. (1989) Campbell-like integration of heterologous plasmid DNA into the chromosome of *Lactococcus lactis* subsp. *lactis. Appl. Environ. Microbiol.* **55**, 394–400.

Leenhouts, K.J., Kok J., and Venema, G. (1990) Stability of integrated plasmids in the chromosome of *Lactococcus lactis. Appl. Environ. Microbiol.* **56**, 2726–2735.

Leenhouts, K.J., Gietema, J., Kok, J. and Venema, G. (1991a) Chromosomal stabilization of the proteinase genes in *Lactococcus lactis. Appl. Environ. Microbiol.* **57**, 2568–2575.

Leenhouts, K.J., Kok, J., and Venema, G. (1991b) Replacement recombination in *Lactococcus lactis. J. Bacteriol.* **173**, 4769–4798.

Leenhouts, K.J., Kok, J. and Venema, G. (1991c) Lactococcal plasmid pWV01 as an integration vector for lactococci. *Appl. Environ. Microbiol.* **57**, 2562–2567.

Lee-Wickner, L.-J. and Chassy, B.M. (1984) Production and regeneration of *Lactobacillus casei* protoplasts. *Appl. Environ. Microbiol.* **48**, 994–1000.

Lin, J.H.-C. and Savage, D.C. (1986) Genetic transformation of rifampicin resistance in *Lactobacillus acidophilus. J. Gen. Microbiol.* **132**, 2107–2111.

Loureiro dos Santos, A.L. and Chopin, A. (1987) Shotgun cloning in *Streptococcus lactis. FEMS Microbiol. Lett.* **42**, 209–212.

Luchansky, J.B., Kleeman, E.G., Raya, R.R. and Klaenhammer, T.R. (1989) Genetic transfer systems for delivery of plasmid DNA to *Lactobacillus acidophilus* ADH: conjugation, electroporation and transduction. *J. Dairy Sci.* **72**, 1408–1417.

Luchansky, J.B., Muriana, P.M. and Klaenhammer, T.R. (1988) Application of electroporation for transfer of plasmid DNA into *Lactobacillus, Leuconostoc, Listeria, Pediococcus, Bacillus, Staphylococcus, Enterococcus* and *Propionibacterium. Mol. Microbiol.* **2**, 637–646.

Maeda, S. and Gasson, M.J. (1986) Cloning, expression and location of the *Streptococcus lactis* gene for phospho ß-D-galactosidase. *J. Gen. Microbiol.* **132**, 331–340.

Maguin, E., Duwat, P., Hege, T., Ehrlich, D. and Gruss, A. (1992) New thermosensitive plasmid for gram-positive bacteria. *J. Bacteriol.* **174**, 5633–5638.

McCarthy, D.M., Lin, J.H.-C., Rinckel, L.A. and Savage, D.C. (1988) Genetic transformation in *Lactobacillus* sp. strain 100-33 of the capacity to colonize the non-secreting gastric epithelium in mice. *Appl. Environ. Microbiol.* **54**, 416–422.

McIntyre, D.A. and Harlander, S.K. (1989a) Genetic transformation of intact *Lactococcus lactis* supsp. *lactis* by high voltage electroporation. *Appl. Environ. Microbiol.* **55**, 604–610.

McIntyre, D.A. and Harlander, S.K. (1989b) Improved electroporation efficiency of intact *Lactococcus lactis* subsp. *lactis* cells grown in defined media. *Appl. Environ. Microbiol.* **55**, 2621–2626.

McKay, L.L., Cords, B.R. and Baldwin, K.A. (1973) Transduction of lactose metabolism in *Streptococcus lactis* C2. *J. Bacteriol.* **115**, 810–815.

McKay, L.L. and Baldwin, K.A. (1974) Simultaneous loss of proteinase- and lactose-utilizing enzyme activities in *Streptococcus lactis* and reversal of loss by transduction. *Appl. Environ. Microbiol.* **28**, 342–346.

McKay, L.L. and Baldwin, K.A. (1978) Stabilization of lactose metabolism in *Streptococcus lactis* C2. *Appl. Environ. Microbiol.* **36**, 360–367.

McKay, L.L., Baldwin, K.A. and Efstathiou, J.D. (1976) Transductional evidence for plasmid linkage of lactose metabolism in *Streptococcus lactis* C2. *Appl. Environ. Microbiol.* **32**, 45–52.

McKay, L.L., Baldwin, K.A. and Walsh, P.M. (1980) Conjugal transfer of genetic information in group N streptococci. *Appl. Environ. Microbiol.* **40**, 84–91.

Mercenier, A. and Chassy, B.M. (1988) Strategies for the development of bacterial transformation systems. *Biochimie* **70**, 503–517.

Mercenier, A. and Lemoine, Y. (1989) Genetics of *Streptococcus thermophilus*: A Review. *J. Dairy Sci.* **72**, 3444–3454.

Mercenier, A., Robert, C., Romero, D.A., Costellino, I., Slos, P. and Lemoine, Y. (1988) Development of an efficient spheroplast transformation procedure for *S. thermophilus*: the use of transfection to define a regeneration medium. *Biochimie* **70**, 567–577.

Mercenier, A., Slos, P., Fallen, M. and Lecocq, J.P. (1988) Plasmid transduction in *Streptococcus thermophilus*. *Mol. Gen. Genet.* **212**, 386–389.

Moller-Madsen, A.A. and Jensen, H. (1962) Transformation of *Streptococcus lactis*. In: *Contributions to the XVIth International Dairy Congress*, Copenhagen, Vol.B 255.

Molskness, T.A., Sandine, W.E. and Elliker, L.R. (1974) Characterization of Lac⁺ transductants of *Streptococcus lactis*. *Appl. Environ. Microbiol.* **28**, 753–758.

Morelli, L., Cocconcelli, P.S., Bottazzi, V., Damiani, G., Ferretti, L. and Sgaramella, V. (1987) *Lactobacillus* protoplast transformation. *Plasmid* **17**, 73–75.

Murphy, E. (1989) Transposable elements in Gram-positive bacteria. In: *Mobile DNA* (eds Berg, D.E., and Howe, M.M.) American Society for Microbiology, Washington, D.C., p.269–288.

Neumann, E., Schaefer-Ridder, M., Wang, Y. and Hofschneider, P. (1982) Gene transfer into mouse myeloma cells by electroporation in high electric fields. *EMBO J.* **1**, 841–845.

Neve, H., Geis, A. and Teuber, M. (1984) Conjugal transfer and characterization of bacteriocin plasmids in group N (lactic acid) streptococci. *J. Bacteriol.* **157**, 833–838.

Novel, M., Huan, X.F. and Novel, G. (1990) Cloning of a chromosomal fragment from *Lactococcus lactis* subsp. *lactis* partially complementing *Escherichia coli recA* functions. *FEMS. Microbiol. Lett.* **72**, 309–314.

Okamoto, T., Fujita, Y. and Irie, R. (1983) Fusion of protoplasts of *Streptococcus lactis*. *Agric. Biol. Chem.* **47**, 2675–2676.

Okamoto, T., Fujita, Y. and Irie, R. (1985) Interspecific protoplast fusion between *Streptococcus cremoris* and *Streptococcus lactis*. *Agric. Biol. Chem.* **49**, 1371–1376.

Otts, D.R. and Day, D.F. (1987) Optimization of protoplast formation and regeneration in *Leuconostoc mesenteroides*. *Appl. Environ. Microbiol.* **53**, 1694–1695.

Petzel, J. and McKay, L.L. (1992) Molecular characterization of the integration of the lactose plasmid from *Lactococcus lactis* subsp. *cremoris* SK11 into the chromosome of *L. lactis* subsp. *lactis*. *Appl. Environ. Microbiol.* **58**, 125–131.

Polzin, K.A. and Shimizu–Kadota, M. (1987) Identification of a new insertion element similar to gram-negative IS26, on the lactose plasmid of *Streptococcus lactis* ML3. *J. Bacteriol.* **169**, 5481–5488.

Polzin, K.M. and McKay, L.L. (1991) Identification, DNA sequence, distribution of IS*981*, a new high-copy-number insertion sequence in lactococci. *Appl. Environ. Microbiol.* **57**, 734–743.

Polzin, K.M. and McKay, L.L. (1992) Development of a lactococcal integration vector by using IS*981* and a temperature-sensitive lactococcal replication region. *Appl. Environ. Microbiol.* **58**, 476–484.

Posno, M., Leer, R.J., van Luijk, N., van Giezen, M.J.F., Hewelmans, P.T.H.M., Lokman, B.C. and Pouwels, P.H. (1991) Incompatibility of *Lactobacillus* vectors with replicons derived from small cryptic *Lactobacillus* plasmids and segregational instability of the introduced vectors. *Appl. Environ. Microbiol.* **57**, 1822–1828.

Powell, I.B., Achen, M.G., Hillier, A. and Davidson, B.E. (1988) A simple and rapid method for genetic transformation of lactic streptococci by electroporation. *Appl. Environ. Microbiol.* **54**, 655–660.

Pucci, M.J., Monteschio, M.E. and Vedamutha, E.R. (1987) Conjugal transfer in *Leuconostoc* spp: intergeneric and intrageneric transfer of plasmid-encoded antibiotic resistance determinants. *FEMS Microbiol. Lett.* **46**, 7.

Rauch, P.J.G. and De Vos, W.M. (1992) Characterization of the novel nisin-sucrose conjugative transposon Tn*5276* and its insertion in *Lactococcus lactis*. *J. Bacteriol.* **174**, 1280–1287.

Raya, R.R. and Klaenhammer, T.R. (1992) High-frequency plasmid transduction by *Lactobacillus gasseri* bacteriophage øadh. *Appl. Environ. Microbiol.* **58**, 187–193.

Raya, R.R., Kleeman, E.G., Luchansky, J.B. and Klaenhammer, T.R. (1989) Characterization of the temperate bacteriophage φadh and plasmid transduction in *Lactobacillus acidophilus* ADH. *Appl. Environ. Microbiol.* **55**, 2206–2213.

Renault, P.P. and Heslot, H. (1987) Selection of *Streptococcus lactis* mutants defective in malolactic fermentation. *Appl. Environ. Microbiol.* **53**, 320–324.

Reniero, R., Cocconcelli, P., Bottazzi V. and Morelli, L. (1992) High frequency of conjugation in *Lactobacillus* mediated by an aggregation-promoting factor. *J. Gen. Microbiol.* **138**, 763–768.

Romero, A., Slos, P., Castellina, R.L., Castelinna, I. and Mercenier, A. (1987) Conjugative mobilization as an alternative vector delivery system for lactic streptococci. *Appl. Environ. Microbiol.* **53**, 2405–2413.

Romero, D.A. and Klaenhammer, T.R. (1990) Characterization of insertion sequence IS*946*, an iso-IS*SI* element, isolated from the conjugative lactococcal plasmid pTR2030. *J. Bacteriol.* **172**, 4151–4160.

Romero, D.A. and Klaenhammer, T.R. (1991) Construction of an IS*946*-based composite transposon in *Lactococcus lactis* subsp. *lactis*. *J. Bacteriol.* **173**, 7599–7606.

Romero, D.A. and Klaenhammer, T.R. (1992) IS*946*-mediated integration of heterologous DNA into the genome of *Lactococcus lactis* subsp. *lactis*. *Appl. Environ. Microbiol.* **58**, 699–702.

Salama, M., Sandine, W.E. and Giovannoni, S. (1991) Development and application of oligonu-cleotide probes for identification of *Lactococcus lactis* subsp. *cremoris*. *Appl. Environ. Microbiol.* **57**, 1313–1318.

Sanders, M.E. and Nicholson, M.A. (1987) A method for genetic transformation of non-protoplasted *Streptococcus lactis*. *Appl. Environ. Microbiol.* **53**, 1730–1736.

Sandine, W.E., Elliker, P.R., Allen, L.K. and Brown, W.C. (1962) Genetic exchange and variability in lactic streptococcus starter organisms. *J. Dairy Sci.* **45**, 1266–1271.

Schäfer, A., Jahns, A., Geis, A. and Teuber, M. (1991) Distribution of the IS elements IS*SI* and IS*904* in lactococci. *FEMS Microbiol. Lett.* **80**, 311–318.

Scheirlinck, T., Mahillon, J., Joos, H., Dhaese, P. and Michiels, F. (1989) Integration and expression of αamylase and endoglucanase genes in the *Lactobacillus plantarum* chromosome. *Appl. Environ. Microbiol.* **55**, 2130–2137.

Scherwitz, K.M., Baldwin, K.A. and McKay, L.L. (1983) Plasmid linkage of a bacteriocin-like substance in *Streptococcus lactis*. subsp. *diacetylactis* strain WM4: transferability to *Streptococcus lactis*. *Appl. Environ. Microbiol.* **45**, 1506–1512.

Sechaud, L., Cluzel, P.-J., Rousseau, M., Baumgartner, A. and Accolas, J.-P. (1988) Bacteriophages of lactobacilli. *Biochimie* **70**, 401–410.

Shimizu-Kadota, M. and Kudo, S. (1984) Liposome-mediated transfection of *Lactobacillus casei* spheroplasts. *Agric. Biol. Chem.* **48**, 1105–1107.

Shimizu-Kadota, M., Kirki, M., Hirokawa, H. and Tsuchida, N., (1985) ISLI: a new transposable element in *Lactobacillus casei*. *Mol. Gen. Genet.* **200**, 193–198.

Shimuzu-Kadota, M., Flickinger, J.L. and Chassy, B.M. (1988) Evidence that *Lactobacillus casei* insertion element ISLI has a narrow host range. *J. Bacteriol.* **170**, 4976–4978.

Shivarova, N., Forster, W., Jacob, H.-E. and Grigorova, R. (1983) Microbiological implications of electric field effects. VII. Stimulation of plasmid transformation of *Bacillus cereus* protoplasts by electric field pulses. *Z. Allg. Mikrobiol.* **23**, 595–599.

Shrago, A.W., Chassy, B.M. and Dobrgosym, W.J. (1986) Conjugal plasmid transfer (pAMβ1) in *Lactobacillus plantarum*. *Appl. Environ. Microbiol.* **52**, 574–576.

Simon, D., Rouault, A. and Chopin, M.-C. (1985) Protoplast transformation of group N streptococci with cryptic plasmids. *FEMS Microbiol. Lett.* **26**, 239–241.

Simon, D., Rouault, A. and Chopin, M.-C. (1986) High-efficiency transformation of *Streptococcus lactis* protoplasts by plasmid DNA. *Appl. Environ. Microbiol.* **52**, 394–395.

Smith, M.D. and Clewell, D.B. (1984) Return of *Streptococcus faecalis* DNA cloned in *Escherichia coli* to its original host via transformation of *Streptococcus sanguis* followed by conjugative mobilization. *J. Bacteriol.* **160**, 1109–1114.

Snook, R.J., McKay, L.L. and Ahlstrand, G.G. (1981) Transduction of lactose metabolism by *Streptococcus cremoris* C3 temperate phage. *Appl. Environ. Microbiol.* **42**, 897–903.

Somkuti, G.A. and Steinberg, D.H. (1988) Genetic transformation of *Streptococcus thermophilus* by electroporation. *Biochimie* **70**, 579–585.

Steele, J.L. and McKay, L.L. (1986) Partial characterization of the genetic basis for sucrose metabolism and nisin production in *Streptococcus lactis*. *Appl. Environ. Microbiol.* **51**, 57–64.

Steele, J.L. and McKay, L.L. (1989) Conjugal transfer of genetic material by *Lactococcus lactis* subsp. *lactis* 110067. *Plasmid* **22**, 32–43.

Steele, J.L., Polzin, K.M. and McKay L.L. (1989) Characterization of the genetic element coding for lactose metabolism in *Lactococcus lactis* subsp. *lactis* KP3. *Plasmid* **22**, 44–51.

Swindell, S., Underwood, H.M. and Gasson, M.J. (1993) Construction and analysis of a plasmid-free *L. lactis* subsp. *lactis* strain carrying chromosomally integrated lactose genes. *J. Gen. Microbiol.* (in press)

Tanaskanen, E.L., Tulloch, D.L., Hillier, A.J. and Davidson, B.E. (1990) Pulsed-field gel electrophoresis of *Sma*I digests of lactococcal genomic DNA, a novel method of strain identification. *Appl. Environ. Microbiol.* **56**, 3105–3111.

Terzaghi, B.E. and Sandine, W.E. (1975) Improved medium for lactic streptococci and their bacteriophages. *Appl. Microbiol.* **29**, 807–813.

Thompson, J., Nguyen, N.G., Sackett, D.L. and Donkershoot, J.A. (1991) Transposon-encoded sucrose metabolism in *Lactococcus lactis. J. Biol. Chem.* **266**, 14573–14579.

Tohyama, K., Sakurai, T. and Arai, H. (1971) Transduction by temperate phage PLS-1 in *Lactobacillus salivarius. Jpn. J. Bacteriol.* **26**, 482–487.

Trautwetter, A., Ritzenthaler, P., Alatossava, T. and Mata-Gilsinger, M. (1986) Physical and genetic characterization of the genome of *Lactobacillus lactis* bacteriophage LL-H. *J. Virol.* **59**, 551–555.

Trieu-Cout, P., Carlier, C., Martin, P. and Courvalin, P. (1987) Plasmid transfer by conjugation from *Escherichia coli* to gram-positive bacteria. *FEMS Microbiol. Lett.* **48**, 289–294.

Tsai, H.J. and Sandine, W.E. (1987) Conjugal transfer of nisin plasmid genes from *Streptococcus lactis* 7962 to *Leuconostoc dextranicum* 181. *Appl. Environ. Microbiol.* **53**, 352–357.

Tulloch, D.L., Finch, L.R., Hillier, A.J. and Davidson, B.E. (1991) Physical map of the chromosome of *Lactococcus lactis* subsp. *lactis* DL11 and localization of six putative rRNA operons. *J. Bacteriol.* **173**, 2768–2775.

Van der Lelie, D., van der Vossen, J.M.B.M. and Venema, G. (1988) Effect of plasmid incompatibility on DNA transfer to *Streptococcus cremoris. Appl. Environ. Microbiol.* **54**, 865–871.

Van der Lelie, D., Wörten, H.A.B., Bron, S., Oskam, L. and Venema, G. (1990) Conjugal mobilization of streptococcal plasmid pMV158 between strains of *Lactococcus lactis* subsp. *lactis. J. Bacteriol.* **172**, 47–52.

Van der Lelie, D., Chavarri, F., Venema, G. and Gasson, M.J. (1991) Identification of a new genetic determinant for cell aggregation associated with lactose plasmid transfer in *Lactococcus lactis. Appl. Environ. Microbiol.* **57**, 201–206.

Van der Vossen, J.M.B.M., Kok, J., Van der Lelie, D. and Venema, G. (1988) Liposome-enhanced transformation of *Streptococcus lactis* and plasmid transfer by intergeneric protoplast fusion of *Streptococcus lactis* and *Bacillus subtilis. FEMS Microbiol. Lett.* **49**, 323–329.

Vescovo, M., Morelli, L., Bottazzi, A. and Gasson, M.J. (1983) Conjugal transfer of broad host range plasmid pAMβ1 into enteric species of lactic acid bacteria. *Appl. Environ. Microbiol.* **46**, 753–755.

Vescovo, M., Morelli, L., Cocconcelli, P.S. and Bottazzi, V. (1984) Protoplast formation, regeneration and plasmid curing in *Lactobacillus reuteri. FEMS Microbiol. Lett.* **23**, 333–334.

Von Wright, A., Taimisto, A.M. and Sivela, S. (1985) Effect of Ca^{2+} ions on plasmid transformation of *Streptococcus lactis* protoplasts. *Appl. Environ. Microbiol.* **50**, 1100–1102.

Walsh, P.M. and McKay L.L. (1981) Recombinant plasmids associated with cell aggregation and high frequency conjugation of *Streptococcus lactis* ML3. *J. Bacteriol.* **146**, 937–944.

Walsh, P.M. and McKay, L.L. (1982) Restriction endonuclease analysis of the lactose plasmid in *Streptococcus lactis* ML3 and two recombinant lactose plasmids. *Appl. Environ. Microbiol.* **43**, 1006–1010.

Ward, L.J.H. and Jarvis, A.W. (1991) Rapid electroporation-mediated plasmid transfer between *Lactococcus lactis* and *Escherichia coli* without the need for plasmid preparation. *Lett. Appl. Microbiol.* **13**, 278–280.

Watanabe, K., Hayashida, M., Nakashima, Y. and Hayashi, S. (1987) Preparation and regeneration of bacteriophage PL-1 enzyme-induced *Lactobacillus casei* protoplasts. *Appl. Environ. Microbiol.* **53**, 2686–2688.

West, C.A., and Warner, P.J. (1985) Plasmid profiles and transfer of plasmid-encoded antibiotic resistance in *Lactobacillus plantarum. Appl. Environ. Microbiol.* **50**, 1319–1321.

Woskow, S.A. and Kondo, J.K. (1987) Effect of proteolytic enzymes on transfection and transformation of *Streptococcus lactis* protoplasts. *Appl. Environ. Microbiol.* **53**, 2583–2587.

Wyckoff, H.A., Sandine, W.E. and Kondo, J.K. (1991) Transformation of dairy *Leuconostoc* using plasmid vectors from *Bacillus, Escherichia,* and *Lactococcus* hosts. *J. Dairy Sci.* **74**, 1454–1460.

Youngman, P., Juber, P., Perkins, J.B., Sondman, K., Igo, M. and Lovick, R. (1985) New ways to study developmental genes in spore-forming bacteria. *Science* **228**, 285–291.

Youngman, P.J. (1987) Plasmid vectors for recovering and exploiting Tn917 transpositions in *Bacillus* and other gram positive bacteria. In *Plasmids: A Practical Approach* (Hardy, K.G. ed), pp. 79–103. IRL Press, Washington D.C., U.S.A.

Zinder, N.D. and Lederberg, J. (1952) Genetic exchange in *Salmonella. J. Bacteriol.* **64**, 679–699.

Zink, A., Klein, J.R. and Plapp, R. (1991) Transformation of *Lactobacillus delbruckii* ssp. *lactis* by electroporation and cloning of origins of replication by use of a positive selection vector. *FEMS Microbiol. Lett.* **78**, 207–212.

2 Gene cloning and expression systems in Lactococci

W.M. DE VOS and G.F.M. SIMONS

2.1 Introduction

The genus *Lactococcus* includes five species, viz. *L. garviae, L. raffinolactis, L. plantarum, L. piscium,* and *L. lactis* (Schleifer *et al.*, 1985; Williams *et al.*, 1990). *L. lactis* represents the best characterized species, that formerly was designated lactic or group N *Streptococcus*. Three *L. lactis* subspecies have been described, viz. *L. lactis* subsp. *lactis, L. lactis* subsp. *cremoris*, and *L. lactis* subsp. *hordniae*. Genetic interest has exclusively been focused on *L. lactis* subsp. *lactis* and *L. lactis* subsp. *cremoris*, since strains of these subspecies are used as starter cultures in industrial dairy fermentations. Differentiation between these commercially important subspecies was based on the properties of *L. lactis* subsp. *lactis* to metabolize arginine, to grow at temperatures above 37°C, and to tolerate salt (Schleifer *et al.*, 1985), However, this phenotypic classification has recently been redefined using molecular markers including restriction fragment length polymorphism (Godon *et al.*, 1992) and 16S rRNA sequence analysis (Salama *et al.* 1991, Klijn *et al.*, 1991). Since not all lactococcal strains have been subject to this molecular analysis, no attempts are made here to differentiate *L. lactis* on the subspecies level. Nevertheless, the described data exclusively relate to results obtained with strains of *L. lactis* subsp. *cremoris, L. lactis* subsp. *lactis* or the citrate-utilizing variant *L. lactis* subsp. *lactis* biovar. *diacetylactis*.

Apart from the need for lactococcal starter cultures with stable and novel properties, there is a variety of other reasons why the implementation of gene technology in *L. lactis* has received considerable attention in recent years. The first relates to the wealth of extrachromosomal elements, viz. plasmids and bacteriophages, that were found in *L. lactis*. These provided not only an excellent substrate for further studies aimed at understanding their biology but also allowed the development of plasmid vectors and induced the need to develop plasmid-free laboratory strains. In addition, these and other *L. lactis* strains can easily be grown and handled in the laboratory while serving in physiological or biochemical studies as a model for anaerobic, Gram-positive bacteria with a simple metabolism. Moreover, there is the observation that many of the genetic tools developed for *L. lactis* can be applied in other lactic acid bacteria, in many cases with no or only minor modifications. This especially holds for the promiscuous plasmid replicons that are functional in many lactic acid bacteria and other hosts. Yet another reason is the possibility of using *L. lactis* as a production host for proteins or primary and secondary metabolites. This is particularly relevant for the production of compounds that are unique for these food-grade bacteria, such as the antimicrobial peptide nisin, the flavour compound diacetyl, or enzymes involved in the generation of cheese flavour (de Vos, 1992). Finally, *L. lactis* is a food-grade Gram-positive bacterium that has a number of additional properties allowing its use in the production of foreign proteins, especially those that are applied in the food industry or otherwise have to be consumed. Lactococci can be grown at industrial scale on cheap, whey-based media using existing fermentation technology, and, moreover, they have the capacity to secrete proteins, allowing surface expression or extracellular production of heterologous enzymes or proteins.

This chapter provides an overview of the present systems that have allowed genetic, metabolic and protein engineering in *L. lactis* (de Vos, 1992). In addition, those aspects of the molecular biology and genetics of *L. lactis* that are relevant for the development of new systems in this industrial microorganism, which has served as a model for studies with related lactic acid bacteria and other Gram-positive bacteria are summarized (for earlier reviews see de Vos, 1987; de Vos *et al.*, 1989a; Kok, 1991; van der Guchte *et al.*, 1992a).

2.2 Replicative gene cloning

Cloning systems by which the gene of interest remains extrachromosomal are here designated as replicative gene cloning systems to distinguish them from the integrative cloning systems designed for integration into the *L. lactis* chromosome. All known lactococcal replicative cloning systems have exclusively been based on plasmids and no vectors derived from bacteriophage replicons have so far been developed in spite of the abundance of different bacteriophages in *L. lactis* (see chapter 3).

The first genetic studies of lactococci revealed that all strains contained various, up to ten, different plasmids some of which encoded important metabolic properties (McKay, 1983). These findings initiated considerable interest in the plasmid biology of *L. lactis*. Various aspects of this actively studied field of research that are relevant for the development and use of plasmids as cloning vectors are discussed below.

To allow for a rapid analysis of cloned genes in *L. lactis* and accurately follow their fate, the lactococcal host strains should be devoid of endogenous plasmids. Various plasmid-free laboratory strains have been developed using successive rounds of plasmid curing (de Vos, 1987). Those used in the gene cloning studies described below include IL1403 (Chopin *et al.*, 1984) and the related strains LM0230 (Efstathiou and McKay, 1977; McKay *et al.*, 1980) and MG1363 (Gasson, 1983). Chromosomal maps are now available for the well-studied strains IL1403 and MG1363 that represent the genotypically different *L. lactis* subsp. *lactis* and *L. lactis* subsp. *cremoris*, respectively (see chapter 1).

In spite of the wealth of plasmid DNA in *L. lactis* strains, none of the identified plasmids encoded antibiotic resistances and many plasmids turned out to be cryptic (McKay, 1983). This delayed the initial use of endogenous plasmids in vector development, especially since it appeared that vectors developed for related Gram-positive bacteria also replicated and expressed their antibiotic resistance markers in *L. lactis*. As a consequence, two fundamentally different classes of vectors may be distinguished that are being used successfully in *L. lactis*, those based on heterologous replicons and those based on genetically marked lactococcal replicons. Representatives of these two distinct groups of vectors are listed in Table 2.1. Their relevant properties and their first use in productive gene cloning experiments in *L. lactis* are described below.

Table 2.1 Properties of cloning vectors used in lactococci. The size and the copy numbers of the vectors in *L. lactis* are shown (low: approximately 5–10 copies per cell; high: more than approximately 50 copies per cell). Resistances to chloramphenicol (Cm^R), erythromycin (Em^R), kanamycin (Km^R), spectinomycin (Sp^R) are indicated

Vector	Replicon	Size markers (kb)	Copy number	References
pGB301	pIP501	9.8 Cm^R Em^R	Low	Behnke *et al.* (1981)
pSA3	pIP501 pACYC184	10.2 Cm^R Em^R	Low	Dao and Ferreti (1986)
pHV1301	pAMβ1	13.0 Em^R	Low	Chopin *et al.* (1986)
pIL252	pAMβ1	4.7 Em^R	Low	Simon and Chopin (1988)
pIL253	pAMβ1	5.0 Em^R	High	Simon and Chopin (1988)
pVA838	pVA380-1 pACYC184	9.2 Cm^R Em^R	Low	Macrina *et al.* (1982)
pDL278	pVA380-1 pUC19	6.9 Sp^R	High	LeBlanc and Lee (1991)
pGK12	pWV01	4.4 Cm^R Em^R	Low	Kok *et al.* (1984)
pGL3	pWV01	5.0 Cm^R Km^R	Low	van der Lelie *et al.* (1989)
pGKV2	pWV01	5.0 Cm^R Em^R	Low	van der Vossen *et al.* (1985)
pNZ12	pSH71	4.3 Cm^R Km^R	High	de Vos (1986a–c)
pNZ121	pSH71	4.3 Cm^R Km^R	Low	de Vos (1986a–c)
pNZ123	pSH71	2.8 Cm^R	High .	de Vos *et al.* (1992)
pCK1	pSH71	5.5 Cm^R Km^R	High	Anderson and Gasson (1985)
pFX1	pDI25	5.5 Cm^R	High	Xu *et al.* (1990)
pFX3	pDI25	4.5 Cm^R	High	Xu *et al.* (1991)
pCI374	pCI305	4.9 Cm^R	Low	Hayes *et al.* (1990)

2.2.1 *Streptococcal plasmid vectors*

Vectors designed for use in streptococci were the first to be applied in lactococcal gene cloning experiments and include derivatives of the 30 kb pIP501 from *Streptococcus algalactiae* (Horodniceanu *et al.*, 1976), the 26.5 kb pAMβ1 originally detected in *Enterococcus faecalis* (Clewell *et al.*, 1974), and the small 4.2 kb cryptic *Streptococcus ferus* plasmid pVA380-1 (Macrina *et al.*, 1980).

The related plasmids pIP501 and pAMβ1 belong to a family of self-transmissible streptococcal and enterococcal plasmids with a size of 24–34 kb that code for resistance to macrolides, lincosamides and streptogramin B (MLS) and in some cases other antibiotics (Brantl *et al.*, 1990). It was not unexpected that derivatives of these plasmids could replicate and express their MLS resistance gene in *L. lactis* since the conjugal transfer of pAMβ1 into lactococci had already been demonstrated (Gasson and Davies, 1980). In fact those plasmids can replicate in many Gram-positive bacteria including all genera of lactic acid bacteria (see chapter 1). Therefore, vectors derived from those plasmids are expected to have a broad host-range and in some cases may have retained the capacity to be mobilized.

The 9.8 kb pIP501-derivative pGB301 (Behnke *et al.*, 1981) was used as vector in the primary description of a successful cloning experiment in lactic acid bacteria, the cloning of fragments containing the *lac* genes in *L. lactis* LM0230 (Kondo and McKay, 1984). A related vector, the 10.2 kb pSA3 (Dao and Ferretti, 1985) based on the same pIP501 replicon and fused to the *Escherichia coli* vector pACYC184, was used in the cloning of the nisin-resistance gene of *L. lactis* DRC3 in strain LM0230 (Froseth *et al.*, 1988). Plasmid pSA3 had the advantage of replicating in both Gram-positive and Gram-negative bacteria, a property that is shared by the pIP501-derived vector pAM401 (10.4 kb; Wirth *et al.*, 1986) used for the cloning of the NCDO 712 *thyA* gene in MG1363 (Ross *et al.*, 1990) and the pVA380-1 derivative pVA838 that also contains pACYC184 (9.2 kb; Macrina *et al.*, 1982). The latter was used for the cloning of the *Streptococcus thermophilus* β-galactosidase gene in a *L. lactis* LM0230-derivative (Romero *et al.*, 1987; Mercenier, 1990).

Whilst the derivatives of the conjugative plasmid pIP501 are not self-transmissible, it has been shown that pSA3 and pVA838 can be mobilized by pVA797, a derivative of pVA380-1 containing the transfer and mobilization functions of pIP501, allowing further dissemination of cloned genes by mating into poorly or non-transformable strains (Romero *et al.*, 1987). Details of this mobilization system that first was developed for *Enterococcus faecalis* (Smith and Clewell, 1984) and its practical impact is discussed in chapter 1.

All the pIP501-derived vectors are relatively large, contain only few unique restriction sites that can be used to insert DNA, and have a low copy number in *L. lactis* (Table 2.1). The same holds for the plasmids based on the non-selftransmissible pVA380-1 that are able to replicate in streptococci, enterococci and *L. lactis* (LeBlanc and Lee, 1991). Improved derivatives based on pVA380-1 include pTG222, containing a large polylinker sequence (Romero *et al.*, 1987), and the 6.9 kb pDL276, which contains a polylinker sequence flanked by *E. coli* transcription terminators and the pUC19 replication origin (Dunny *et al.*, 1991). The latter vector has a high copy number in *L. lactis* but codes for kanamycin-resistance, a marker that is poorly selective in lactococci due to a high level natural resistance (see below). Replacement of this resistance marker by a spectinomycin-resistance gene resulted in the 6.6 kb pDL278 which could be transformed and stably maintained in *L. lactis* (LeBlanc and Lee, 1991).

A series of useful cloning vectors has been developed based on pAMβ1 that eliminate some of the disadvantages of the plasmids based on pIP501. Early cloning experiments already showed the applicability of the non-self-transmissible deletion derivative pHV1301 (13.0 kb), that was used for cloning UV-resistance in *L. lactis* IL1403 (Chopin *et al.*, 1986). A smaller cloning derivative was derived from pHV1301 and equipped with a polylinker sequence. The copy number of the resulting 4.7 kb vector, designated pIL252, is low and the vector shows high segregational instability (see section 2.2.3; Simon and Chopin, 1988). However, the pAMβ1 replicon has the capacity to support replication of large inserts as was shown by the cloning in strain IL1403 of a 30 kb fragment from pNP40 from *L. lactis* DRC3 encoding nisin resistance using pIL252 (Simon and Chopin, 1988) while also a 22 kb fragment containing Tn*916* and host DNA was cloned using pIL204, a plasmid from which pIL252 has been derived (Loureiro dos Santos and Chopin, 1987). An important addition to these cloning vectors was the construction of the 4.8 kb vector pIL253 (Figure 2.1) based on a high copy number derivative of pAMβ1 that had been isolated in *Bacillus subtilis* (Lereclus *et al.*, 1984). The polylinker-containing vector pIL253 has a high copy number in *L. lactis* and, as a consequence, a somewhat higher segregational stability (Simon and Chopin, 1988). This vector has been used for the cloning and expression of the *L. lactis* SK11 *prtM* gene in MG1363 (Vos *et al.*, 1989a). This study also showed that pIL253 was compatible with vectors based on the lactococcal replicon pSH71, allowing the first *trans* complementation experiments in *L. lactis*. In a later study it appeared that the promoter activity of the replication region of pIL253 could be used to drive expression of the promoterless *lacF* gene in *L. lactis* YP2-5 (de Vos *et al*, 1990). This high copy number derivative of pAMβ1 may also accommodate inserts of large size since the 22 kb fragment of Tn*916* could be cloned and maintained in pIL206, a plasmid from which pIL253 has been derived (Simon and Chopin, 1988). Recently, plasmids pIL252 and pIL253 served as the basis for the construction of a series of low and high copy number derivatives with extended polylinker sequences (O' Sullivan *et al.*, 1993).

Although it was assumed for a long time that the routine plasmid vectors developed for *B. subtilis* that were based on small *Staphylococcus aureus* plasmids did not replicate in *L. lactis* (Kok *et al.*, 1984; Chopin *et al.*, 1989), there have been reports that the replicons of the staphylococcal plasmids pUB110, pE194 and pC194 have some residual activity in *L. lactis* MG1363 and other lactococcal strains. This was concluded from the physical presence of these plasmids in *L. lactis* after transfer by protoplast fusion with *B. subtilis* (Baigori *et al.*, 1988) and the detection of replication intermediates originating from chromosomally inserted copies of these plasmids (Leenhouts *et al.*, 1990). Some of the staphylococcal plasmids have been found to replicate in various lactobacilli, although efficient replication of pUB110 in *Lactobacillus casei* needed some modification (see section 2.2.3; Shimuza-Kadota *et al.*, 1991). It is possible that those plasmids can also be used for gene cloning in *L. lactis*.

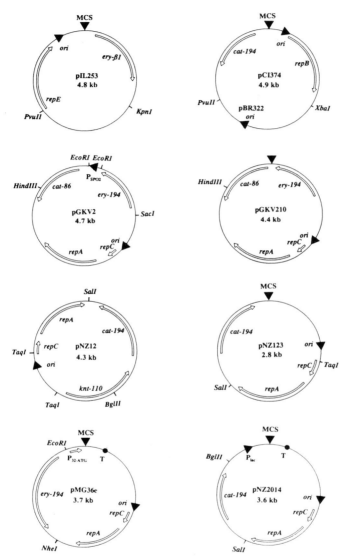

Figure 2.1 Physical and genetic maps of various well-known cloning and expression vectors used in *L. lactis* (see text for a discussion).

A recent example includes the Tn*917* delivery vectors based on pE194ts, a temperature-sensitive derivative of pE194, which was found to replicate in *L. lactis* MG1363 depending on the antibiotic selection (Israelson and Hansen, 1993).

2.2.2 Lactococcal plasmid vectors

Two approaches have been used to convert the abundantly present lactococcal replicons into useful cloning vectors: genetic marking of replicons and the use of a replication probe plasmid for the selection of replicons. In many cases the initially generated vectors were improved by constructing smaller replicating derivatives, detailed characterization of their replicons, and introduction of polylinkers or other dedicated sequences. Here the main vectors are described that have been used successfully in the cloning of genes in lactococci.

The construction of vectors based on lactococcal replicons was greatly facilitated by the finding that the small 2.1 kb lactococcal cryptic plasmid pWV01 (Otto et al., 1982) could be introduced in protoplasts of B. subtilis 8G5 and maintained at a low copy number in this heterologous host (Vosman and Venema, 1983). This allowed the genetic marking of pWV01 by antibiotic resistance genes in B. subtilis, at the time that a transformation system for L. lactis was still being developed and improved (Kok et al., 1984). The resulting vector, pGK12, was marked with two antibiotic resistance genes from S. aureus plasmids that were known to be expressed in several Gram-positive and Gram-negative bacteria, the chloramphenicol acetyl transferase gene from pC194 and the erythromycin-resistance gene from pE194, designated here cat-194 and ery-194, respectively. Importantly, pGK12 could be introduced into protoplasts of L. lactis MG1363 and expressed both resistance markers (Kok et al., 1984). Unexpectedly, the pWV01 replicon was also functional in the E. coli strain BHB2600 and its derived vectors could replicate in this host with a high copy number. In contrast, the copy number of pGK12 in B. subtilis and more importantly also in L. lactis was low (Kok et al., 1984). The first reported use of a pGK12-derivative, the 4.8 kb pGKV2 (van der Vossen et al., 1985; Figure 2.1), was the cloning and expression of the proteinase gene from strain Wg2 first in B. subtilis PSL1 and, subsequently, in L. lactis MG1363, demonstrating the advantage of using an intermediate host in gene cloning (Kok et al., 1985). In following studies derivatives of pGKV2, such as pGKV210 (4.5 kb, see Figure 2.1; van der Vossen et al., 1987), or other vectors based on pWV01 such as pMG24 (3.2 kb; van Belkum et al., 1989) or pMG36 and pMG36e (3.7 and 3.6 kb, respectively, see Figure 2.1; van de Guchte et al., 1989) were developed and used for the cloning and expression of bacteriocin or lysozyme genes (van Belkum et al., 1989; van de Guchte et al., 1989).

The observation that pWV01 derivatives had a low copy number in L. lactis limited its application potential and stimulated studies aimed at developing high copy number plasmid vectors. In the first replicon screening approach, the 6 kb vector pNZ10 was constructed containing the E. coli vector pAT153 (Twigg and Sheratt, 1980 and a portable cassette with two antibiotic resistance genes from staphylococcal plasmids, the cat-194 gene from pC194 and the kanamycin nucleotidyl transferase (knt-110) gene from pUB110, that were readily expressed in B. subtilis (de Vos, 1986a; de Vos, 1986b; de Vos and Simons,

1987). Restriction fragments from different lactococcal plasmids were ligated in shot-gun experiments to the antibiotic-resistance gene cassette and introduced into protoplasts of *B. subtilis* 1G33. Subsequently, plasmid DNA was isolated from antibiotic-resistant *B. subtilis* transformants and introduced into protoplasts of *L. lactis* MG1363. Finally, transformants in the latter host were analysed for the presence of plasmid DNA and a high and a low copy number plasmid, designated pNZ12 and pNZ121, respectively, were further analysed. Both plasmids had a size of 4.1 kb and appeared to contain the two largest *Taq*I fragments of the 2.1 kb plasmid pSH71 from *L. lactis* NCDO 712 (see Figure 2.1). In addition, they contained a unique *Sal*I site that was used for the insertion of foreign sequences, but differed in the orientation of the antibiotic-resistance gene cassette. Both pNZ12 and pNZ121 showed a broad host-range and replicated in *B. subtilis* 1G33, *S. aureus* RN403, and *E. coli* MC1061. The vector pNZ121 and its 3.5 kb derivative pNZ122, from which the *knt-110* gene was deleted, were first used in the cloning and expression the proteinase gene of *L. lactis* SK11 in MG1363 (de Vos, 1986a; de Vos *et al.*, 1989b). In a later study pNZ121 was provided with a polylinker site resulting in pNZ17, that was used to express in *L. lactis* MG1363 a gene fusion of part of the proteinase (*prtP*) gene and the *E. coli* lacZ gene, demonstrating the capacity of *L. lactis* to produce heterologous proteins (de Vos and Simons, 1988). Both vectors pNZ12 and pNZ121 were used in the cloning of the phospho-β-galactosidase gene from strain NCDO 712 in *L. lactis* MG1363 that showed the effect of copy number on the expression of cloned genes (de Vos and Gasson, 1989; de Vos, unpublished results). Based on pNZ12 and pNZ122 derivatives with polylinkers were developed such as pNZ18/pNZ19 and the small and versatile pNZ123 (Figure 2.1), respectively, that were instrumental in the cloning of many lactococcal and heterologous genes (de Vos, 1987; Simons and de Vos, 1987; de Vos, 1992; van Asseldonk *et al.*, 1993a) In a parallel study with a similar objective, genetically marked derivatives of pSH71 were constructed with identical broad host-range replication properties (Anderson and Gasson, 1985). One of the resulting vectors, the 5.5 kb pCK1, that also contained the *S. aureus* cat-194 and *knt-110* genes, was used for the cloning and partial expression of the *L. lactis* 712 proteinase gene in strain MG1363 (Gasson *et al.*, 1987). Derivatives of pCK1 were equipped with the *E. coli* gene fragment for the *lacZα* peptide with a multiple cloning site, resulting in pTG262 (5.6 kb), which was used for the cloning of a *Bacillus* α-amylase gene in *L. lactis* (A. Mercenier and M. O'Regan, personal communication; van Asseldonk *et al.*, in preparation). In an analagous approach the pSH71 replicon was marked with the staphylococcal *cat-194* and *ery-194* genes, resulting in the 4.9 kb pVS2 (von Wright *et al.*, 1987). This vector was used to clone in a derivative of strain MG1363 a 9 kb chromosomal *L. lactis* DNA fragment that later appeared to encode an oligopeptide transport system (von Wright *et al.*, 1987; see also chapter 4).

Recently, the replication region of the 5.5 kb cryptic plasmid pDI25 from *L.*

lactis 5136 was genetically marked by the *cat-194* gene of pC194 resulting in the construction of the 5.5 kb pFX1 and its polylinker-containing derivative, the 4.5 kb pFX3. These vectors were used to clone in *L. lactis* 4125 the proteinase gene from the *L. lactis* strain H1 (Xu *et al.*, 1990) and the *L. lactis* 4560 gene for tagatose-1,6-diphosphate aldolase, respectively (Xu *et al.*, 1991). Although aimed at providing a vector with different replication properties than the pGK-, pNZ- or pCK-vectors, the pFX-series of vectors showed also replication in some *E. coli* strains and was later found to contain a replicon that was highly related to that of the almost identical plasmids pWV01 and pSH71 (see section 2.2.3).

The promiscuous, heterogramic replication properties of the vectors based on the replicons of pWV01/pSH71 and related plasmids have received considerable attention. A very relevant property is their wide host range that allows the use of these vectors in all genera of the lactic acid bacteria (see section 2.2.3). Although this generalization may hold in the majority of cases, heterogeneity between strains has been noted as in the case of the *L. casei* strain MSK248 that could not be transformed with the pGK12-derivative pGKV11 (Shimuza-Kadota *et al.*, 1991).

A completely new set of lactococcal vectors was derived from the 8.7 kb cryptic plasmid pCI305 from strain *L. lactis* UC317. The minimal replicon of this plasmid was identified in *L. lactis* MG1363 by a similar replicon screening approach as described above, using the *E. coli* vector pUC19 provided with the pC194 *cat-194* gene (Hayes *et al.*, 1990). The smallest derivative, the 4.6 kb pC1374 (Figure 2.1) was used for the cloning of the lactococcal *pepN* gene in *L. lactis* MG1363 (I.J. van Alen-Boerrigter and W.M. de Vos, unpublished observations). Although the copy number of these vectors is relatively low they have two important properties that make them attractive candidates for future developments. First their host range seems to be limited to *L. lactis* (Hayes *et al.*, 1990 and 1991). This apparent drawback for the manipulations could be an advantage in developing self-contained vectors for applications in which the genetic modification is not to be disseminated. Second, they have a mode of replication that differs from that of the other developed lactococcal vectors (Hayes *et al.*, 1991; see below). This explains their compatibility with vectors based on the pSH71 replicon (van Alen-Boerrigter and de Vos, unpublished results). Recently, vectors based on the cryptic plasmid pWV02 (Otto *et al.*, 1982) have been reported (Kiewit *et al.*, 1993a). Interestingly, their replicon is highly similar to that of pCI305.

A replicon screening strategy was also used to clone replicons from other lactococcal plasmids. Using the 6.0 kb pSA34, a deletion derivative of the vector pSA3 from which the streptococcal replication region had been removed, a 17 kb plasmid-derived DNA fragment encoding replicon activity and a restriction-modification system from *L. lactis* KH was cloned in strain LM0230 (Schultz and Sanders, 1990). In a variant of this approach a nisin resistance gene was used as a marker in the cloning of a replicon from a cryptic lactococcal plasmid

(Froseth and McKay, 1991). However, further applications of these potentially useful lactococcal vectors have so far not been described.

2.2.3 Plasmid replication and stability

Various replicons that served as the basis for constructing the well-known vectors described above have been sequenced and their mode of replication has been studied in detail. Based on sequence similarity, mode of replication, and host range these can be grouped into four different families of replicons: pAMβ1/pIP501; pVA380-1; pSH71/pWV01; and pCI305/pWV02. Here the salient features of those replicons are described that are relevant for their use in gene cloning in lactococci and other lactic acid bacteria. Specific attention will be given to the mode of replication and its impact on plasmid stability.

There are two basically different modes of plasmid replication that are named after the characteristic structure of their replication intermediates, the theta and sigma modes (Viret et al., 1991). In the theta mode the sites for priming of leading- and lagging-strand synthesis are located close to another within the origin of replication. During replication via the theta mode, that can be unidirectional or bidirectional, both DNA strands remains covalently closed, except during the resolution of daughter molecules. During the sigma mode of replication the priming events for leading and lagging strand synthesis are separated in time and space. For the initiation of replication one of the DNA strands is nicked at a position termed the plus origin by the replication protein. Subsequently, a single-stranded intermediate is formed during displacement synthesis that is eventually converted into a double-stranded molecule after priming of the lagging strand synthesis at the minus origin. This mode of replication is also designated rolling circle replication (RCR) a term that will be used here. An important feature that distinguishes both replication modes is the fact that the RCR-type plasmids accumulate detectable single-stranded intermediates depending on the efficiency of the lagging strand initiation process (te Riele et al., 1986; Gruss and Ehrlich, 1989).

Plasmid replication is intimately linked to plasmid stability. Plasmid instability includes both segregational instability, the loss of entire plasmid molecules, and structural instability, the deletion of specific plasmid sequences. From studies in B. subtilis it is known that RCR-type plasmids often show a high degree of structural and segregational instability. This has been attributed both at the formation of single-stranded DNA intermediates and the formation of linear high molecular weight plasmid multimers (HMW) that are a consequence of the mode of replication of RCR-type plasmids (Gruss and Ehrlich, 1988 and 1989; Leonhardt and Alonso, 1991; Viret et al., 1991). Plasmid multimerization by homologous recombination contributes to segregational instability in theta-type plasmids and requires site-specific recombination systems to resolve plasmid multimers into monomers. In all systems the copy number of monomers is an important factor in segregational stability (Summers and Sheratt, 1984).

In a series of elegant experiments it has been demonstrated that the pAMβ1-derivative pIL253 shows an unidirectional theta-type replication (Bruand et al., 1991). The minimal replicon of pAMβ1 has been characterized and includes the repE gene for a 57 kDa replication protein and the sites for leading and lagging strand initiation that are located immediately downstream of the repE gene (Swinfield et al., 1990; Bruand et al., 1991). The RepE protein positively controls replication and the repE gene is preceded by sequences involved in its expression (Le Chatelier et al., 1993). This replication and copy number control region showed high sequence identity with that of pIP501 which also has a highly similar counterpart in pSM19035 (Brantl and Behnke, 1992; Brantl et al., 1990). Recently, it was shown that pIP501 also shows a unidirectional theta-mode of replication, has a similar replication protein encoded by the repR gene and an origin of replication within or immediately downstream of the repR gene (Le Chatelier et al., 1993).

The stability of pAMβ1-derived vectors has been the subject of various studies. The low-copy number derivative pIL252 is relatively unstable in L. lactis with an average plasmid loss of more than 10^{-2} per generation while that of the high copy number derivative has a higher segregational stability (loss of 10^{-4} per generation; Simon and Chopin, 1988). Recently, it was found that the repE gene of pAMβ1 is followed by the res gene for a 23.9 kDa resolvase that reduces plasmid multimerization and increases segregational stability in B. subtilis by reducing the average plasmid loss per generation approximately 20-fold (Swinfield et al., 1990). Both pIL252 and pIL253 lack the res gene, also designated orfH. Insertion of the res gene in pIL252 significantly increased its stability (Kiewit et al., 1993b) and a similar effect may be expected in pIL253 or other derivatives. It is likely that pAMβ1 carries other yet unidentified stabilization determinants (Swinfield et al., 1990 that could prove useful in future vector development. A derivative of pIL252 carrying the res gene was used to address its segregational stability in L. lactis after insertion of heterologous inserts (Kiewit et al., 1993b). These studies showed that inserts up to a size of 16.8 kb did not alter the segregational stability (plasmid loss of less than 10^{-4} per generation) or the copy number (about 15 per chromosome). Structural instability has not been reported for cloning of DNA fragments in vectors based on the pAMβ1/pIP501 family of replicons. In contrast, large inserts (up to 30 kb) could be maintained in L. lactis with the use of those vectors (Simon and Chopin, 1988), which illustrates their utility in gene cloning experiments and especially for the preparation of gene libraries.

Thus, vectors containing the well-characterized pAMβ1/pIP501 family of replicons, show a theta-type replication, high structural stability, and when equipped with the res gene also high segregational stability in L. lactis. In addition, they have promiscuous replication properties within Gram-positive bacteria and can be maintained in Bacillus, Clostridium, Lactococcus, Streptococcus, Enterococcus, Staphylococcus, Lactobacillus, Leuconostoc, Pediococcus (see Brantl and Behnke, 1992).

The replicon functions of the other sequenced broad host-range Gram-positive plasmid pVA838-1 have also been studied. Its 2.5 kb minimal replicon contains a 714 bp ORF that presumably encodes *trans*-acting function (LeBlanc and Lee, 1991). Although the mode of replication of pVA838-1 has not yet been studied, its segregational stability was found to be high in *L. lactis* for the derivative pDL278 and a presumed stability locus was identified in the pVA838-1 minimal replicon (LeBlanc and Lee, 1991).

Plasmid pSH71 was the first of the lactococcal plasmids to be sequenced in order to define the high copy-number pNZ-vectors that were derived from it (de Vos, 1986c; de Vos and Simons, 1987). Its replication properties were characterized in detail by deletion and complementation studies (de Vos *et al.*, 1989c). It is one of the smallest lactococcal plasmids with a size of 2060 nucleotides. Its 1.8 kb minimal replicon contains the *rep*-operon comprising two genes, the *repC* gene encoding a 6 kDa repressor involved in replication control, and the *repA* gene for a 27 kDa replication protein that was found to be rate limiting in replication (see section 2.5.3 and Figure 2.9; de Vos *et al.*, 1989c). Plasmid pSH71 is a RCR-type plasmid and its derivatives, such as pNZ12, produce single-stranded DNA intermediates in *E. coli*, *B. subtilis* and *L. lactis* (de Vos *et al.*, 1989c; Hayes *et al.*, 1991). The direction of replication at the plus origin is the same as the transcription of the *repCA* operon. Considerable sequence similarity for the *rep* genes and the plus origin was found between pSH71 and other RCR-type plasmids (Gruss and Ehrlich, 1989) that include the well-studied *Staphylococcus aureus* pE194 (Horinouchi and Weisblum, 1992; Villafane *et al.*, 1987) and *Streptococcus agalactiae* pLS1 (Lacks *et al.*, 1986; de la Campa *et al.*, 1990). These pSH71 Rep proteins have been overproduced, purified, and their interaction with the pSH71 DNA was studied by gel mobility shift experiments and footprinting studies (de Vos and Mulders, unpublished results). The RepA protein was found to interact with a restriction fragment that contains the putative plus origin followed by a 11 bp sequence that was three times tandemly repeated. The RepC protein is a repressor that is involved in plasmid copy number control (see section 2.5.3; de Vos *et al.*, 1989c). Hybridization studies with labelled pSH71 DNA showed that its replicon is disseminated widely and present in one out of ten industrial strains and even more frequently in laboratory strains of *L. lactis* (de Vos and de Haard, unpublished observations). This was substantiated by sequence analysis of several lactococcal plasmids. The 2178 bp pWV01 used for the construction of the pGK-vectors differed from pSH71 only in a few nucleotides and the presence of a direct repeat (Leenhouts *et al.*, 1991c). This confirmed previous hybridization and mapping data suggesting that pWV01 and pSH71 were highly similar (de Vos, 1986b). In addition, the replicon of 5.5 kb pDI25 that was used in the construction of the pFX-plasmids was found to be similar to that of pSH71/pWV01 (Xu *et al.*, 1991). It appeared that RCR-type plasmids similar to the pSH71/pWV01 family are also found in other lactic acid bacteria and include pLS1 (4.4 kb) from *Streptococcus agalactiae* (Lacks *et al.*, 1986), pLB4 (3.5 kb) and pA1 (2.8 kb) from two different strains of *Lactobacillus plantarum* (Bates and Gilbert, 1989; Vujcic and Topisirovic, 1993), and pLC2 (2.5 kb) from *Lactobacillus curvatus* (Klein *et al.*, 1993). RCR-plasmids appear to be abundant in lactic bacteria since apart from this pSH71/pWVO1 family various small RCR-type plasmids belonging to other replicon families have been characterized in *Lactobacillus* species (see chapter 6).

The segregational stability of vectors based on the pSH71/pWV01 replicons appeared to be high in *L. lactis* (plasmid loss of less than 10^{-5} per generation) and other Gram-positive hosts but low in *E. coli* (loss of more than 10^{-2} per generation) in spite of their high copy number in this host (Kok *et al.*, 1984; de Vos and Simons, 1987). In a systematic study the effect of heterologous inserts in a pWV01-derivative was determined on the segregational stability and compared to that in a pIL252-derivative (Kiewit *et al.*, 1993b). All theta-type plasmids were stably maintained but not all RCR-plasmids. The RCR-derivatives carrying no or small inserts were as stably maintained as the theta-type plasmids while RCR-derivatives with inserts larger than approximately 8 kb showed a size-dependent instability. All RCR plasmids with heterologous inserts produced HMW DNA, that could amount to up to 40% of total cellular DNA with the inserts above 8 kb, yielding an average plasmid loss of less than 10^{-2} per generation. It is likely that this affects the growth rate and hence the segregational stability although effects on the control of the circular plasmid copy number can not be excluded. These results are in agreement with previous observations that several cloned genes, such as a 7.5 kb fragment containing the SK11 *prtP* and *prtM* genes, could be stably maintained on pNZ-vectors in *L. lactis* MG1363 (loss per generation of less than 10^{-4}; de Vos, unpublished observations). Since in most industrial food fermentations the total number of generations is limited (up to 25 generations) it could be argued that segregational instability is only a problem for RCR-type plasmids with large insert sizes. In addition, the original pWV01 derivative used in this study did not produce HMW DNA in spite of the presence of heterologous antibiotic resistance genes that made up about half of the plasmid (Kiewit *et al.*, 1993b). Therefore, the sequence of the insert could also affect the formation of HMW DNA and hence segregational instability. Structural instability in RCR-plasmids in *L. lactis* has not been studied systematically. However, a great variety of homologous and heterologous genes (see section 2.7) has been successfully cloned and expressed using RCR-type vectors. Therefore, it remains to be determined whether the high structural instability with RCR-type plasmids observed in *B. subtilis* is also to be encountered in *L. lactis*.

The broad host-range replication properties of the pSH71/pWV01 family of plasmids have been studied in several hosts. Most *E. coli* K12 strains could be transformed efficiently with those lactococcal plasmids and served as intermediate hosts in many cloning experiments (de Vos, 1987). However, strongly reduced transformation frequencies were observed in *E. coli recA* strains that appeared to accumulate increased amounts of single-stranded replication intermediates (de Vos and Ventris, unpublished results). This may be a consequence of the mechanism used to initiate lagging strand synthesis, which for the pSH71/pWV01 plasmids was found to differ from that of other RCR plasmids. No consensus minus origins was detected in these lactococcal plasmids and the conversion of single-stranded intermediates was found to be independent of the host-encoded RNA polymerase in *L. lactis* (Leenhouts *et al.*, 1991c). In contrast, in *B. subtilis* it was found that RNA polymerase was involved in the conversion of single-stranded intermediates into double stranded DNA, indicating that in different hosts distinct mechanisms exist for lagging strand initiation. Nevertheless, this does not seem to have a great impact on the host range of the pSH71/pWV01 replicon since it has been reported that the pGK- and pNZ-series of vectors, in addition to their heterogramic replication in *E. coli*, are capable of efficient transformation and replication in all genera that constitute the lactic acid bacteria. In addition, they replicate in various other genera of the low-GC Gram-positive bacteria, and their present host range includes *Bacillus, Carnobacterium, Clostridium, Lactobacillus, Lactococcus, Leuconostoc, Listeria, Pediococcus, Propionibacterium, Staphylococcus*, and *Streptococcus* (Kok *et al.*, 1984; de Vos, 1987; de Vos *et al.*, 1989a; Luchansky *et al.*, 1988; Kok, 1991; M.E. Stiles, personal communication).

Theta-type plasmids are also widespread in lactic acid bacteria and the prototype of this expanding family of replicons is the 8.7 kb *L. lactis* plasmid pCI305. The 1.6 kb minimal replicon of pCI305 was delineated, sequenced and its replication was analysed in *L. lactis* (Hayes *et al.*, 1991). The minimal replicon contained the *repB* gene encoding a *trans*-acting 46 kDa replication protein, and a 0.3 kb *cis*-acting AT-rich region designated *repA*, containing a 22 bp sequence that was tandemly repeated three and one-half times. No single-stranded DNA was found to be generated from pCI305 in *L. lactis* under conditions that allowed the ready detection of RCR-type intermediates Hayes *et al.*, 1991; Hayes and de Vos, unpublished results). Hybridization studies showed this type of replicon to be present in many *L. lactis* strains (Hayes *et al.*, 1991). This was substantiated in subsequent sequencing studies that showed highly similar replicons to be present in the 8 kb citrate plasmid pSL2 in strain BU2 (Jahns *et al.*, 1991), the 47 kb lactose plasmid pSK11L of strain SK11 (Horng *et al.*, 1991), the 40 kb lactose-proteinase plasmid pUCL22 from strain CNRZ270 (Frère *et al.*, 1993), and the cryptic 3 kb pWV02 from strain Wg2 (Kiewit *et al.*, 1993a). The mode of replication of these replicon families was studied using a similar approach as used for the analysis of pAMβ1 replication (Bruand *et al.*, 1991) and it was established that pWV02 and pUCL22 followed a theta-type replication in *L. lactis* (Kiewit *et al.*, 1993a; Frere *et al.*, 1993). A remarkable aspect is that more than one of these theta-type plasmids may reside in a single strain of *L. lactis* indicating that they are compatible (Kiewit *et al.*, 1993a; Frère *et al.*, 1993). This is of special interest for the use of vectors based on the pCI305/pWV02 family of replicons that may be compatible with endogenous lactococcal plasmids. Comparison of the sequences of the identified replicons showed that they all contained the tandem repeats identified in the *repA* locus of pCI305 but that the sequence of the repeats varied slightly. By analogy with IncQ *E. coli* plasmids it was suggested that these repeats have a role in plasmid incompatibility and may be the target for the replication protein (Hayes *et al.*, 1991). If so, the sequence heterogeneity in these repeats may underlie the compatibility of these highly similar replicons. In conclusion, the pCI305/pWV02 family of replicons shows a theta-type of replication and is widely distributed in *L. lactis*. The recent finding that members of this family are also present in other lactic acid bacteria (Frère *et al.*, 1993), holds a promise for the further development of these narrow host-range replicons as vectors for lactic acid bacteria.

2.3 Integrative gene cloning

Integrative gene cloning results in the specific integration of a sequence into the chromosome and is an expanding area of research in lactococci. Chromosomal insertion could prevent segregational instability of plasmids in the absence of appropriate selective pressure that could cause problems during large scale or continuous fermentations. This applies both to naturally occurring and recombinant plasmids since in all cases there is a need for stable industrial strains. In addition, by integrative gene cloning it is possible to construct strains with disrupted or deleted genes. Finally, there is a growing interest in the lactococcal chromosome. Integrative gene cloning is a helpful tool in analysing the structure, function and location of chromosomal genes, the control of chromosomal gene expressions, and chromosome organization and mapping. While the general procedures and methodologies used to achieve this are described in chapter 1, we focus here on the vectors that allow integration and amplification of genes in the chromosome and expression of integrated genes.

2.3.1 Integration vectors

Because of their inability to replicate efficiently in lactococcal strains integration vectors have been constructed based on known *E. coli* replicons as those from pBR322 (pUC19) or pSC101, and those of plasmids pTB19, or pE194 and pUB110 originally isolated from *B. subtilis* or *Staphylococcus aureus*, respectively (Leenhouts *et al.*, 1989; Chopin *et al.*, 1989; Leenhouts *et al.*, 1990). The same selection markers were used as in replicative gene cloning and included the *cat-194* gene from pC194 and the *ery-194* gene from pE194 (Table 2.2). In those initial studies randomly isolated DNA from the chromosome or from a temperate bacteriophage with a size of about 1.5 kb served as homology region. Integration occurred at low frequency and a few transformants were found per μg of the integration plasmid DNA under conditions that replicating plasmids showed $10^5 - 10^6$ transformants per μg.

Table 2.2 Plasmids used for the construction of lactococcal integration vectors, the host used for their replication, and their selection markers

Plasmid	Host	Selection marker(s)	Reference
pBR322	*E. coli*	*cat-194*	Leenhouts *et al.*, 1989
pSC101	*E. coli*	*ery-194*	Leenhouts *et al.*, 1990
pUC19	*E. coli*	*cat-194*	Simons *et al.*, 1993
pACYC184	*E. coli*	*cat-194*	Simons *et al.*, unpublished
pE194	*B. subtilis*	*ery-194*	Chopin *et al.*, 1989
pUB110	*B. subtilis*	*ery-194 cat-194*	Leenhouts *et al.*, 1990
pTB19	*B. subtilis*	*ery-194*	Leenhouts *et al.*, 1990
pWV01 Ts	*L. lactis*	*ery-194*	Maguin *et al.*, 1992

Further studies showed that integration had occurred via homologous single-cross-over recombination but that the final configuration was dependent on the resistance marker and the used replicon (see section 2.3.2). From a later study it appeared that homologous integration can occur along the entire chromosome and that the single cross-over frequency is logarithmically proportional to the length of the homology region and frequencies varied from approximately 10^{-4} with a homology region of as small as 360 bp to more than 10^{-2} with fragments of 2.5 kb or larger (Biswas *et al.*, 1993).

Using similar non-replicative vectors it was shown that replacement recombination is also feasible in *L. lactis* and can be used to insertionally inactivate chromosomal genes (Leenhouts *et al.*, 1991a). Replacement recombination requires two recombination events to take place. These events between the integration plasmid and the chromosome may occur simultaneously (double cross-over event) or consecutively (single cross-over followed by loop-out deletion). In the first example of replacement recombination the *ery-194* gene was inserted in the *pepXP* gene of *L. lactis* MG1363 (Leenhouts *et al.*, 1991a). Using a similar approach but with other (the *ery-β1* gene from pAMβ1) or no antibiotic resistance markers, mutations were introduced in the *pepN* gene of *L. lactis* MG1363 (van Alen-Boerrigter and de Vos, unpublished results), and the *lacR* and *lacG* genes of MG5276, containing a single copy of the *lac* operon (van Rooijen *et al.*, 1993b; Simons *et al.*, 1993). Replacement recombination also allows deletions to be introduced in chromosomal genes, as was shown for the *nisA* and *lacF* genes (Kuipers *et al.*, 1993a; de Vos *et al.*, unpublished observations). Mutants obtained by replacement recombination or gene inactivation will play an important role in the further analysis of chromosomal genes. Therefore, efficient systems are being developed based on conditionally replicating plasmids. A thermosensitive (Ts) lactose plasmid detected in strain SK11 could be used to promote plasmid integration by homologous recombination in *L. lactis* (Feirtag *et al.*, 1991; Horng *et al.*, 1991). In addition, a Ts derivative was isolated from the broad host-range pGK12 (Maguin *et al.*, 1992). This Ts plasmid, pVE6002, is nonreplicative in *L. lactis* at temperatures above 35°C as a consequence of four mutations in the *repA* gene of the pWV01 replicon. The use of this plasmid as a delivery vector is described in chapter 1. Recently, it was shown that highly efficient replacement recombination can be realized in *L. lactis* by making use of the thermosensitive RCR of this Ts plasmid (Biswas *et al.*, 1993). The excision step during loop out deletion was highly stimulated at permissive temperatures that allowed RCR of the integrated Ts plasmid. The use of this broad host-range Ts plasmid as a delivery vector and for the stimulation of excision during loop out deletion has great potential for the efficient production of mutations in *L. lactis* and other lactic acid bacteria and eliminates the need for high transformation frequencies.

2.3.2 Chromosomal gene amplification and stability

Chromosomal gene amplification is a promising tool to increase the copy number of the gene of interest and obtain overexpression. Since it relies on single cross-over recombination the amplified structure has potential to resolve, resulting in strain instability and loss of amplification. Therefore, specific attention will be given here to factors affecting the stability of the amplified structures that include the selective pressure and growth conditions, the locus of insertion, and integration vector.

By selecting for an increased expression of the integrated sequences it is possible to obtain chromosomal gene amplification in *L. lactis*. This is most easily done by using increased resistance to antibiotics. However, the nature of the resistance gene used and the antibiotic applied appeared to be critical in this approach. Resistance to selective concentrations of antibiotics were found with the single copies of the *ery-194* gene from pE194, the *ery-β1* gene from pAMβ1, and the *tetM* gene from Tn*1545* (Chopin *et al.*, 1989; van Alen-Boerrigter and de Vos, unpublished results; Biswas *et al.*, 1993). In contrast, a single copy of the *cat-194* gene from pC194 did not result in resistance to chloramphenicol above background level (Simons *et al.*, 1993). Metabolic markers could also be used to promote amplification. An unexpected high and stable chromosomal amplification of pNZ305, a replicating derivative of pIL253 carrying the *lacF* gene, was observed in the LacF-deficient *L. lactis* strain YP2-5 (de Vos, 1989). Although the mechanism of integration remains to be elucidated, it is likely that amplification was due to the need for multiple copies of the *lacF* gene in order to support rapid growth on lactose.

The first report on gene amplification by homologous recombination in *L. lactis* involved a single copy insertion of the *ery-194* gene linked to a homology region that was amplified tandemly by using increasing concentrations of the erythromycin analogue clindamycin (Chopin *et al.*, 1989). This antibiotic does not induce the *ery-194* gene and therefore erythromycin also was added at an inducing concentration. A six- to nine-fold amplification was obtained but these amplified sequences were unstable in the absence of selective pressure (loss of approximately 10^{-2} per generation). Using a similar approach a *usp45*-α-amylase gene fusion linked to the *ery-194* gene was amplified up to three-fold in the chromosomal *usp45* locus resulting in an increased secretion of the α-amylase (van Asseldonk *et al.*, 1993c).

Since a single copy of the *cat-194* gene is not sufficient to provide resistance to selective concentrations (5 μg/ml) of chloramphenicol, multicopy integrations up to 15 tandem copies per chromosome are selected when using this marker (Leenhouts *et al.*, 1989; Simons *et al.*, 1993). These multicopy integrations may be either caused by the direct integration of plasmid multimers or by the amplification of single copy insertions. Under non-selective conditions the amplifications in an undefined chromosomal locus were gradually lost although a single copy remained stably integrated in the chromosome (Leenhouts *et al.*, 1989). Such instability however could be dependent on the location of the insertion and the growth conditions as was observed when insertions were introduced in the chromosomal *lac* genes of strain MG5267 (van Rooijen *et al.*, 1992). This was realized with the integration vector pNZ932, a pUC19-derivative carrying the *cat-194* gene, containing homologous fragments of the *lac* operon (Simons *et al.*, 1993). Multicopy insertions immediately downstream from an intact copy of the *lacE* gene were stably maintained during growth on media containing glucose as sole energy source. During growth on lactose, however, a high degree of instability was observed as a consequence of a reduced growth rate. This was attributed to a polar effect on the expression of the distal *lacG* gene that is essential for growth on lactose. In contrast, multicopy (11-fold) insertions downstream of a functional *lacG* gene, resulting in an additional amount of approximately 50 kb of foreign sequences, did not affect the growth rate and were maintained stably both on lactose- and glucose-containing media (Simons *et al.*, 1993). This indicates that multicopy insertions may be stably maintained and have potential for increasing gene expression.

The nature of the used replicon also affects the stability of the amplified structures as was found when pUB110 was used as an integration vector. Amplified plasmid copies of pUB110 derivatives were unstable and lost completely upon non-selective growth. Plasmid pUB110 is a RCR plasmid and has residual activity in *L. lactis* since it was found to produce single-stranded DNA. The instability found with integrated pUB110-derivatives may be caused by stimulation of recombination analogous to that promoted by replicon activation of the integrated Ts plasmid in *L. lactis* (Biswas *et al.*, 1993). This also explains the high instability of amplified structures obtained with pE194, that presumably shows some activity in *L. lactis* (Chopin *et al.*, 1989; see above) Therefore, in order to achieve stable chromosomal amplification the use of RCR plasmids that show residual activity in *L. lactis* should be avoided.

2.3.3 Integrative gene expression

Systematic studies on the expression of stable chromosomally integrated genes should involve the effect of the locus used for integration, read-through transcription at the integration site, and level of amplification. Although such studies have not been done in *L. lactis* there are various generalizations to be made based on the integration studies reported so far.

From the first study on gene amplification by homologous recombination it appeared that an increase in amplification level resulted in an increased level of resistance to clindamycin (Chopin *et al.*, 1989). However, the expression level of the amplified *ery-194* gene was not quantified. In addition, in another study no amplification was obtained with clindamycin, which was attributed to strain differences (Leenhouts *et al.*, 1990).

In a more recent study, stable insertion and amplification was obtained of the proteinase (*prtP* and *prtM*) genes by using a pTB19-derivative carrying the *ery-194* gene as an integration vector (Leenhouts *et al.*, 1991b). The amplification level varied between different isolates of the integrants and two cultures were obtained with an amplification level of about 2 and 8 copies of the *prt* genes per chromosome. Although it is difficult to estimate the amount of the cell-envelope located proteinase (see chapter 4) it appeared that the culture with the highest amplification level produced approximately 11 times more proteinase than the other culture, indicating that there is a non-linear correlation between the number of integrated copies and level of gene expression. Recently, various copies of an *usp45*-α-amylase gene fusion were stably integrated in the *usp45* or *lacX* locus of the isogenic strains MG1363 or MG5267 (van Asseldonk *et al.*, 1993c). Integration of the gene fusion flanked by sequences known to terminate transcription in either the *usp45* or the *lacX* locus did not affect expression and showed a linear relation between the number of integrated copies (one or four) and the production of the α-amylase. However, when the integration was realized in such a way that the upstream sequences of the *usp45* promoter were reconstituted, overexpression was obtained with just a single copy integration and only increased marginally after amplification to two or three copies. These results indicate that there is no detectable effect of the locus on the expression of this reporter gene fusion and that integrative gene expression can be used for the overproduction of a heterologous protein.

In conclusion, the amplification level obtained with chromosomal insertions in *L. lactis* does not exceed 15 copies per chromosome which is considerably less than that obtained with high copy number plasmids. However, overexpression can be achieved even with single copy integrations in regions with strong transcriptional activity.

2.4 Gene expression signals

Studies on gene expression in lactococci and other lactic acid bacteria have received considerable attention in recent years and useful information has been collected since previous reviews (de Vos, 1987; de Vos *et al.*, 1989a; van de Guchte *et al.*, 1992a). Here the main features of lactococcal transcription and translation are summarized.

2.4.1 Vectors for the selection of expression signals

In initial approaches to study gene expression in *L. lactis* various selection vectors were developed containing promoterless reporter genes. These vectors were used to isolate sequences that activated the reporter genes but proved also instrumental in analysing complex regulatory systems as will be detailed below. In most cases the isolated sequences contain signals involved in initiation of both transcription and translation since these are often closely linked to each other. In addition, vectors in which a promoter sequence has been introduced can be subsequently used to select for fragments carrying a terminator of transcription.

Two classes of reporter genes have been used. The first class allows dominant selection and includes genes for antibiotic resistance or that provide complementation of auxotrophic mutations. The other class of reporter genes codes for a detectable phenotype that allows macroscopic identification of colonies. These reporter genes were inserted in selected cloning vectors described above and in most cases equipped with versatile multiple cloning sites. Here we describe the current *L. lactis* vectors that enable the isolation of expression sequences and allow quantification of promoter efficiency in lactococci. The various vectors, their reporter genes, and the nature of their replicon are listed in Table 2.3.

As antibiotic resistance reporter genes two promoterless chloramphenicol acetyltransferase (*cat*) genes have been used that had shown their utility in other Gram-positive bacteria, the *cat-86* gene from *Bacillus pumilis* (Williams *et al.*, 1981) and the *cat-194* of *Staphylococcus aureus* (Band *et al.*, 1983). Although both *cat* genes are homologous, it is difficult to compare the efficiency of promoters cloned in front of one of the two *cat* genes on the basis of resistance to chloramphenicol. This is not only due to differences in copy numbers of the used vectors that are based on various replicons but can also be attributed to differences in the cat genes and their gene products. In addition, the expression of the *cat-86* gene is regulated by translation attenuation in contrast to the non-inducible *cat-194* gene (Lovett, 1990). Since the inducibility may limit the possibilities of the *cat-86* reporter gene, a vector was developed based on a mutated version of the *cat-86* gene which is constitutively expressed when activated by the insertion of a promoter sequence (Bojovic, *et al.*, 1991). Notably, it was found that in *L. lactis* the constitutively expressed *cat-86* gene has four times higher activity than the inducible version of the same gene when both have the same promoter inserted upstream of the *cat-86* gene.

A major disadvantage of the promoter screening vectors based on the *cat-86* or *cat-194* reporter genes is the difficulties in correlating the resistance level to promoter strength in *L. lactis* due to the low level of chloramphenicol resistance and CAT activities as compared to other hosts. This was found in comparative studies in which the same promoters were analysed in *L. lactis, B. subtilis* and *E. coli* followed by correction of the obtained data for differences in copy number of the used vectors (van der Vossen *et al.*, 1987; Simons, unpublished results).

Table 2.3 Main features of vectors used for the isolation and chatacterization of expression and secretion signals. * Indicates non-inducible mutant

	Reporter gene	Selection marker	Replicon	Reference
Promoter selection vector				
pGKV210	cat-86	ery-194	pWV01	van der Vossen et al., 1985
NZ220	cat-86	knt-110	pSH71	de Vos and Simons, 1987
pMU1328	cat-194*	ery-194	pVA380-1	Achen et al., 1986
pKTH1750	cat-86	ery-194	pSH71	Koivula et al., 1991a
pBV5030	cat-86*	ery-194	pWV01	Bojovic et al., 1991
pNZ336	lacG	cat-194	pSH71	Simons et al., 1990a
pKSB8	luxAB	cat-194	pSH71	Ahmad and Stewart, 1991
pNZ272	gusA	cat-194	pSH71	Platteeuw et al., 1993
Translational fusion vector				
pFX3,4,5,6	lacZ	cat-194	pDI125	Xu et al., 1991
pNZ262	prtP-lacZ	cat-194	pSH71	de Vos and Simons, 1988
pMG57	P32-lacZ	ery-194	pWV01	van de Guchte et al., 1992a
Terminator selection vector				
pGKV11	P_{SPO2}-cat-86	ery-194	pWV01	van der Vossen et al., 1985
pGKV259	P_{59}-cat-86	ery-194	pWV01	van der Vossen et al., 1985
Signal sequence selection vector				
pGA14	amyL	ery-194	pWV01	Perez-Martinez et al., 1992
pGB14	bla	ery-194	pWV01	Perez-Martinez et al., 1992
pKTH33	bla	cat-194	pSH71	Sibakov et al., 1991

Plausible explanations for these observations include a low translational efficiency of the *cat* genes or a short half-life of the CAT enzyme in *L. lactis*.

A good correlation between expression level and enzyme activity would be expected from a well-expressed, homologous reporter gene encoding a readily detectable enzymatic activity. Therefore, a promoter screening vector was constructed based on the promoterless lactococcal *lacG* gene encoding phospho-β-galactosidase that already had been used to show the effect of gene dosage in *L. lactis* (de Vos and Gasson, 1989; de Vos, unpublished observations). The developed vector, pNZ336, also contains translational stop codons in all three reading frames between the multiple cloning site and the *lacG* gene (Simons *et al.*, 1990a). As a result, initiation of translation of the *lacG* gene occurs in all constructs at the same position independent of the sequence of the inserted promoter fragment. Complementation of a LacG-deficient host using lactose-indicator plates should be possible but the hitherto obtained *lacG* deletion strains that were constructed by replacement recombination, showed undesired levels of a cryptic phospho-β-glucosidase activity (Simons *et al.*, 1993). Alternatively, it is possible to directly assay phospho-β-galactosidase activity using 5-bromo-4-chloro-3-indolyl-β-D-galactopyranoside phosphate (X-gal phosphate) but this chromogenic substrate is not yet commercially available (W. Hengstenberg, personal communication).

The well-studied *E. coli lacZ* gene for β-galactosidase was also incorporated in expression-selection vectors after its expression was demonstrated in *L. lactis* (de Vos and Simons, 1988). This resulted in a set of gene fusion vectors that allowed the isolation and study of transcription-translation fusions (Xu *et al.*, 1991). Other vectors based on the *lacZ* gene were developed to analyse the efficiency of translational coupling and the effect of mRNA secondary structure on translation initiation (van de Guchte *et al.*, 1991; van de Guchte *et al.*, 1992b).

Other useful reporter genes include those based on the *Vibrio fischeri luxAB* genes that when expressed in *L. lactis* have shown to result in bioluminescent lactococci upon addition of exogenous dodecanol (Ahmad and Stewart, 1991). In addition, these *lux* genes have also been used for the analysis of the inducible *lac* promoter in *L. lactis* (van Rooijen *et al.*, 1991; Eaton *et al.*, 1993a). Finally, a recently developed promoter screening vector is based on the promoterless *E. coli* β-glucuronidase (*gusA*) gene (Platteeuw *et al.*, 1993). This vector pNZ272 has been used to isolate chromosomal promoters by screening for blue colonies on plates containing the chromogenic substrate 5-bromo-4-chloro-3-indolyl-β-D-glucuronic acid (X-gluc). In addition, because of the broad host-range replication properties of the used replicon of pSH71 it was possible to compare the efficiency of three characterized promoters from the bacteriophage, plasmid and chromosome, in *L. lactis* with that in *Lactobacillus casei, Lactobacillus plantarum* and *Leuconostoc lactis* (Platteeuw *et al.*, 1993). These studies revealed a considerable difference in promoter efficiency between the various lactic acid bacteria suggesting that dedicated promoters should be used when optimizing expression in lactic acid bacteria.

2.4.2 Transcription initiation and termination

From all studies on lactococcal gene expression, transcription initiation has received most attention. This process is now being studied from the two possible perspectives. On the one hand promoters are identified and analysed as will be discussed below and on the other hand studies have been initiated to characterize proteins that determine the specificity of transcription initiation. Recently, the major sigma (σ) factor from *L. lactis* has been characterized by the cloning and sequencing of its gene (Araya *et al.*, 1993). This σ factor binds to the core RNA polymerase and makes sequence-specific contacts with the promoter. The *rpoD* gene for the major σ factor of *L. lactis* was found to encode a 38.8 kDa protein with high similarity to the vegetative *B. subtilis* σ^{43} factor and the C-terminal part of the *E. coli* σ^{70} factors. Remarkably, the lactococcal *rpoD* gene could partially complement an *E. coli rpoD* Ts mutant. Based on these results it was proposed to designate the RpoD protein the *L. lactis* σ^{39} factor (Araya *et al.*, 1993).

Presently more than 40 promoters have been characterized by sequence analysis and mapping of the transcription initiation site by primer extension or S1-nuclease mapping (Figure 2.2). These promoters have been derived from random screening experiments or from defined genes on chromosomal, bacteriophage or plasmid DNA. Almost half of the characterized promoter sequences have been isolated in random cloning experiments with the vectors based on the promoterless *cat* genes (Figure 2.2A). The function or regulation of these promoters is not known and therefore it is difficult to conclude on their significance and contribution to gene expression. The only available information is their location as deduced from the source of DNA used to isolate them. Chromosomal promoters (Figure 2.2A) have been isolated from *L. lactis* Wg2 (P21, P23, P32, P44, and P59) and *L. lactis* MG1363 (P1, P2 and P21). Bacteriophage promoters (Figure 2.2B) have been derived from DNA of the temperate phage BK5-T (Pal, Pf2, Pa3, Pg2, and Pf1) and virulent phage φ SK11G (P11G). Sequences with transcriptional activity from strain Wg2 were selected directly in *L. lactis* whereas the others were initially selected in *E. coli*, *B. subtilis* or *S. sanguis* and subsequently transformed into *L. lactis*.

The identification of a rapidly growing number of lactococcal genes allowed the characterization of their transcriptional organization. While most initial attention has primarily been focused on plasmid-located genes most of the presently defined promoters have been derived from chromosomal genes (Figure 2.2C–2.2G). Several of the defined promoters seem to be regulated and will be discussed below. For others there is presently no information available about their regulation.

Comparison of the listed promoter sequences (Figure 2.2) shows that in all cases the transcriptional initiation site, usually a purine base, is preceded within 4–10 bp by an AT-rich hexanucleotide sequence. Alignment of these promoter sequences with respect to the conserved −10 region shows that most of the lactococcal promoters contain the canonical −35 and −10 consensus sequences found for the promoters recognized by the *E. coli* σ^{70} and *B. subtilis* σ^{43} transcription factors (Harley and Reynolds, 1987; Moran, 1989). Although the sequences that are bound by the main lactococcal transcription factor have not been identified yet, these results strongly suggest that the consensus promoter recognized by the *L. lactis* σ^{39} transcription factor is characterized by a −35 TTGACA and −10 TATAAT sequence that is spaced by 17 nucleotides. Sequence conservation is higher in the −10 region than in the −35 region. This especially holds for the positively controlled promoters that also show a larger spacing between the canonical sequences and the *usp45* promoter that will be discussed below (Figure 2.2F and 2.2G).

Approximately half of the promoter sequences contain a TG doublet one bpupstream of the −10 region, confirming previous observations with a limited number of lactococcal promoters (de Vos, 1987; van de Guchte *et al.*, 1992a).

PROMOTER	SEQUENCE	Spacing	Reference

A. Undefined chromosomal promoters

P21	AGTTTTTTCTTGACAGAAGAAGGCGAAAAATGGTATTATATTAGGTACT	18	van der Vossen et al., 1987
P23	ACCTAAGACTGATGACAAAAGAGCAAATTTGATAAAATAGTATTAGAAT	17	van der Vossen et al., 1987
P32	TGAGCTTGGACTAGAAAAAACTTCACAAAATGCTATACTAGTAGGTAAT	22	van der Vossen et al., 1987
P44	AGCTAAACTCTTGTTTACTTGATTTTATGTTAAAATAATTAATGAGTGTA	18	van der Vossen et al., 1987
P59	ACATTAAATTCTTGACAGGGAGAGATAGGTTTGATAGAATATAATAGTTGT	17	van der Vossen et al., 1987
P1	GAAAGAAAGACTTGCATTTGTTGTTGAAAAAATGCTAAAATACATAAGTCCG	17	Koivula et al., 1991a
P2	GAGTTTTTTCTTGACAGAAGATGGCGAAAAAATGGTATTATATCTAGGTACT	18	Koivula et al., 1991a
P21	TCAAGTTGACCTTGAAAAAAAACTGAAAATCTGTTATCATAAATAATGGAC	17	Koivula et al., 1991a

B. Undefined bacteriophage promoters

Pa1	TCGAATTTTCTTCCATATTTCAAAGAATCAGTTACTATCTAACGATCAT	18	Lakshmidevi et al., 1990
Pf2	TATATGAAAAATGACTGATGTAACTCCGTTGACTGATGATGCAGAAG	17	Lakshmidevi et al., 1990
Pa3	ATTCATTATTTTTATAATCCTCACTAGTTATACATATAGTATTTGGGTTTT	18	Lakshmidevi et al., 1990
Pg2	TATGGAAAAATACAGACAAGCAAACTAAGGAGGGTATATTGAATGACCGAC	16	Lakshmidevi et al., 1990
Pf1	AAAATGTAAGATTGGAGTTACTAAAACAGTAACTTACTCCACTGGAGGTA	17	Lakshmidevi et al., 1990
P11G	TATGGCCTATTTTAGACAGAATCAGCTTCTTGTTAAAATGGGAGAGCAAC	15	Platteeuw et al., unpublished

C. Defined promoters

usp45	TCATAAAGAAATATTAAGGTGGGGTAGGAATAGTATAATGTTTATTCA	14	van Asseldonk et al., 1990
infA	AAAGAAAGACTTGCATTTGTTGTTGAAAAAATGCTAAAAATACATAAGTCCG	17	Koivula et al. 1991b
pepN	TTCTTTGTACTCGAAATTTTCTATTCAATTGATAATAATATTATTAATAC	17	Tan et al., 1992
gap	GGTAGTTTATGTTTGCAAATTTTAAAAAAGTGTTAAAATAAAAGAGTAAG	16	van Asseldonk et al., unpublished
als	GTCAAATAATTGTAAAAGGTTCTATTATCTGATAAAATGATTGTGAAGT	17	Marugg et al., unpublished
rrnA	CAGTTAATTCTTGACAAGTTTAGTTAGGTTTGATAGAATATAATAGTTGT	17	Chiaruttini and Millet, 1993
sacB	GTGGCAAACGCTTGACATATATCAAAAAAATGATAAAATAACTTCTGTAA	16	Rauch and de Vos, 1992b
sacA	CTTATTTATATTGATTTTTTTTATAAAAAACGTTTATCATAAATATATATA	17	Rauch and de Vos, 1992b
lct	TCATTAGTTATTGCATTTTACTAATCGAAGGAGTATAAAGATTTCGAAGG	17	Piard et al., 1993
trpA	TTTTAAATCCTTGACAAGTCTCATAAAAAGTTTTTACAATTATTTTCAACA	17	Bardowski et al., 1992; Chopin, 1993

D. Defined and negatively controlled promoters

lacA	ACAAAAATAGTTGCGTTTTGTTGAATGTTTGATATCATATAAACAAAGA	17	van Rooijen et al., 1992
repC	AGCTTTATTGTGTTTTTATGATTACAAAGTGATACACTAATTTTATAAA	17	van Rooijen and de Vos, unpublished
dnaJ	GCTAATTTTTTTGCCAAAAATGAAAAAACGTGGTAAAATAGTGCTATTGA	17	van Asseldonk et al., 1993b

E. Defined and putative negatively controlled promoters

prtP-S	TAAAATTTCGTTGAATTGTTCTTCAATAGTATATAATAATAGTATA	16	Vos et al., 1989b; Marugg et al., unpublished
prtP-W	TAAAATTTCGTTGAATTGTTCTTCAATAGTATATAATAATAGTATA	16	van der Vossen et al., 1992
prtM-S	AACCCTACGCTTGATGTAGTTAAGATTATATTATTAATTAATATTATA	17	Marugg et al., unpublished
prtM-W	AACCCTACGCTTGATGTAGTTAAGATTATATTATTAATTAATATACTA	17	van der Vossen et al., 1992
hisC	AAAATAAATAATAACATAACTATTATAATTCAGATAATTA	17	Godon et al., 1992; Chopin, 1993
leuL	TATTATTTTATTGACAATTTAAAATATTAAGAGTATTATAATGTAAATTA	17	Godon et al., 1992; Chopin, 1993
ilvD	TGACAGATTATGTATTTTCATTTTTTAGTGATAAAATAGCTCTATGTA	17	Godon et al., 1992; Chopin, 1993

F. Defined and positively controlled promoters

nisA	AACGGCTCTGATTAAATTCTGAAGTTTGTTAGATACAATGATTTCGTTCG	20	Kuipers et al., 1993a
lacR	TTAATTTTTTGTTTTTTTTTTTATTTGTTTTTTTAAAAAATAGATAACACCG	19	van Rooijen and de Vos, 1990

G. Defined and possibly positively controlled promoters

nisR	ATTGATAGATATATTTCTTCAGAATGAATGGTATAATGAAGTAATGAG	22	Kuipers et al, unpublished
lcnMa	ACATTTGTTAACGAGTTTATTTTTATATAATCTATAATAGATTTATAAA	23	van Belkum et al., 1991
lcnA	ACATTTGTTAACGAGTTTATTTTTATATAATCTATAATAGATTTATAAA	23	van Belkum et al., 1991

Figure 2.2 Lactococcal transcription initiation sites. Transcriptional initiation sites, conserved TG-dinucleoside and putative –35 and –10 hexanucleotides are underlined. The spacing between the putative –35 and –10 sequences is also presented. Promoters are arranged in eight groups according to their isolation strategy, location, or (possible) control. Sequences are aligned with respect to the conserved –10 sequences or transcriptional initiation sites.

This feature is weakly conserved in *E. coli* but also found in promoters of other Gram-positive bacteria (Moran, 1989).

As reported previously, the DNA region upstream of the −35 region of lactococcal promoter sequences is characterized by a higher A+T content (up to 78%) than the average value reported for *L. lactis* DNA (62.8%). The A+T rich region at position −322 to −76 of the *lac* promoter has been shown to contribute more than tenfold to the efficiency of transcription initiation in *L. lactis* but not so in *E. coli* (van Rooijen *et al.*, 1992). A similar effect of the upstream A+T rich region was recently also observed for the *usp45* promoter, that shows an unusual spacing between the canonical hexanucleotides. The *usp45* promoter was studied in fusions of the *usp45* gene with the structural part of the *amyS* gene of *B. stearothermophilus* (van Asseldonk *et al.*, 1993c). A six-fold higher level of expression was obtained in plasmids when the promoter region was extended with a 121 bp upstream region (position −247 to −36). Similar observations have been reported for other bacterial promoters and suggest that the regions upstream of the promoter contribute to the efficiency of transcription initiation in *L. lactis*. It has been demonstrated that curving of A+T-rich sequences could contribute to an elevated expression of the downstream located DNA region (Bracco *et al.*, 1991; Gartenberg and Crothers, 1991). Indeed, a high curvature score could be calculated for the 121 nt upstream of the −35 promoter region of the *usp45* gene which supports this possibility (van Asseldonk *et al.*, 1993c).

Vectors containing a transcriptionally activated reporter gene can be used to select and analyse sequences that terminate transcription (Table 2.3). This approach has been used to show that the rho-independent terminator of the *B. licheniformis* penicillinase gene is functional in *L. lactis* (van der Vossen *et al.*, 1985) and to show that the presumed *prtP* terminator has only limited efficiency in *L. lactis* (van der Vossen *et al.*, 1992). In addition, many genes and operons contain typical GC-rich palindromes followed by T-rich sequences at their 3' end suggesting that rho-independent terminators are common in *L. lactis*. In some cases such as that of the *L. lactis pepN* gene (Tan *et al.*, 1992; Strøman, 1992), the size of the transcript and the mapped transcription initiation site supports the functionality of such a terminator *in vivo*. However, no lactococcal terminators have so far been characterized by S1 mapping. In the *L. lactis rrn* operon a putative transcriptional terminator located downstream of the distal tRNA[asn] gene was found to be inefficient in arresting transcription (Chiaruttini and Millet, 1993). Whether this is due to the presence of several consensus antitermination signals in this *rrn* operon that conform to the *E. coli* boxA and boxB remains to be established.

2.4.3 Translation initiation and codon usage

Base pairing between a purine-rich sequence upstream of the initiation codon in the mRNA and a complementary sequence near the 3' end of 16S rRNA is required for the initiation of translation. Many putative translation initiation sites for *L. lactis* genes have been postulated. However, at present the amino terminal peptide sequence of *L. lactis* gene products and the nucleotide sequence of their corresponding genes have only been determined in a few cases and those are shown in Figure 2.3. In most cases the AUG triplet is used as translational start codon. Only LacF initiates with a GUG triplet. As often observed in bacterial systems the N-terminal methionine is cleaved off in some cases. In the region preceding the AUG or GUG codon, Shine and Dalgarno (SD) sequences (Shine and Dalgarno, 1974) could be detected that showed complementarity to the 3' end of *L. lactis* 16S rRNA which has the sequence 3'-CUUUCCUCC-5' (Chiaruttini and Millet, 1993). The free energies (ΔG) of these SD sequences ranges from −8.4 to −17.8 kcal/mol with a spacing between the SD region and the initiation codon of 5–9 nucleotides. The ΔG values found for the *L. lactis* genes described in Figure 2.3 are intermediate between those observed in *E. coli* and *B. subtilis* where values are found of −9 to −12 and −14 to −19 kcal/mol, respectively (McLaughlin *et al.*, 1981). Putative translation initiation sites for more than 20 cloned lactococcal genes have been reported. In all cases AUG is used as initiation codon, although it is known that *L. lactis* also can initiate translation at the initiation codons UUG and GUG from the *Bacillus cat-86* and *nprE* genes, respectively (van de Guchte *et al.*, 1992a). The configuration of the SD sequences and their complementarity to the 3' end of the known lactococcal 16 S rRNA is similar to those shown in Figure 2.3. An exception is formed by the SD sequence of the *usp45* gene for the major extracellular protein from *L. lactis* that shows only limited similarity to the 3' end of the 16S rRNA (ΔG value of only −4.6 kcal/mol) and is spaced by 7 nucleotides from the AUG initiation site (van Asseldonk *et al.*, 1990). A second SD region for the *usp45* gene has been postulated with a higher complementarity to the rRNA (ΔG value of −14.4 kcal/mol), however the spacing to the AUG initiation codon amounts to 21 nucleotides. Site directed mutagenesis experiments indicated that both SD sequences function equally well in *L. lactis* suggesting that deviations from the consensus translational initiation are tolerated (van Asseldonk *et al.*, unpublished observations).

In various lactococcal operons overlapping genes are found suggesting that translational coupling may occur. By changing the distance between the 5' coding region of ORF32 and the *E. coli lacZ* gene it was found that translational coupling could function in *L. lactis* (van de Guchte *et al.*, 1991). The highest coupling efficiency was found with a configuration containing the sequence ATGA in which the stop (TGA) and start (ATG) codons were overlapping, resulting in an approximately three-fold increase of translational efficiency. Evidence that translational coupling contributes to controlling expression of naturally occurring genes in lactic acid bacteria has been provided by the analysis of the β-galactosidase in *Leuconostoc lactis* that is encoded by two partially overlapping genes, *lacL* and *lacM* (David *et al.*, 1992).

The codon usage in *L. lactis* has been the subject of various studies, since the availability of charged tRNAs contributes to the efficiency expression (Sharp

Gene	Relevant Gene and Protein Sequence	ΔG	Reference
nisA	ATAAATTATAAGGAGGCACTCAAAATGAGTACAAAAGATTTTAACTTGGATTTG ******* M S T K D F N L D L	-17.8	[a]
lacF	AAACAUAAAACGGAGGAUAUUGUUGUGAACAGAGAAGAGAUGACUCUCAAUGGG ***** M N R E E M T L L G	-14.4	[b]
pepXP	CAUGUUUAUUACGGAGGAUUUAAAAUGCGCUUUAACCAUUUUUCAAUUGUUGAC ***** M R F N H F S I V D	-14.4	[c]
lacG	CUUUUUUUGAAAGGACUUACACUUAUGACUAAAAACACUUCCUAAAGAUUUUAUU ******* M T K T L P K D F I	-14.0	[d]
pepO	CUUUUAUUAAAAAAGGAGUUUGAUAUGACAAGGAUUCAAGAUGAUUUAUUCGCU ******* M T R I Q D D L F A	-12.8	[e]
pepN	AAUACUGAAUAUUUAGGAGAAGAUAUGGCUGUAAAACGUUUAAUUGAAACUUUU ***** M A V K R L I E T F	-11.6	[f]

sacA	AAUUUAGUAAAUGAGGAAAAAAAAAUGAAAUGGUCUACCAAACAACGAUAUCGU
	       **** M K W S T K Q R Y R
	                     Q
		-9.4	[g]

repA	UUGUUUUUGAUAAGGUAAUAUAUCAUGGCUAUUAAAAAUACUAAAGCUAGAAAU **** M A I K N T K A R N	-8.4	[h]
lacA	AGUACCACGAUUAAGGUAUAACCAAUGAUUCUGACAGUCACACUAAAUCCUUCA **** M I L T V T L N P S	-8.4	[i]
16S rRNA	3' UCUUUCCUCC-5'		[j]

Figure 2.3 Lactococcal translational initiation sites. [a] van der Meer *et al.*, 1993b; [b] de Vos *et al.*, 1990; [c] Mayo *et al.*, 1991; [d] de Vos and Gasson 1989; [e] Mierau *et al.*, 1993; [f] Tan *et al.*, 1992; [g] Rauch and de Vos, 1992b; [h] de Vos, unpublished results; [i] van Rooijen *et al.*, 1991; C.H. Moore, personal communication; [j] Chiaruttini and Millet, 1993.

and Li, 1987; Kurland, 1991; Andersson and Kurland, 1990). A compilation of the codon usage of 23 *L. lactis* genes has been reported (van de Guchte *et al.*, 1992a). Recently, this was extended to include 74 genes derived from the lactococcal chromosome, plasmids or bacteriophages (Chopin, 1993). Trends in lactococcal codon usage were noted and included a strong preference for the CGU (Arg) and AUU (Ile) codons and a selection against AGG (Arg) and CUA (Leu) codons. Eccentric codon usage was noted for several genes that are usually highly expressed in bacteria and include the *rpmG, rpmJ* and *infA* genes, that code for ribosomal proteins and a translation initiation factor, respectively, the *citP* encoding the citrate permease, and several *lac* genes involved in the degradation of lactose (see section 2.5.1).

2.5 Control of gene expression

Parallel with the growing interest in gene expression and organization, studies on regulation of gene expression in *L. lactis* are emerging. Since these are essential for the development of inducible expression systems, several of these systems will be described here with emphasis on those that are controlled at the transcriptional level by known regulators.

2.5.1 Lactose utilization and its control

The most important function of lactococci in industrial dairy fermentations is the fermentative conversion of lactose into lactic acid. The metabolic conversions involved in this process have been established and include the lactose phospho-transferase system (PTS), the tagatose-6-phosphate pathway and the glycolytic pathway (Figure 2.4). It has been known for decades that the key enzymes in the utilization of lactose are induced during growth on lactose and often are located on plasmids (McKay, 1982). In recent years, the mechanism by which the utilization of lactose is controlled has been elucidated at the molecular level resulting in the first model for gene organization and expression in *L. lactis*. Those studies that have led to the identification of *cis* and *trans* acting factors in the control of lactococcal gene expression are summarized here.

The *lac* genes for the lactose-specific enzymes of the PTS (LacEF), the phospho-β-galactosidase (LacG) and the enzymes of the tagatose-6-phosphate pathway (LacABCD) have been located on the indigenous 23.7 kb plasmid pMG820 in *L. lactis* strain MG1820 (de Vos and Gasson, 1989; de Vos *et al.*, 1990; van Rooijen *et al.*, 1991). They are organized in a 7.8 kb operon with the gene order *lacABCDFEGX* that is followed by an *iso*-IS*SI* element (Figure 2.5). The function of the distal *lacX* gene is presently unknown but seems to be dispensable for growth on lactose (Simons *et al.*, 1993). The *lac* genes are transcribed as two 6 and 8 kb polycistronic transcripts of the *lacABCDFE* and *lacABCDFEGX* genes, respectively. An inverted repeat is located between the *lacE* and *lacG* genes and could function as an intercistronic terminator. Transcription of the *lac* operon was found to be regulated by the product of the divergently transcribed 0.8 kb *lacR* gene (van Rooijen and de Vos, 1990). The deduced amino acid sequence of the 28 kDa LacR protein appeared to be highly similar to that of several *E. coli* repressors, now known as the DeoR family of repressors.

Northern blot analysis showed that the transcription of the *lac* operon is induced up to tenfold during growth on lactose. In contrast, transcription of the *lacR* gene is similarly induced during growth on glucose. The *lac* promoter is located in a back-to-back configuration with the promoter of the divergently transcribed *lacR* gene, encoding the LacR repressor (Figure 2.6). Identical transcription start sites are used under induced (lactose) and non-induced (glucose) conditions (van Rooijen and de Vos, 1990; van Rooijen *et al.*, 1992).

In subsequent genetic studies using LacR-overproducing and LacR-deficient strains it was shown that LacR is the repressor of the *lac* operon and presumably activates its own transcription (van Rooijen and de Vos, 1990; van Rooijen *et al.*, 1992). This was confirmed in physical studies using purified LacR that

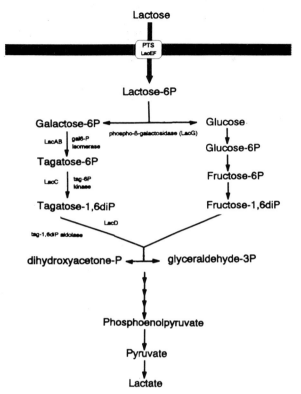

Figure 2.4 Pathway of lactose utilization by industrial *L. lactis* strains. The enzymes encoded by the *lac* operon are indicated.

showed the presence of two operators with different affinities for lacR (van Rooijen and de Vos, 1993): one high affinity operator *lacO1*, which shows a marked bidirectional symmetry, is located in the *lac* promoter (Figure 2.6), while the other operator *lacO2*, with a three-fold lower affinity, located between the *lacR* promoter and the start of the *lacR* gene.

Protein engineering studies with the LacR repressor that was overproduced, purified and found to form dimers, provided evidence for the presence of two domains in the family of DeoR repressors. One is located at the N-terminus and includes a helix-turn-helix motif involved in DNA binding (van Rooijen *et al.*, 1993c). The other includes a number of conserved charged residues at the C-terminal end that were shown to provide the response to the inducer, which in the case of LacR was found to be tagatose-6-phosphate (van Rooijen *et al.*, 1993a; van Rooijen and de Vos, 1993).

Combination of the physical and genetic results has led to a model for the regulation of *lac* gene expression that includes the following three stages.

First, binding of the LacR repressor to *lacO1* during growth on glucose activates transcription of the *lacR* promoter. The *lacR* promoter has an unusual

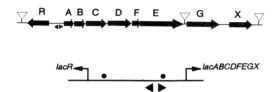

Figure 2.5 Genetic and transcriptional organization of the pMG820 *lac* operon. The intercistronic promoter region is enlarged. The black arrowheads indicate the transcription initiation sites, the white triangles the transcription terminators, and the black dots the position of *lacO2* (left) and *lacO1* (right).

Figure 2.6 Nucleotide sequence of the lac promoter region. The canonical sequences and transcriptional initiation sites of the *lac* operon and *lacR* promoter are indicated as are the nucleotides protected from DNase digestion by LacR (van Rooijen and de Vos, 1993).

spacing between the −10 and −35 canonical sequences observed more often for activated promoters (Figure 2.2; Hoopes and McClure, 1987). In addition, the location of *lacO1* and the *lacR* transcriptional start site coincides with the distance commonly observed for a transcription activator (Collado-Vides *et al.*, 1991).

Second, binding of LacR to *lacO2* at increasing LacR concentrations during growth on glucose results in repression of the *lacR* gene and *lac* operon expression. Since *lacO2* has a lower affinity for LacR than *lacO1*, it will be bound at increasing LacR concentrations. *In vivo* studies using transcriptional fusions of the *lac* promoter to the promoterless *cat-86* gene have shown that repression during growth on glucose requires the presence of both *lacO1* and *lacO2* (van Rooijen *et al.*, 1992). This may be attributed to the formation of a DNA loop between these operators as has been described for other regulatory systems (Matthews, 1992).

Third, binding of the LacR repressor to tagatose-6-phosphate during growth on lactose results in dissociation of the LacR-operator complex concomitant with the induction of the *lac* operon expression. Previously, it had been found that growth on lactose and galactose induced the *lac* operon expression (McKay, 1983). Tagatose-6-phosphate but no other phosphorylated intermediates formed during catabolism of lactose or galactose were found to prevent LacR binding to the *lacO1O2*. In addition, LacR mutants obtained by site specific mutation of the charged conserved residues of the inducer binding domain of LacR prevented induction *in vivo* and response to tagatose-6-phosphate in vitro (van Rooijen *et al.*, 1993a). These results strongly indicate that the actual inducer of the *lac* operon expression is tagatose-6-phosphate and not galactose-6-phosphate as assumed earlier (McKay, 1983).

Recently, it was observed that in a *L. lactis* strain, in which the *lacR* gene had been deleted by replacement recombination, a second transcriptional control system was operating that reacted to glucose (van Rooijen *et al.*, 1993b). A tetradeca nucleotide sequence that conforms to the consensus sequence involved in catabolite repression in other Gram-positive bacteria has been identified at +12 to +25 in the *lac* promoter region. It is likely that this type of global control also contributes to control of gene expression of the *lac* operon in *L. lactis*. Similar consensus sequences have been found near the transcriptional initiation sites of genes that respond to glucose, including the *sacB* gene of *L. lactis* and the *lacS* gene of *Streptococcus thermophilus*, suggesting that this glucose repression system is widespread in lactic acid bacteria (Figure 2.7).

The molecular characterization of the *lac* genes detailed above has provided insight into two different mechanisms of transcriptional regulation in *L. lactis*: a specific one mediated by the LacR repressor and one that is likely to be more general and is involved in catabolite repression. In addition, the structural genes for lactose transport and degradation have been identified and characterized. These results have been instrumental in developing the first systems for food-grade selection and induction of gene expression in lactic acid bacteria that are widely used now (see sections 2.7 and 2.8).

```
                    TGWAANCGNTNWCA
                    oooooooooooooo
      TATCTATAAACAAAGAAATGATGAAAACGTTATCTTGAACAT     lacA
                    ooooooooo oooo
      AAAAAATGATAAAATAACTTCTGTAAGCGAAATCATTTTATT     sacB
                    ooooooooo oooo
      GACTAACCAATTTTCATATAATGTAAACGTATTCAAATAATA     lacS
```

Figure 2.7 Conserved glucose responsive elements in the promoter region of the *L. lactis lac* operon (van Rooijen *et al.*, 1993b), *L. lactis sacB* (Rauch and de Vos, 1992b), and *S. thermophilus lacS* (Poolman *et al.*, 1989). Transcription start sites are indicated.

2.5.2 Heat shock response

It has been well established that virtually all bacteria show a heat shock response when exposed to temperatures above the normal growth temperature. The heat shock response is characterized by the induction of the synthesis of heat shock proteins (HSPs) including DnaJ, DnaK, and GroEL. Although their actual function is not known, evidence for their role as molecular chaperones is accumulating suggesting that they assist in protein folding. Heat shock response in *L. lactis* has been studied and several HSPs have been identified (Whitaker and Batt, 1991). Presently, the genes for *dnaJ*, *dnaK* and *groESL* have been identified and characterized (van Asseldonk *et al.*, 1993b; Eaton *et al.*, 1993b; Kim and Batt, 1993). They belong to different transcriptional units that are all induced to various levels during heat shock. The promoter of the *dnaJ* gene has been characterized (Figure 2.2) and found to be preceded by an inverted repeat that contains a conserved motif in a number of Gram-positive bacteria (Figure 2.8; van Asseldonk *et al.*, 1993b). Deletion of these sequences resulted in constitutive expression of the *dnaJ* gene as determined by using a promoterless *usp45-amyS* gene fusion as a reporter system, indicating that the

```
dnaJ             AATTAGCACTCTTATAAAAAGAGTGCTAATT
                   ********            *********
dnaK/grpE        TGTTAGCACTTAGAGGTAAAGAGTGCTAAAA
                   *********           *********
groES/groEL      TGTTAGCACTCGTTTAATAAGAGTGCTAAAA
```

Figure 2.8   Conserved heat shock motif in the heat shock genes *dnaJ* (van Asseldonk *et al.*, 1993b), *dnaK* (Eascton *et al.*, 1993b), and *groELS* (Kim and Batt, 1993).

heat shock response of the *dnaJ* gene is repressible. Using these signals the first heat induced secretion system has been developed for *L. lactis* (see above). Interestingly, the promoter regions of the operons that include the *dnaK* and *groELS* genes also contain this conserved motif although it is probably located downstream from the transcriptional initiation start (Figure 2.8). These results indicate that heat-inducible promoters share a conserved sequence motif at their promoter region. Further research is required to determine whether the lactococcal heat shock promoters also react to other stress conditions.

2.5.3 *Negative regulation of transcription*

Apart from the promoters of the *lac* operon (section 2.5.1) and the *dnaJ* gene (section 2.5.2) various other repressible promoters have been found (Figure 2.2D) and some will be discussed here. The first repressor-operator system characterized in lactic acid bacteria was that operating in the copy number control of the broad host-range replicon of the *L. lactis* plasmid pSH71 (de Vos *et al.*, 1989c). The 6 kDa repressor encoded by the *repC* gene contains a helix-turn-helix motif and was found to interact with an operator sequence located around the transcription initiation site of the *repCA* promoter. In this way it controls its own transcription and a steady state level of RepC and RepA is ensured in the cell (Figure 2.9). Since the replication protein RepA is interacting with the plus origin of replication (see section 2.2.3) and is rate limiting for replication, a stable plasmid copy number is maintained. A schematic model for this control system is provided in Figure 2.9.

Promoters that are likely to be under negative control include the Tn*5276*-located promoters of the *sac*-operon controlled by the product of the *sacR* gene containing an N-terminal helix-turn-helix motif (Rauch and de Vos, 1992a and 1992b). In addition, it is likely that the promoters of the *prt* genes from SK11

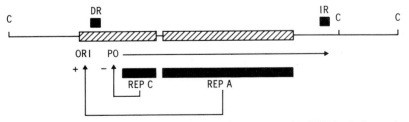

Figure 2.9   Copy number control of the replicon of the *L. lactis* plasmid pSH71 by the interaction of the repressor RepC and the promoter-operator of the *repCA* operon. See text for further explanation.

and Wg2 are also under negative control and induced by specific medium components (Bruinenberg *et al.*, 1992; Marrug and de Vos, unpublished results). These *prt* promoters contain several direct repeats and a region with bidirectional symmetry that is most prominent in that of the SK11 *prtP* gene (de Vos *et al.*, 1989a). Finally, it has been suggested that the initiation of transcription of the *his* and *leu-ilv* operons is controlled at the transcriptional level by a repressor. These operons for histidine and branched chain amino acid biosynthesis have been recently characterized and are induced by histidine and isoleucine, respectively (Delorme *et al.*, 1992; Godon *et al.*, 1992; Chopin, 1993). Apart from this negative control these biosynthetic operons seem to be also controlled by an attenuation mechanism (Chopin, 1993)

2.5.4 Positive regulation of transcription

Only few promoters have been identified that could be under positive control. The activation of the pMG820-located *lacR* promoter has been described above and is under autogenous control of LacR that binds to the upstream-located *lacO1* (Figure 2.6; van Rooijen and de Vos, 1990). The two Tn*5276*-located *nis* promoters are probably also positively regulated. The *nisA* promoter is most likely activated by a two component system activator, the product of the *nisR* gene (van der Meer *et al.*, 1993a). This could also apply to the *nisR* promoter which then should also be autogenously regulated (Kuipers *et al.*, unpublished observations). All these promoters show aberrant −35 sequences that are preceded by inverted repeated sequences, and a large spacing between the putative −35 and −10 sequences. These features are shared by the plasmid-derived promoters from the *lcn* operons. It is therefore tempting to speculate that the *lcn* promoters are under positive control. However, the participation of minor σ factors can not be excluded. Minor σ factors have not yet been reported in lactococci but are an important component in the regulation of gene expression in the environmental response in *E. coli* or *B. subtilis* (Harley and Reynolds, 1987; Moran, 1989). Therefore, it is to be expected that this class of molecules that determine RNA polymerase specificity could also contribute to differential gene expression in *L. lactis*.

2.6 Protein secretion

In general, protein secretion in bacteria can be divided into two main mechanisms. Sec-dependent transport is the general secretory pathway by which proteins with a consensus N-terminal signal sequence are translocated. For this translocation system a variety of proteins are required including SecA, SecY and SecE. SecY protein is an essential component of the protein export machinery in *E. coli* and appears to interact with the signal sequence of secretory proteins but it also shows interactions with at least two other components of the protein

A

Protein	Function	Leader Sequence	Reference
Ss1	unknown	M-K-K-I-L-I-I-G-L-L-I-G-S-I-A-L-G↑I	[a]
Ss30	unknown	M-K-I-W-T-K-L-G-L-S-L-V-G-L-T-L-T-A-C-G↑S	[a]
Ss38	unknown	M-K-K-I-L-I-T-T-L-A-L-A-L-S-L-G-A↑C	[a]
Ss45	unknown	M-K-K-K-I-L-I-I-A-L-I-I-I-A-A-M-S-A-I-F-I-S-A↑K	[a]
Ss80	unknown	M-K-K-I-F-I-A-L-M-A-S-V-S-L-F-T-L-A-A↑C	[a]

B

Protein	Function	Leader Sequence	Reference
PrtP	proteinase	M-Q-R-K-K-K-G-L-S-I-L-L-A-G-T-V-A-L-G-A-L-A-V-L-P-V-G-E-I-Q-A-K-A↑A	[b]
NisP	proteinase	M-K-K-I-L-G-F-I-V-C-S-L-G-L-S-A-T-V-H-G↑E	[c]
Usp45	unknown	M-K-K-K-I-I-S-A-I-L-M-S-T-V-I-L-S-A-A-A-P-L-S-G-V-Y-A↑D	[d]
PrtM	maturation	M-K-K-K-M-R-L-K-V-L-L-A-S-T-A-T-A-L-L-L-S-G↑C	[e]
NisI	immunity	M-R-Y-L-I-L-I-V-A-L-I-G-I-T-G-L-S-G↑C	[f]

C

Protein	Function	Leader Sequence	Reference
NisA/NisZ	lantibiotic	M-S-T-K-D-F-N-L-D-L-V-S-V-S-K-D-S-G-A-S-P-R↑I	[g]
Lct	lantibiotic	M-K-N-L-F-N-L-L-Q-E-V-T-E-S-E-D-L-I-L-G-A↑K	[h]
LcnA	bacteriocin	M-K-N-Q-L-N-F-N-I-V-S-D-E-L-S-E-A-N-G-G↑K	[i]
LcnB	bacteriocin	M-K-N-Q-L-N-F-N-I-V-S-D-E-L-A-E-V-N-G-G↑S	[j]
LcnMa	bacteriocin	M-K-N-Q-L-N-F-E-I-L-S-D-E-L-Q-G-I-N-G-G↑I	[k]

Figure 2.10 Lactococcal leader sequences with defined cleavage sites. A: sequences obtained by screening using signal sequence selection vectors; B: sec-dependent signal sequences derived from defined genes; C: sec-independent leader sequences derived from defined genes. [a] Shibakov et al., 1991; [b] Kok et al., 1985; Vos et al., 1989b [c] van der Meer et al., 1993b; [d] van Asseldonk et al., 1990; [e] Vos et al., 1989; [f] Kuipers et al., 1993a; [g] Buchman et al., 1988; Mulders et al., 1991 [h] Piard et al., 1993; [i] Holo et al., 1991; van Belkum et al., 1991; [j] van Belkum et al., 1991; [k] van Belkum et al., 1992.

translocation system: the peripheral membrane protein SecA and the integral membrane protein SecE. Recently, the lactococcal *secY* gene has been characterized and found to encode a product with considerable similarity to the *E. coli* and *B. subtilis* SecY proteins (Koivula *et al.*, 1991b). This strongly suggests that the sec-dependent pathway is functional in *L. lactis*. The other transport system known in various bacteria is not dependent on the Sec proteins and involves dedicated transmembrane translocators that belong to the family of the ABC-proteins, such as the well-characterized *E. coli* haemolysin translocator HlyB (Pugsley, 1991, 1993). These translocators have an intracellularly located nucleotide binding site suggesting that transport is associated with the hydrolysis of ATP. The proteins that are transported via the sec-independent secretory pathway do not contain the consensus N-terminal signal sequence found in the sec-dependent pathway. At present two translocators that belong to this family have been identified in *L. lactis*, LcnB and NisT (Stoddard *et al.*, 1992; Engelke *et al.*, 1992; Kuipers *et al.*, 1993a). Each of these proteins is assumed to be involved in the specific translocation of a leader peptide-containing bacteriocin. Since no sec-deficient *L. lactis* mutants have been described so far, it is – strictly speaking – not appropriate to describe the lacto-coccal secretion machinery as sec-dependent and sec-independent. However, these terms are used here by analogy with the better characterized systems to indicate the fundamental differences between these secretory pathways. Figure 2.10 summarizes the leader sequences described so far in lactic acid bacteria.

2.6.1 Export-signal selection vectors

By analogy to their successful use in *B. subtilis* export-signal selection vectors have been designed for application in *L. lactis*. Vectors based on replicons belonging to the pSH71 or pWVO1 family have been equipped with the *amyL* or *bla* genes for the α-amylase of *B. licheniformis* or the *E. coli* TEM β-lactamase, respectively (Table 2.3). Cloning in the correct reading frame of DNA fragments carrying a promoter, ribosome binding site, start codon and signal sequence should result in synthesis and secretion of these easily assayable enzymes. Since the export-signal selection vectors replicate also in *E. coli* the first selection of secretion signals was done in this host. Sequences directing the secretion of the reporter enzyme in *E. coli* were subsequently analysed in *L. lactis*. Various lacto-coccal sequences have been isolated that directed the secretion of the used reporter enzymes. However, it is difficult to conclude on the efficiency of the isolated sequences since the contribution of the signals involved in initiating gene expression is not known (Shbakov *et al.*, 1991a; Perez-Martinez *et al.*, 1992). A few of the isolated sequences have been characterized and were found to encode signal-type sequences with variable size (Figure 2.10A; Shbakov *et al.*, 1991a). However, it remains to be demonstrated whether these sequences actually are derived from proteins that are secreted via the sec-dependent pathway.

2.6.2 Sec-dependent secretion

Up to now five large proteins are known to be secreted by *L. lactis*. They all contain signal sequences that conform to the consensus tripartite structure proposed for sec-dependent signal sequences (Figure 2.10B; von Heijne, 1986).

Two are serine proteinases and the largest is the plasmid-encoded 200 kDa serine proteinase PrtP from the strains Wg2, SK11 and NCDO 763 (Kok *et al.*, 1988; Vos *et al.*, 1989b; Kiwaki *et al.*, 1989). This proteinase is the key enzyme in the cascade of proteolytic reactions that allow lactococci to grow in milk (see chapter 4). Recently the serine proteinase NisP, encoded by Tn*5726*, was characterized and found to be involved in the cleavage of the leader peptide of nisin (van der Meer *et al.*, 1993a). Both proteins are cell envelope-located because of the presence of a C-terminal located membrane anchor, that conforms to the stop transfer sequences found in other Gram-positive bacteria (Fischetti *et al.*,1990). While those serine proteinases are encoded by mobile elements and do not belong to the standard make up of all lactococcal cells, it has been found that all *L. lactis* strains produce another secreted protein of approximately 45 kDa, represented by Usp45 of strain MG1363 (van Asseldonk *et al.*, 1990). The signal sequences vary in size between 19 and 33 amino acid residues but all show a consensus leader peptidase I cleavage site. In the leader region of Usp45 two additional signal peptidase I cleavage sites could be postulated. In frame fusions with the *B. stearothermophilus amyS* gene have demonstrated that amino terminal processing of Usp45 can be excluded and that the *usp45* gene encodes a signal peptide of 27 amino acids (van Asseldonk *et al.*, 1993a).

Two other proteins, PrtM and NisI, that are found outside the lactococcal cells contain the consensus sequence L-S-G-C characteristic of lipoproteins that is recognized by a signal peptidase II (Pugsley, 1993). PrtM has been shown to be involved in maturation of the cell envelope located proteinase PrtP (Haandrikman *et al.*, 1989; Vos *et al.*, 1989a) and is now considered to be an extracellular chaperone (see chapter 5). Labelling studies with ^{3}H-palmitic acid and immuno electron microscopy using antibodies against purified PrtM have demonstrated that PrtM is indeed a lipoprotein and is covalently bound to the membrane (Haandrikman *et al.*, 1991). The location of the other secreted protein of this class, the Tn*5276* encoded NisI has not been studied but based on its function, which is to confer immunity to nisin, and sequence conservation, it is assumed to be a lipoprotein (Kuipers *et al.*, 1993a).

2.6.3 Sec-independent secretion

Translocation that is mediated by a dedicated translocator has been described for the plasmid-encoded lactococcins of *L. lactis* that are made with an N-terminal leader peptide. Secretion of lactococcin A, encoded by the *lcnA* gene, was found to depend on the presence of two genes, the *lcnB* gene encoding an ABC-type translocator of the HlyB family (Pugsley, 1991), and *lcnC* for a protein with deduced similarity to the HlyC protein involved in the translocation of haemolysin (Stoddard *et al.*, 1992). It is assumed that other lactococcins, LcnB and LcnM1, are also translocated via this or a similar system. The leader peptides of those bacteriocins show high similarity and belong to the family of bacteriocins that contain the G-G sequence at position −1 and −2 (Figure 2.10C; see chapter 3). In contrast, the lantibiotics lacticin 481, nisin A and nisin Z, have N-terminal extensions that do not conform to this consensus and show similarity to other leader sequences found in lantibiotics (Figure 2.10C; van der Meer *et al.*, 1993b). It is likely that nisin is translocated via NisT, an ABC-type translocator of the HlyB family (Engelke *et al.*, 1992; Kuipers *et al.*, 1993a). Proteolytic cleavage of the leader of nisin was found to be independent of translocation and appeared to be mediated by the serine proteinase NisP (van der Meer *et al.*, 1993a).

2.7 Expression and secretion systems

The development of efficient gene transfer methods and cloning vectors and the isolation of transcription, translation and secretion signals has allowed the investigation of heterologous gene expression in lactococci. Several interesting and useful heterologous genes have been expressed in lactococci. An overview of the expression of those heterologous genes is provided in Table 2.4. It is important to note that in some cases heterologous expression and secretion signals proved to be recognized by the lactococcal host. However, in various cases, especially for the expression of eukaryotic genes or to increase the expression level, use has been made of specifically designed expression or secretion vectors that will be summarized below.

2.7.1 Expression vectors

Several expression vectors for lactococci have been constructed and are summarized in Table 2.5. All contain the pSH71/pWVO1 replicon, a resistance marker and a promoter. In some cases they also contain efficient translation initiation signals derived from lactococcal genes. The first vector, pGKV2, contained the *B. subtilis* SPO2 promoter that was shown to function in *L. lactis* although with limited efficiency (van der Vossen *et al.*, 1987). Another vector contained the lactococcal promoter A but its efficiency has not been established (Gasson *et al.*, 1987). The expression signals of other expression vectors have been derived from the SK11 proteinase *prtP* gene, the chromosomal P32 promoter, and the well-studied *lac* operon promoter.

Previous studies had revealed that the *prtP-lacZ* gene fusion in pNZ262 (see Table 2.3) showed high levels of β-galactosidase in *E. coli* (up to 5% total protein) and to a lesser extent also in *L. lactis* (de Vos and Simons, 1988). Therefore, the transcription and translation initiation signals of the *prtP* gene

Table 2.4 Heterologous genes expressed *L. lactis*

Genes	Source (plasmid)	Vector(s)	Secreted	Reference
Cm^R; *cat-194*	*S.aureus* (pC194)	pGK12	–	Kok *et al.*, 1984
Cm^R; *cat-86*	*B.pumilis*	pGKV210	–	van der Vossen *et al.*, 1985
Em^R; *ery-194*	*S.aureus* (pE194)	pGK12	–	Kok *et al.*, 1984
Em^R; *ery-β1*	*E.faecalis* (pAMβ1)	pIL253	–	Simon and Chopin, 1988
Km^R; *knt-110*	*S.aureus* (pUB110)	pNZ12	–	de Vos and Simons, 1987
Km^R; *aphA3*	*S.faecalis* (pJH1)	pMG36	–	van de Guchte *et al.*, 1989
Tc^R; *tetM*	*E.faecalis* (Tn1545)	pVE6002	–	Biswal *et al.*, 1993
Tc^R; *tetL*	*E.faecalis* (pJH1)	pNZ280	–	Platteeuw *et al.*, 1993
β-galactosidase; *lacZ*	*E.coli*	pNZ262	–	de Vos and Simons, 1988
Egg white lysozyme	Hen	pIL253	–	van de Guchte, *et al.*, 1992b
T4 lysozyme	*E.coli* phage T4	pMG36e	–	van de Guchte, *et al.*, 1992b
λ lysozyme	*E.coli* phage λ	pMG36e	–	van de Guchte, *et al.*, 1992b
β-glucuronidase; *gusA*	*E.coli*	pNZ123	–	Platteeuw *et al.*, 1993
β-galactosidase	*C.acetobutylicum*	pFX1	–	Pillidge and Pearce, 1991
Luciferase; *luxAB*	*V.fischerii*	pCK17	–	Ahmad and Stewart, 1991
Prochymosin	Bovine	pNZ18	–	Simons *et al.*, 1988; 1992
T7 RNA polymerase	*E. coli*	pIL253	–	Wells *et al.*, 1993
Prochymosin	Bovine	pKM1363	Yes	Klessen *et al.*, 1988
Prochymosin	Bovine	pNZ18	Yes	Simons *et al.*, 1988;1992
Streptodornase; *sdc*	*S.equisimilis*	pDG13	Yes	Wolinowska *et al.*, 1991
Neutral protease; *nprE*	*B.subtilis*	pMG36e	Yes	van de Guchte *et al.*, 1990
β-lactamase; *bla*	*E.coli*	pVS2	Yes	Sibakov *et al.*, 1991
α-amylase; *amyS*	*B.stearotherm.*	pNZ123	Yes	van Asseldonk *et al.*, 1993a
α-amylase; *amyL*	*B.licheniformis*	pGA14	Yes	Perez-Martinez *et al.*, 1991
α-galactosidase	*C.tetragonoloba*	pGK13	Yes	Haandrikman, 1990
Surface protein	*S.mutans*	pSA3	Yes	Iwaki *et al.*, 1990
Phage lysin	*L.monocytogenes*	pCK17	Yes	Gasson, unpublished results
Tetanus toxin fragment C	*C.tetanus*	pLET1	–	Wells *et al.*, 1993

Table 2.5 Main features of expression and secretion vectors

Expression vector	Expression signal	Selection marker	Replicon	Reference
pGKV2	P$_{SPO2}$	*ery-194*	pWV01	van der Vossen *et al.*, 1985
pCK965	P$_A$	*cat-194*	pSH71	Gasson *et al.*, 1987
pNZ337	P$_{prtP}$	*cat-194*	pSH71	Simons *et al.*, 1990
pMG36	P$_{32}$	*aphA3*	pWV01	van de Guchte *et al.*, 1989
pMG36e	P$_{32}$	*ery-194*	pWV01	van de Guchte *et al.*, 1989
pNZ3004	P$_{lac}$	*ery-194*	pWV01	van Rooijen *et al.*, 1992
pNZ3005	P$_{lac}$, *lacR*	*ery-194*	pWV01	van Rooijen *et al.*, 1992
pNZ2014	P$_{lac}$	*cat-194*	pSH71	Platteeuw and de Vos, unpublished; Kuipers *et al.*, 1992
pLET1	P$_{T7}$	*cat-194*	pSH71	Wells *et al.*, 1993
pILPol	P$_{lac}$	*ery-194*	pAMβ1	Wells *et al.*, 1993

Secretion vector	Secretion signal	Selection marker	Replicon	Reference
pNZ150	PrtP	*cat-194*	pSH71	Simons *et al.*, unpublished results
pNZ100	Usp45	*cat-194*	pSH71	Simons *et al.*, unpublished results

and its first 8 codons were inserted in the promoter-selection vector pNZ336, resulting in the expression vector pNZ337 (Simons *et al.*,1990a). Using a similar vector the bovine prochymosin gene was expressed, resulting in the intracellular production of inactive chymosin containing an N-terminal extension (see below).

The expression signals of pMG36 and pMG36e include the P32 promoter (see Figure 2.2) followed by the SD region, initiation codon and 12 residues of the P32 open reading frame in which a polylinker sequence has been inserted (van de Guchte *et al.*, 1989). Because the *ery-194* gene is a better marker, pMG36e is the preferred expression vector (Figure 2.1). Gene fusions were constructed with the P32 open reading frame and the coding region of the hen egg white lysozyme or the lysozyme genes of the *E. coli* bacteriophages T4 and λ (van de Guchte *et al.*, 1992b). Low expression levels and no lysozyme activity were found in *L. lactis* after cloning the eukaryotic hen egg white lysozyme into the expression plasmid pMG36. However, intracellular production of the bacteriophage T4 lysozyme was obtained and lysozyme activity could be detected in cell lysates of *L. lactis* strain IL1403. In derivatives of pMG36e in which the bacteriophage λ R or T4 lysozyme genes were translationally coupled to ORF32, highest expression level was obtained (van de Guchte *et al.*, 1992b).

Using the expression vector pMG36e the *B. subtilis* neutral protease (*nprE*) gene has also been expressed (van de Guchte *et al.*, 1990). The *nprE* gene-specific translation initiation signals (including its GUG initiation codon) and sec-dependent secretion signal were used and appeared to be recognized in *L. lactis* since the mature form of the neutral protease was found in the culture supernatant.

The lactococcal *lac* promoter (see 2.5.1) can be used in lactose-deficient *L. lactis* strains as a strong constitutive promoter. This application of the *lac* promoter was demonstrated by the overproduction of nisin and engineered derivatives (Kuipers *et al.*, 1992). In lactococcal strains that are equipped with the *lac* genes, either on a plasmid or integrated into the chromosome, the *lac* promoter can be used for lactose-inducible gene expression. This concept was first shown in transcriptional fusions of various regions of the *lac* promoter and the *cat-86* gene in *L. lactis* MG5276 containing a single chromosomal copy of the *lac* operon (van Rooijen *et al.*, 1992). A 3-fold induction of transcription was obtained by growth on lactose using pNZ3004 and the induction efficiency could be increased to approximately 6-fold by including a copy of the lacR gene such as in plasmid pNZ3005, although this decreased the absolute expression level several times. Based on these results the expression vector pNZ2014 (Figure 2.1) was constructed that also contained the terminator of the *pepN* gene. This expression vector has been used for the inducible overexpression of various genes in *L. lactis* strains containing a chromosomal copy of the *lac* genes (Platteeuw and de Vos, unpublished observations). Recently, the *lac* promoter has been used to develop a regulated gene expression system for *L. lactis* based on the *E. coli* T7 promoter system (Wells *et al.*, 1993). For this purpose the T7 polymerase gene has been placed under control of a *lac* promoter fragment also containing the *lacR* gene on pIL277, a pAMβ1 derived vector, resulting in the plasmid pILPol. In addition, based on a derivative of pCK17 the expression vector pLET1 was constructed that contains the T7 promoter of gene 10 followed by its translation initiation region, a multiple cloning site, and the gene 10 transcription terminator. Both plasmids contain different markers and are compatible since they are based on the pAMβ1 and pSH71 replicons (Table 2.1). By introducing pLET1 and pILpol in *L. lactis* strain MG1820 containing the lactose plasmid pMG820, an inducible expression system was constructed that was used to express substantial quantities (up to 22% of soluble protein) of the tetanus toxin fragment C (Wells *et al.*, 1993).

2.7.2. Secretion vectors

For the sec-dependent secretion of heterologous proteins into the medium of *L. lactis* cells two secretion vectors, designated pNZ150 and pNZ100 (Table 2.5), have been constructed. They are based on the regulatory sequences of the *prtP* or *usp45* gene including the 33 or 27 residue signal coding sequence, respectively. In frame fusions with heterologous genes it is possible by using a unique restriction enzyme site, *Eag*I or *Pvu*II respectively, too overlap the coding sequence of the last residue of the signal peptide. The usefulness of these secretion vectors in the secretion of α-amylase and bovine prochymosin is discussed here.

The *B. stearothermophilus amyS* gene was used to test the efficiency of the secretion vectors pNZ100p and pNZ50u in *L. lactis*. The resulting plasmids pNZα15 and pNZ10α5 were introduced in *L. lactis* MG1363 and were found to induce the secretion of α-amylase that could be visualized by the production of large haloes around their colonies on agar plates after staining with iodine (van Asseldonk *et al.*, 1993a). The α-amylase activity in the supernatant of strain MG1363 harbouring pNZ10α5 was about 0.6 U/ml using amylose azure as a substrate. However, still a substantial amount (about 50% of the total α-amylase production) of unprocessed and inactive α-amylase remained intracellularly. Nevertheless, *L. lactis* MG1363 harbouring pNZ10α5 was able to grow rapidly on a medium containing starch as sole energy source (van Asseldonk *et al.*, 1993c).

The leader peptide preceding the amino-terminus of mature Usp45 contains two additional putative signal peptidase cleavage sites after Ala[19] and Ala[20] (van Asseldonk *et al.*, 1990). To test whether a shorter Usp45 leader peptide such as residue 1–19 and 1–20 would also be functional in directing α-amylase secretion to the culture supernatant, the coding regions for residue 21–27 and residue 20–27 were deleted in pNZ10α5. In *L. lactis*, the shorter leader peptides were unable to direct α-amylase secretion and unprocessed α-amylase accumulated intracellularly. In contrast, the shorter leader peptides directed efficient α-amylase secretion into the periplasm of *E. coli* (van Asseldonk *et al.*, 1993a). These data confirmed that *L. lactis* as other Gram-positive bacteria require a longer signal peptide length as compared to *E. coli* (von Heijne and Abrahamsen, 1989).

The *usp45* signal sequence was also used to generate the first inducible secretion system, in which an *usp45-amyS* fusion was placed under control of the heat shock promoter of the *dnaJ* gene. Thermoinduction induces the secretion approximately three-fold (van Asseldonk *et al.*, 1993b). In similar constructions with the *usp45* signal sequence an inducible secretion system was developed based on the combination of the T7 and lac expression system described in section 2.7.1 (J. Wells, personal communication).

Bovine chymosin is synthesized as a 381-aa precursor, preprochymosin, which is processed during secretion to form the 365-aa prochymosin, that is autocatalytically activated at low pH to the active 323-aa protein, chymosin. With the use of secretion vectors, the production of extracellular prochymosin has been realized using secretion vectors in various eukaryotic microbial hosts, such as yeast and fungi. However, in *E. coli* only the production of an intra-cellular insoluble prochymosin has been reported, that can be isolated only with the use of a protein denaturant (Marston *et al.*, 1984). Initial attempts to secrete prochymosin in *L. lactis* MG1363 by using the expression signals of the streptokinase gene of *S. equisimilis* revealed a rapid proteolytic degradation of the prochymosin-streptokinase fusion protein (Klessen *et al.*, 1988). This is most likely due to the use of the tripartite streptokinase gene fusion vector rather than the *L. lactis* host since the used strain MG1363 does not produce any proteinases and in another study the intact production of prochymosin was reported in this host. In the latter study the transcription and translation initiation signals of the *prtP* gene and various parts of its coding sequence were fused to the coding region of the bovine prochymosin gene (Simons *et al.*, 1992; de Vos *et al.*, 1989a). the most relevant ones are those in which the prochymosin gene was fused to the coding sequence of the first 7, 33 and 62 residues of PrtP, respec-tively. The levels of prochymosin mRNA of the three constructs and of *prtP* nRNA were compared by means of Northern blot analysis. The mRNA species found were in agreement with the expected size and the amount of mRNA produced by the three gene fusions equals the amount of *prtP* mRNA in the wild-type *L. lactis* strain. Intra- and extracellular prochymosin production was measured by means of immuno blots. In *L. lactis* MG1363 harbouring the 7 residue PrtP-prochymosin fusion cassette only intracellular prochymosin synthesis could be detected, as expected. This material could not be activated to active chymosin at low pH. The 33 residue PrtP-prochymosin fusion cassette directed the synthesis of extracellular prochymosin, with the size expected for prochymosin. However, a substantial amount of unprocessed prochymosin still remained intracellularly. This could be avoided by using a leader peptide of 62 residues, by which the majority of the synthesized prochymosin was secreted. The size of this secreted prochymosin suggested that the 33 residue signal peptide has been cleaved off and that a 29 (62 minus 33) residue PrtP-prochy-mosin fusion protein is released into the medium.

The observation that for efficient secretion a leader region is required which is about 30 residues longer than the signal peptide could be due to the presence of several basic residues at the N-terminus of prochymosin. It has been found that for the secretion of a mutant leader peptidase to the periplasmic space of *E. coli* cells the presence of positively charged residues immediately downstream of the signal sequence adversely affects secretion (Anderson and von Heijne, 1991).

Recently, the PrtP and Usp45 signal sequences have also been used to construct a series of secretion vectors in which transcription was dependent on the T7 polymerase system. Efficient secretion of the tetanus toxin fragment B was obtained in *L. lactis*, indicating the potential of these well-characterized secretion signals (J. Wells, personal communication).

The developed secretion vectors also offer the possibility to obtain surface expression in *L. lactis* by providing the protein of interest with a C-terminal membrane anchor derived from PrtP or NisP. In particular this is an interesting way to expose new surface antigens in *L. lactis* or other lactic acid bacteria.

No secretion vectors have been described based on the sec-independent pathway. However, it has shown to be possible to secrete heterologous or new peptides from lactococci via the nisin transport system, most likely a sec-independent pathways. This appeared from protein engineering studies in which mutations were introduced in the leader- and pro-region of nisin and by the fusion of sequences encoding the *B. subtilis* subtilin leader to those for the nisin pro-region (Kuipers *et al.*, 1992; van der Meer *et al.*, 1993b; Kuipers *et al.*, 1993b). Secretion was observed of various mutant forms of nisin including those containing homologous or heterologous N-terminal extensions. It is expected that this sec-independent pathway opens new avenues for the secretion of heterologous peptides while also the translocation of larger proteins a priori cannot be excluded.

2.8 Food-grade systems

The term food-grade has been coined to designate systems that are ultimately to be applied in the food industry or result in products to be used in foods (de Vos, 1989). Although ill-defined and without legal status, this term is used here to qualify gene cloning and expression systems that do not rely on the presence of antibiotic-resistance markers. The current novel food directives rule out the presence of antibiotic resistance genes in genetically modified lactic acid bacteria or constituents thereof that are used to produce foods. As a consequence there is a strong need for food-grade markers by which genetically modified lactic acid bacteria can be selected, the genetic modification can be maintained stably, and the desired phenotype can be realized. Different requirements could be envisaged to which such a food-grade marker system should comply with. In general, food-grade markers should be well-defined, acceptable in foods, and applicable on a large scale. Preferably, the marker should be dominant. In addition, the marker gene should be derived from the host that is aimed to be genetically modified, at least when the intention is to achieve self-cloning, a special form of genetic modification that has been excluded from the European Directives on the contained use of genetically modified microorganisms (CEC, 1990).

The first food-grade system that was described for lactococci but also can be used in other lactic acid bacteria was based on the complementation of one or more auxotrophic markers (de Vos, 1989; de Vos *et al.*, 1989a). The detailed characterization of the *L. lactis lac* operon (see section 2.5.1) allowed for the identification of the *lacF* gene that was found to complement *L. lactis* strain YP2-5 containing one of the first described mutations in the chromosomal *lac* operon (Park and McKay, 1982). The mutated *lacF* gene in *L. lactis* YP2-5 was characterized by sequence analysis and found to contain a missense mutation resulting in the substitution of a glycine for a glutamate residue at position 18 (de Vos *et al.*, 1990). Expression of the *lacF* gene by the vector-located promoter in pNZ305 resulted in complementation of the lactose-deficient phenotype of strain YP2-5 and cells containing this plasmid could be selected easily on lactose indicator plates. In addition, the plasmid could be maintained stably by growing its host in media containing lactose. Since lactose is a cheap sugar present in various industrial media based on whey, the use of lactose as a means to maintain selective pressure offers large scale application. Finally, the *lacF* gene has a very small size and could be cloned on a 0.4-kb DNA fragment allowing easy manipulation (de Vos *et al.*, 1990).

A second food-grade marker system has been based on the ability of some *L. lactis* strains to grow in the presence of nisin. This nisin-resistance determinant that was first reported for *L. lactis* strain DRC3 and later also found in other *L. lactis* strains such as 10.084, could be isolated on defined restriction fragments and cloned and expressed in *L. lactis* (Simon and Chopin, 1988; Froseth *et al.*, 1988; von Wright *et al.*, 1991). These studies also showed that the nisin-resistant strains did not produce nisin and from subsequent sequence analysis it appeared that the nisin-resistance determinant *nsr* could encode a hydrophobic protein that has no relation with the recently described nisin immunity determinant NisI (Froseth *et al.*, 1991; see above). *L. lactis* cells expressing the *nsr* gene could be selected by specific plating methods and the potential of the *nsr* gene in providing a food grade marker was illustrated by the cloning of a replication origin (Froseth *et al.*, 1992).

Complementation based on the cloned lactococcal *thyA* gene for thymidilate synthase was also described as a way to select and maintain foreign sequences in *L. lactis* and other bacteria (Ross *et al.*, 1990a; Ross *et al.*, 1990b). However, no *L. lactis* strains deficient in the *thyA* gene have been described so far. Other possible dominant food-grade markers for *L. lactis* and other hosts have been proposed including the α-galactosidase gene from the plant *Cyamopsis tetragonoloba*, the *Kluyveromyces marxianus* inulase gene, the bacteriocin production and/or immunity genes from *Pediococcus acidilacti* PAC 1.0, or the sucrose sucrose genes from *Pediococcus pentasaceus* (Leenhouts *et al.*, 1992).

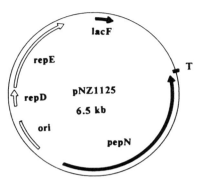

Figure 2.11   Physical and genetic map of pNZ1125 containing the food-grade *lacF* marker and used to overexpress the *pepN* gene for the debittering aminopeptidase PepN.

Although these genes have all been cloned and partially characterized, no experimental results were provided on the use of those genes as food-grade markers in *L. lactis*. However, recently it was shown that the sucrose genes of *P. pentasaceus* could be expressed in *L. lactis* (C. Leenhouts, personal communication). Many *L. lactis* strains are unable to ferment sucrose, although the nisin producing strains of *L. lactis* are known to contain sucrose genes (Rauch and de Vos, 1992a; Rauch and de Vos, 1992b). Therefore, selection for sucrose utilization mediated by cloned pediococcal gene fragments could be a simple, although not homologous way, to stabilize genes in *L. lactis*. Other examples for food-grade markers could be envisaged in *L. lactis* and other lactic acid bacteria but have not been thoroughly tested. These include selection based on bacteriophage insensitivity mechanisms that are widespread in lactococci (chapter 3), complementation of essential genes such as those in certain biosynthetic operons some of which are well characterized now (see section 2.5), immunity to bacteriocins (see chapter 5), or on suppressor mutations.

The applicability of the food-grade *lacF* complementation system was tested in *L. lactis* by replacing the *ery-β1* of pNZ305 (de Vos *et al.*, 1990a) by the lactococcal *pepN* gene encoding a debittering aminopeptidase (Tan *et al.*, 1993; Figure 2.11). Previous studies had shown that the *pepN* gene could be overexpressed in *L. lactis* using a non food-grade vector (van Alen-Boerrigter *et al.*, 1991). This is the first example of a industrially useful and completely sequenced plasmid that can be stably maintained in *L. lactis*. In addition, the resulting strain has been obtained by self cloning and has been excluded from the existing biosafety legislation (de Vos, 1992).

Because of the exceptional position that self cloning has within the CEC Directives on contained use of genetically modified microorganisms, it is envisaged that lactococci or other lactic acid bacteria improved by self cloning may be among the first to be introduced into the market place.

References

Achen, M.G., Davidson, B.E. and Hillier, A.J. (1986) Construction of plasmid vectors for the detection of streptococcal promoters. *Gene* **45**, 45–49.

Ahmad, K.A. and Stewart, G.S.A.B. (1991) The production of bioluminiscent lactic acid bacteria suitable for the rapid assessment of starter culture activity in milk. *J. Appl. Bacteriol.* **70**, 113–120.

Anderson, P.H. and Gasson, M.J. (1985) High copy number plasmid vectors for use in lactic streptococci. *FEMS Microbiol. Lett.* **30**, 193–196.

Anderson, S.G.E. and Kurland, C.G. (1990) Codon preferemnces in free-living microorganisims. *Microbiol. Rev.* **54**, 198–210.

Anderson, H., von Heijne, G. (1991) A thirty-residue long "export initiation domain" adjacent to the signal sequence is critical for protein translocation across the inner membrane of *Escherichia coli*. *Proc. Natl. Acad Sci. USA* **88**, 9751–9754

Araya, T., Ishinashi, N., Shimamura, S., Tanaka, K. and Takahashi, H. (1993) genetic and molecular analysis of the *rpoD* gene from *Lactococcus lactis*. *Biosc. Biotech. Biochem.* **57**, 88–92.

Baigori, M., Sesma, F., de Ruiz Holdago, A.P. and de Mendoza, D. (1988) Transfer of plasmids between *Bacillus subtilis* and *Streptococcus lactis*. *Appl. Environ. Microbiol.* **54**, 1309–1311.

Band, L., Yansura, D.G. and Henner, D.J. (1983) Construction of a vector for cloning promotors in *Bacillus subtilis*. *Gene* **26**, 313–315.

Bates, E.E.M. and Gilbert, H.J. (1989) Characterization of a cryptic plasmid from *Lactobacillus plantarum*. *Gene* **85**, 253–258.

Bardowski, J., Ehrlich, S.D. and Chopin, A. (1992) Trytophan biosynthesis genes in *lactococcus lactis* subsp. *lactis*. *J. Bacteriol.* **174**, 6563–6570.

Behnke, H., Gilmore, M.S. and Ferretti, J.J. (1981) Plasmid pGB301, a new multiple resistance strepto-coccal cloning of a gentamicin/kanamycin resistance determinant. *Mol. Gen. Genet.* **182**, 414–421.

Biswas, P., Gruss, A., Ehrlich, S.D. and Maguin, E. (1993) High-efficiency gene inactivation and replacement system for gram-positive bacteria. *J. Bacteriol.* **175**, 3628–3635.

Bojovic, B., Djordjevic, G. and Topisirovic, L. (1991) Improved vector for promoter screening in lactococci. *Appl. Environ. Microbiol.* **57**, 385–388.

Bracco, L., Kortlarz, D., Kolb, A., Diekmann, S. and Buc, H. (1989) Synthetic curved DNA sequences can act as transcriptional activators in *Eschherichia coli*. *EMBO J.* **8**, 4289–4296.

Brantl, S. and Behnke, D. (1992) Characterization of the minimal origin required for replication of the streptococal plasmid pIP501 in *Bacillus subtilis*. *Mol. Microbiol.* **6**, 3501–3510.

Brantl, S., Behnke, D. and Alonso, J.C. (1990) Molecular analysis of the replication region of the conjugative *Streptococcus agalactiae* plasmid pIP501 in *Bacillus subtilis*. Comparison with plasmids pAMβ1 and pSM19035. *Nucl. Acids Res.* **18**, 4783–4790.

Bruand, C., Ehrlich, S.D. and Janniere, L. (1991) Unidirectional theta replication of the structurally stable *Enterococcus faecalis* plasmid pAMβ1. *EMBO J.* **10**, 2171–2177.

Bruinenberg, P.G., Vos, P. and de Vos, W.M. (1992) Proteinase overproduction in *Lactococcus lactis* strains: Regulation and effect on growth and acidification in milk. *Appl. Environ. Microbiol.* **58**, 78–84.

Buchman, G.W., Banerjee, S. and Hansen, J.N. (1988) Structure, expression and evolution of a gene encoding the precursor of Nisin, a small protein antibiotic. *J. Biol. Chem.* **263**, 16260–16266

CEC (1990) Council Directive on the contained use of genetically modified micoorganisms. 90/220/EEC

Chiaruttini, C. and Millet, M. (1993) Gene organization, primary structure and RNA processing analysis of a ribosomal RNA operon in *Lactococcus lactis*. *J. Mol. Biol.* **230**, 57–76.

Chopin, A. (1993) Organization and regulation of genes for amino acid biosynthesis in lactic acid bacteria. *FEMS Microbiol. Rev.* **12**, 21–39.

Chopin, M.C., Chopin, A., Moillo-Bott, A. and Langella, P. (1984) Two plasmid-determined restriction and modification systems in *Streptococcus lactis*. *Plasmid* **11**, 260–263.

Chopin, M.C., Chopin, A., Rouault, A. and Simon, D. (1986) Cloning in *Streptococcus lactis* of plasmid-mediated UV resistance and effect on prophage stability. *Appl. Environ. Microbiol.* **51**, 233–237.

Chopin, M.C., Chopin, A., Rouault, A. and Galleron, N. (1989) Insertion and amplification of foreign genes in the *Lactococcus lactis* subsp. *lactis* chromosome. *Appl. Environ. Microbiol.* **55**, 1769–1774.

Clewell, D.B., Yagi, Y., Dunny, G.M. and Schultz, S.K. (1974) Characterization of three deoxyribonucleic acid molecules in a strain of *Streptococcus faecalis*: identification of a plasmid determining erythromycin resistance. *J. Bacteriol.* **117**, 283–289.

Collado-Vides, J., Maganasik, B. and Gralla, J.D. (1991) Control site of location and transcriptional regulation in *Escherichia coli*. *Microbiol. Rev.* **55**, 371–394.

Dao, M.L. and Feretti, J.J. (1985) Streptococcus-Escherichia coli shuttle vector pSA3 and its use in the cloning of streptococcal genes. *Appl. Environ. Microbiol.* **49**, 115–119.

David, S., Simons, G. and de Vos, W.M. (1989) Plasmid transformation by electroporation in *Leuconostoc paramesenteroides* and its use in molecular cloning. *Appl. Environ. Microbiol.* **55**, 1483–1489

David, S., Stevens, H., van Riel, M., Simons, G. and de Vos, W.M. (1992) *Leuconostoc lactis* β-galactosidase is encoded by two overlapping genes. *J. Bacteriol.* **174**, 4475–4481.

de la Campa, A.G., de Solar, G.H. and Espinosa, M. (1990) Initiation of replication of plasmid pLS1. The initiator protein *rep*B acts on two distant DNA regions. *J. Mol. Biol.* **213**, 247–262.

de Vos (1986a) Genetic improvement of starter streptococci by the cloning and expression of the gene coding for a non-bitter proteinase. In: *Biomolecular Engineering in the European Community: Achievements of the Research Programme (1982–1986): Final Report* (E. Magnien, ed.) Martinus Nijhoff Publishers, Dordrecht, pp. 465–472.

de Vos, W.M. (1986b) Gene cloning in lactic streptococci. *Neth. Milk Dairy J.* **40**, 141–154.

de Vos, W.M. (1986c) Sequence organization and use in molecular cloning of the cryptic lactococcal plasmid pSH71. *Second Streptococcal Genetics Conference*, Miami Beach, American Society for Microbiology P223.

de Vos, W.M. (1987) Gene cloning and expression in lactic streptococci. *FEMS Microbiol. Rev.* **46**, 281–295.

de Vos, W.M. and Simons, A.F.M. (1987) Method for preparing proteins using transformed lactic acid bacteria. European Patent 0 228 726

de Vos, W.M. and Simons, G. (1988) Moleular cloning of lactose genes in dairy lactic streptococci: the phospho-β-galactosidase and β-galactosidase genes and their expression products. *Biochimie* **70**, 461–473.

de Vos, W.M. (1989) Food grade selection and maintenance of recombinant DNA in lactic acid bacteria. Eur. Patent. Appl. 0 355 036

de Vos, W.M. and Gasson, M.J. (1989) Structure and expression of the *Lactococcus lactis* gene for phospho-β-galactosidase (*lacG*) in *Escherichia coli* and *L. lactis*. *J. Gen. Microbiol.* **135**, 1833–1846.

de Vos, W.M., Vos, P., Simons, G. and David, S. (1989a) Gene organization and expression in mesophilic lactic acid bacteria. *J. Dairy Sci.* **72**, 3398–3409

de Vos, W.M., Vos, P., de Haard, H. and Boerrigter, I. (1989b) Cloning and expression of the *Lactococcus lactis* subsp. *cremoris* SK11 gene encoding an extracelluar serine proteinase. *Gene* **85**, 169–176.

de Vos, W.M., Kuiper, H., Lever, A. and Ventris, J. (1989c) Heterogrammic replication of *Lactococcus lactis* plasmid pSH71 is regulated by a repressor-operator control circuit. *American Society for Microbiology, Annual Meeting,* New Orleans, abstract H276.

de Vos, W.M., Boerrigter, I., van Rooijen, R.J., Reiche, B. and Hengstenberg, W. (1990) Characterization of the lactose-specific enzymes of the phosphotransferase system in *Lactococcus lactis. J. Biol. Chem.* **265**, 22554–22560.

de Vos, W.M., Siezen, R.J. and Kuipers, O.P. (1992) Lantibiotics similar to nisin A. PCT Patent Application WO 92/18633

de Vos, W.M. (1992) Engineering of lactic acid bactera for improved production of foods, flavors and additives. In: *Harnessing Biotechnology in the 21st Century*. (R. Lodisch & A. Bose, eds.) American Chemical Society, pp. 524–527.

Delorme, C., Ehrlich, S.D. and Renault, P. (1992) Histidine biosynthesis genes in *Lactococcus lactis* subsp. *lactis. J. Bacteriol.* **174**, 6571–6579.

Dunny, G.M., Lee, L.N. and LeBlanc, D.J. (1991) Improved electroporation and cloning vector system for gram-positive bacteria. *Appl. Environ. Microbiol.* **57**, 1194–1201.

Eaton, T., Shearman, C., and Gasson, M.J. (1993a) The use of bacterial luciterase genes as reporter genes in *Lactococcus*: regulation of the *Lactococcus lactis* subsp. *lactis* lactose genes. *J. Gen. Microbiol.* **139**, 1495–1501.

Eaton, T., Shearman, C. and Gasson, M.J. (1993b) Cloning and sequence analysis of the *dnaK* gene region of *Lactococcus lactis* subsp. *lactis. J. Gen. Microbiol.* **139**, in press.

Efstathiou, J.D. and McKay, L.L. (1977) Inorganic salts resistance associated with a lactose-fermenting plasmid in *Streptococcus lactis*. *J. Bacteriol.* **130**, 257–265.

Engelke, G., Gutowski-Eckel, Z., Hammelman, M. and Entian, K.-D. (1992) Biosynthesis of lantibiotic nisin: genomic organization and membrane location of the NisB protein. *Appl. Environ. Microbiol.* **58**, 3730–3734.

Feirtag, J.M., Petzel, J.P., Pasalodos, E., Baldwin, K.A. and McKay, L.L. (1991) Thermosensitive plasmid replication, temperature sensitive host growth, and chromosomal plasmid integration conferred by *Lactococcus lactis* subsp. *cremoris* lactose plasmids in *Lactococcus lactis* subsp. *lactis*. *Appl.Environ. Microbiol.* **57**, 539–548.

Fischetti, V.A., Pancholi, V., Schneewind, O. (1990) Conservation of a hexapeptide sequence in the anchor region of surface proteins from gram-positive bacteria. *Mol. Microbiol.* **4**, 1603–1605.

Frere, J., Novel, M. and Novel, G. (1993) Identification of the theta-type minimal replicon of the *Lactococcus lactis* CNRZ270 lactose-protease plasmid pUCL22. *Current Microbiology*, **27**, 97–102.

Froseth, B.R., Herman, R.E. and McKay, L.L. (1988) Cloning of a nisin resistance determinant and replication origin on a 7.6-kilobase EcoRI fragment of pNP40 from *Streptococcus lactis* subsp. *lactis* DRC3. *Appl. Environ. Microbiol.* **54**, 2136–2139.

Froseth, B.R. and McKay, L.L. (1991) Development and application of pFM011 as a possible food-grade cloning vector. *J. Dairy Sci.* **74**, 1445–1453.

Gartenberg, M.R. and Crothers, D.M. (1991) Synthetic DNA bending sequences increase the rate of *in vitro* transcription initiation at the *Escherichia coli lac* promoter. *J. Mol. Biol.* **219**, 217–230.

Gasson, M.J. and Davies, F.L. (1980) Conjugal transfer of the drug resistance plasmid pAMβ1 in the lactic streptococci. *FEMS Microbiol. Lett.* **7**, 51–53.

Gasson, M.J., Hill, S.A. and Anderson, P.H. (1987) Molecular genetics of metabolic traits in lactic streptococci. In: *Streptococcal Genetics* (Ferretti, J., and Curtiss, R. eds.) American Society for Microbiology, pp. 242–246.

Gasson, M.J. (1983) Plasmid complements of *Streptococcus lactis* NCDO 712 and other lactic streptococci after protoplast-induced curing. *J. Bacteriol.* **154**, 1–9.

Godon, J.J., Delorme, C., Ehrlich, S.D. and Renault, P. (1992) Divergence of genomic sequences between *Lactococcus lactis* subsp. *lactis* and *Lactococcus lactis* subsp. *cremoris*. *Appl. Environ. Microbiol.* **58**, 4045–4047.

Gruss, S. and Ehrlich, S.D. (1989) The family of highly interrelated single-stranded deoxyribonucleic acid plasmids. *Microbiol. Rev.* **53**, 231–241.

Gruss, A. and Ehrlich, S.D. (1988) Insertion of foreign DNA into plasmids from gram-positive bacteria induces formation of high molecular weight plasmid multimers. *J. Bacteriol.* **170**, 1183–1190.

Haandrikman, A.J., Kok, J., Laan, H., Soemitro, S., Ledeboer, A., Konings, W.N. and Venema, G. (1989) Identification of a gene required for maturation of an extracellular lactococcal serine proteinase. *J. Bacteriol.* **171**, 2789–2794.

Haandrikman, A.J. (1990) Development and use of a broad-host-range vector for the simultaneous analysis of divergent promoters. PhD thesis, University of Groningen, The Netherlands.

Haandrikman, A.J., Kok, J. and Venema, G. (1991) The lactococcal proteinase-maturation protein is a lipoprotein. *J. Bacteriol.* **173**, 4517–4525

Hager, P.W. and Rabinowitz, J.C. (1985) Translational specificity in *Bacillus subtilis*. In: Dubnau, D.A. (ed.), *The Molecular Biology of the Bacilli*. Vol II, pp 1–31, Academic press, Inc., Orlando.

Harley, C.B. and Reynolds, R.P. (1987) Analysis of *E. coli* promoter sequences. *Nucl. Acids Res.* **15**, 2343–2361.

Hayes, F., Daly, C. and Fitzgerald, G.F. (1990) Identification of the minimal replicon of *Lactococcus lactis* subsp. *lactis* UC317 plasmid pCI205. *Appl. Environ. Microbiol.* **56**, 202–209.

Hayes, F., Vos, P., Fitzgerald, G. de Vos, W.M. and Daly, C. (1991) Molecular organization of novel, narrow host-range, lactococcal plasmid pCI305. *Plasmid* **25**, 16–26.

Holo, H., Nilssen, O. and Nes, I.F. (1991) Lactococcin A, a new bacteriocin from *Lactococcus lactis* subsp. *cremoris*: isolation and characterization of the protein and its gene. *J. Bacteriol.* **173**, 3879–3887.

Hoopes, B.C. and McClure, W.R. (1987) Strategies in regulation of transcription initiation. In: Neidhardt, F.L. (ed) *Esherichia coli and Salmonella typhimurium; Cellular and Molecular Biology*. ASM, Washington D.C., pp 1231–40.

Horinouchi, S. and Weisblum, B. (1992) Nucleotide sequence and functional map of pE194, a plasmid that specifies inducible resistance to macrolide, lincosamide, and streptogramin type B antibiotics. *J. Bacteriol.* **150**, 804–814

Horng, J.S., Polzin, K.M. and McKay, L.L. (1991) Replication and temperature-sensitive maintenance functions of lactose plasmid pSK11L from *Lactococcus lactis* subsp. *cremoris. J. Bacteriol.* **173**, 7573–7581.

Horodniceanu, T., Bouanchaud, D., Biet, G. and Chabbert, Y. (1976) R plasmids in *Streptococcus algalactiae* (group B) *Antimicrob. Agents Chemother.* **10**, 795–801.

Israelson, H. and Hansen, E.B. (1993) Insertion of transposon Tn917 derivatives into the *Lactococcus lactis* subsp. *lactis* chromosome. *Appl. Environ. Microbiol.* **59**, 21–26.

Iwaki, M., Okahashi, N., Takahashi, I., Kanamitsu, T., Susiota, Y., Aibara, K. and Koga, T. (1990) Oral immunization with recombinant *Streptococcus lactis* carrying the *Streptococcus mutans* surface protein antigen. *Infect. Immunol.* **58**, 2929–2934.

Jahns, A., Schaefer, A., Geis, A. and Teuber, M. (1991) Identification, cloning and sequencing of the replication region of *Lactococcus lactis* subsp. *lactis* biovar diacetylactis Bu2 citrate plasmid pSL2. *FEMS Micobiol. Lett.* **80**, 253–258.

Kiewit, R., Bron, S. de Jonge, K., Venema, G. and Seegers, J.F.M.L. (1993a) Theta replication of the lactococcal plasmid pWV02. *Molec. Microbiol.*, in press.

Kiewit, R., Kok, J., Seegers, J.F.M.L., Venluna, G. and Bron, S. (1993b) The mode of replication is a major factor in segregational plasmid instability in *Lactococcus lactis. Appl. Environ. Microbiol.* **59**, 358–364.

Kim, S.G. and Batt, C.A. (1993) Cloning and sequencing the *Lactococcus lactis* subsp. *lactis groELS* operon. *Gene* **127**, 121–126.

Kiwaki, M., Ikemura, H., Shimuzu-Kadota, M. and Hirashima, A. (1989) Molecular characterization of a cell-wall associated proteinase gene from *Streptococcus lactis* NCDO 763. *Molec. Microbiol.* **3**, 359–369.

Klein, J.R., Ulrich, C. and Plapp, R. (1993) Characterization and sequence analysis of a small cryptic plasmid from *Lactobacillus curvatus* LTH683 and its use for the construction of new *Lactobacillus* cloning vectores. *Plasmid*, **14**, 14–30.

Klessen, C., Schmidt, H., Ferretti, J.J., Malke, H. (1988) Tripartite streptokinase gene fusion vector for gram-positive and gram-negative prokaryoates. *Mol. Gen. Genet.* **212**, 295–300

Klijn, N., Weerkamp, A.H. and de Vos, W.M. (1991) Identification of mesophilic lactic acid bacteria by using polymerase chain reaction-amplified variable regions of 16S rRNA and specific DNA probes. *Appl. Environ. Microbiol.* **57**, 3390–3393.

Koivula, T., Sibakov, M. and Palva, I. (1991a) Isolation and characterization of *Lactococcus lactis* subsp. *lactis* promoters. *Appl. Environ. Microbiol.* **57**, 333–340.

Koivula, T., Palva, I. and Hemilä, H. (1991b) Nucleotide sequence of the secY gene from *Lactococcus lactis* and identification of conserved regions by comparison of four SecY proteins. *FEBS Lett.* **288**, 114–118.

Kok, J., van der Vossen, J.M.B.M. and Venema, G. (1984) Construction of plasmid cloning vectors for lactic streptococci which also replicate in *Bacillus subtilis* and *Escherichia coli. Appl. Environ. Microbiol.* **48**, 726–731.

Kok, J., van Dijl, J.M., van der Vossen, J.M.B.M. and Venema, G. (1985) Cloning and expression of a *Streptococcus cremoris* proteinase gene in *Bacillus subtilis* and *Streptococcus lactis. Appl. Environ. Microbiol.* **50**, 94–101.

Kok, J., Leenhouts, K.J., Haandrikman, A.J., Ledeboer, A.M. and Venema, G. (1988) Nucleotide sequence of the cell wall proteinase gene of *S. cremoris* Wg2. *Appl. Environ. Microbiol.* **54**, 231–238

Kok J. (1991) Special-purpose vectors for lactococci. In: *Genetics and Molecular Biology of Streptococci, Lactococci and Enterococci*, eds. Dunny, G.M., Cleary, P.P. and McKay, L.L., American Society for Microbiology; pp. 97–108.

Kondo, J.K. and McKay, L.L. (1984) Plasmid transformation of *Streptococcus lactis* protoplasts: optimization and use in molecular cloning. *Appl. Environ. Microbiol.* **48**, 252–259.

Kuipers, O.P., Rollema, H.S., Yap, W.M.G.J., Boot, H.J., Siezen, R.J. and de Vos, W.M. (1992) Engineering dehydrated amino acid residues in the antimicrobial peptide nisin. *J. Biol. Chem.* **267**, 24340–24346.

Kuipers, O.P., Rollema, H.S., de Vos, W.M., and Siezen, R.J. (1993a) Characterization of the nisin gene cluster *nisABTCIPR* of *Lactococcus lactis*: requirement of expression of the *nisA* and *nisI* genes for producer immunity. *Eur. J. Biochem.*, **216**, 281–291.

Kuipers, O.P., Rollema, H.S., de Vos, W.M., and Siezen, R.J. (1993b) Biosynthesis and secretion of a precursor of nisin Z by *Lactococcus lactis*, directed by the leader peptide of subtilin. *FEBS Letters*, **330**, 23–27.

Lacks, S., Lopez, P., Greenberg, B. and Espinoza (1986) Identification and analysis of genes for tetracycline resistance and replication functions in the broad host-range plasmid pLS1. *J. Mol. Biol.* **192**, 753–765.

Laforet, G.A., Kaiser, E.T. and Kendall, D.A. (1989) Signal peptides subsegments are not always functionally interchangable. *J. Biol. Chem.* **264**, 14478–14485.

Lakshmidevi, G., Davidson, B.E. and Hillier, A.J. (1990) Molecular characterization of promoters of the *Lactococcus lactis* subsp. *cremoris* temperate bacteriophage BK5-T and identification of a phage gene implicated in the regulation of promoter activity. *Appl. Environ. Microbiol.* **56**, 934–942.

Le Blanc, D.J. and Lee, L.N. (1991) Replication functions of pVA380–1. In: *Genetics and Molecular Biology of Strepococci, Lactococci and Enterococci*, eds Dunny, G.M., Cleary, P.P. and McKay, L.L., American Society for Microbiology, pp. 235–240.

Le Chatelier, E., Ehrlich, S.D. and Janniere, L. (1993) Biochemical and genetic analysis of the unidirectional theta replication of the *S. algalactiae* plasmid pIP501. *Plasmid* **29**, 50–56.

Leenhouts, K.J., Kok, J. and Venema, G. (1989) Campbell integration of heterologous plasmid DNA into the chromosome of *Lactococcus lactis* subsp. *lactis. Appl. Environ. Microbiol.* **55**, 394–400.

Leenhouts, K.J., Kok, J. and Venema, G. (1990) Stability of integrated plasmids in the chromosome of *Lactococcus lactis* subsp. *lactis. Appl. Environ. Microbiol.* **56**, 2726–2735.

Leenhouts, K.J., Kok, J. and Venema, G. (1991a) Replacement recombination in *Lactococcus lactis.* *J. Bacteriol.* **173**, 4794–4798.

Leenhouts, K.J., Gietema, J., Kok, J. and Venema, G. (1991b) Chromosomal stabilization of the proteinase genes in *Lactococcus lactis. Appl. Environ. Microbiol.* **57**, 2568–2575.

Leenhouts, K.J., Tolner, B., Bron, S., Kok, J., Venema, G. and Seegers, J.F.M.L. (1991c) Nucleotide sequence and characterization of the broad host range plasmid pWVO1. *Plasmid* **26**, 55–66.

Leenhouts, C.J., Marugg, J.D., Verrips, C.T. (1992) A food grade vector. Eur. Patent Appl. 0 487 159

Leonhardt, H. and Alonso, J. (1991) Parameters affecting plasmid instability in *Bacillus subtilis. Gene* **103**, 107–111.

Lereclus, D., Ribier, J., Klier, A., Menou, G. and Lecadet, M. -H. (1984) A transposon-like structure related to the delta-endotoxin gene of *Bacillus thuringiensis. EMBO J.* **3**, 2561–2567.

Loureiro dos Santos, A.D. and Chopin, A. (1987) Shotgun cloning in *Streptococcus lactis. FEMS Microbiol. Lett.* **42**, 209–212.

Lovett, P.S. (1990) Translational attenuation as the regulator of inducible *cat* genes. *J. Bacteriol.* **172**, 1–6.

Luchansky, J.B., Muriana, P.M. and Klaenhammer, T.R. (1988) Application of electroporation for transfer of plasmid DNA to *Lactobacillus, Lactococcus, Leuconostoc, Listeria, Pediococcus, Bacillus, Staphylococcus, Enterococcus*, and *Propionibacterium. Molec. Microbiol.* **2**, 637–646.

Macrina, F.L., Tobian, J.A., Jones, K.R., Evans, R.P. and Clewell, D.B. (1982) A cloning vector able to replicate in *Escherichia coli* and *Streptococcus sanguis. Gene* **19**, 345–353.

Macrina, F.L., Wood, P.H. and Jones, K.R. (1980) Genetic transformation of *Streptococcus sanguis* (Challis) with cryptic plasmids from *Streptococcus ferus. Infect. Immunol.* **28**, 692–699.

Maguin, E., Duwat, P., Hege, T., Ehrlich, S.D. and Gruss, A. (1992) New thermosensitive plasmid for gram-positive bacteria. *J. Bacteriol.* **174**, 5633–5638.

Marston, F.A.O., Lowe, P., Dole, M.T., Schoemaker, J.N., White, S., Angal, S (1984) Purification of calf prochymosin synthesized by *Escherichia coli. Biotechnology* **3**, 800–804

Matthews, K.S. (1992) *Microbiol. Rev.* **56**, 123–136.

Mayo, B., Kok, J., Venema, G.K., Bockelman, W., Teuber, M., Reinke, H. Venema, G. (1991) Molecular cloning and sequence analysis of the X-prolyl dipeptidyl aminopeptidase gene from *Lactococcus lactis* subsp. *lactis. Appl. Environ. Microbiol.* **57**, 38–44.

McKay, L.L., Baldwin, K.A. and Walsh, P. (1980) Conjugal transfer of genetic information in group N streptococci *Appl. Environ. Microbiol.* **40**, 84–91.

McKay, L.L. (1982) in: *Developments in Food Microbiology–1*, Applied Science Publ., London, pp. 153–182.

McKay, L.L. (1983) Functional properties of plasmids in lactic streptococci. *Antonie van Leeuwenhoek* **49**, 259–274.

McLaughlin, J.R., Murray, C.L. and Rabinowitz, J.C. (1981) Unique features in the ribosome binding site of the Gram-positive *Staphylococcus aureus* β-lactamase gene. *J. Biol. Chem.* **256**, 11283–11291.

Mercenier, A. (1990) Molecular genetics of *Streptococcus thermophilus. FEMS Microbiol. Rev.* **87**, 61–79.

Mierau, I., Tan, P.S.T., Haandrikman, A.J., Kok, J., Leenhouts, K.J., Konings, W.N. and Venema, G. (1993) Cloning and sequencing of a gene for a lactococcal endopeptidase, an enzyme with sequence similarity to mammalian enkephalinase. *J. Bacteriol.* **175**, 2087–2096.

Moran, C.P. (1989) Sigma Factors and the regulation of transcription. In: Smith, I., Slepecky, R.A. and Setlow, P. (eds), *Regulation of Prokaryotic Development*, ASM, Washington D.C., pp 167–184.

Moreno, F., Fowler, A.V., Hall, M., Silhavy, T.J., Zabin, I. and Schwartz, M. (1980) A signal sequence is not sufficient to lead β-galactosidase out of the cytoplasm. *Nature* **286**, 356–359.

Mulders, J.W.M., Boerrigter, I.J., Rohema, H.S., Siezen, R.J. and de Vos, W.M. (1991) Identification and Characterization of the lantibiotic nisin Z, a natural nisin variant. *Eur. J. Biochem.* **201**, 581–584.

Nakamura, K., Nakamura, A., Takamatsu, H., Yoshikawa, H. and Yamane, K. (1990) Cloning and characterization of a *Bacillus subtilis* gene homologous to *E. coli* secY. *J. Biochem.* **107**, 603–607.

O'Sullivan, D.J., Hill, C. and Klaenhammer, T.R. (1993) Effect of increasing the copy number of bacteriophage origins, *in trans*, on incoming phage proliferation. *Appl. Environ. Microbiol.* **59**, 2449–2456.

Otto, R., de Vos, W.M. and Gavrieli, J. (1988) Plasmid DNA in *Streptococcus cremoris*: influence of pH on selection in chemostats of a variant lacking a protease plasmid. *Appl. Environ. Microbiol.* **43**, 1272–1277.

Park, Y.H. and McKay, L.L. (1982) Distinct galactose phosphoenol pyruvate-dependent phosphotransferase systems in *Streptococcus lactis. J. Bacteriol.* **149**, 420–427.

Peréz Martinez, G., Smith, H., Bron, S., Kok, J. and Venema, G. (1992) Protein export elements from *Lactococcus lactis. Molec. Gen. Genet.* **234**, 401–411.

Piard, J.C., Kuipers, O.P., Rollema, H.S., Desmazeaud, M.J., and de Vos, W.M. (1993) Structure, organization, and expression of the *lct* gene for the lacticin 481, a novel lantibiotic produced by *Lactococcus lactis. J. Biol. Chem.* **268**, 16361–16368.

Pillidge, C.J. and Pearce, L.E. (1991) Expression of a β-galactosidase gene from *Clostridium acetobutylicum* in *Lactococcus lactis* subsp. *lactis. J. Appl. Bacteriol.* **71**, 78–85.

Platteeuw, C., Simons, G. and de Vos, W.M. (1993) Use of the *Escherichia coli* β-glucuronidase (*gusA*) gene as a reporter gene for analyzing promoters in lactic acid bacteria. *Appl. Environ. Microbiol.*, in press.

Poolman, B., Rayer, T.J., Mainzer, S.E. and Schmdt (1989) Lactose transport system of *Streptococcus thermophilus*: a protein with homology to the melibiose carrier and enzyme III of phosphoenolpyruvate-dependent phosphotransferase systems. *J. Bacteriol.* **171**, 244–253.

Pugsley, A. (1991) Superfamilies of bacterial transport systems with nucleotide binding components. In: *Prokaryotic Structure and Function: a New Perspective*. Mohan, S., Dow, C., and Cole, J.A. (eds), Society for General Microbiology, Cambridge University Press, Cambridge, pp. 223–248.

Pugsley, A. (1993) Complete general secretory pathway in Gram-negative bacteria. *Microbiol. Rev.* **57**, 50–108.

Rauch, P.J.G. and de Vos, W.M. (1992a) Characterization of the novel nisin-sucrose conjugative transposon Tn*5276* and its insertion in Lactococcus lactis. *J. Bacteriol.* **174**, 1280–1287.

Rauch, P.J.G. and de Vos, W.M. (1992b) Transcriptional regulation of the Tn*5276*-located *Lactococcus lactis* sucrose operon and characterization of the SacA gene encoding sucrose–6-phosphate hydrolase. *Gene* **121**, 55–61.

Reznikoff, W.S. and McClure, W.R. (1987) *E. coli* promoters. In: Reznikoff, W. and Gold, L. (eds) *Maximizing Gene Expression*, Butterworth Publ., Stoneham, pp 1–33.

Romero, D.A., Slos, P., Robert, C., Castellino, I. and Mercenier, A. (1987) Conjugative mobilization as an alternative vector delivery system for lactic streptococci. *Appl. Environ. Microbiol.* **53**, 2405–2413.

Ross, P., O'Gara, F. and Condon, S. (1990) Thymidilate synthetase gene from *Lactococcus lactis* as a gene marker: an alternative to antibiotic resistance genes. *Appl. Environ. Microbiol.* **52**, 2164–2169.

Ross, P., O'Gara, F. and Condon, S. (1990) Cloning and characterization of the thymidilate synthetase gene from *Lactococcus lactis. Appl. Environ. Microbiol.* **52**, 2156–2163.

Salama, M., Sandine, W. and Giovannoni, S. (1991) Development and application of oligonucleotide probes for identification of *Lactococcus lactis* subsp. *cremoris. Appl. Environ. Microbiol.* **57**, 1313–1318.

Schleifer, K.-H., Kraus, J., Dvorak, C., Killper-Balz, R., Collins, M.D. and Fischer, W. (1985) Transfer of *Streptococcus lactis* and related streptococci to the genes *Lactococcus* gen. nov. *Syst. Appl. Microbiol.* **6**, 183–195.

Schultz, J.W. and Sanders, M.E. (1990) Cloning of the bacteriophage resistance gene from *Lactococcus lactis* subsp. *cremoris* KH *J. Dairy Sci.* **73**, 2044–2053.

Sharp, P.M. and Li, W.H. (1987) The codon adaptation index – a measure of directional synonymous codon usage bias, and its potential applications. *Nucl. Acids Res.* **15**, 1281–1295.

Shimuza-Kadota, M., Shibahara-Sone, H. and Ishiwa, H. (1991) Shuttle plasmid vectors for *Lactobacillus casei* and *Escherichia coli* with a minus origin. *Appl. Environ. Microbiol.* **57**, 3292–3300.

Shine, J. and Dalgarno, L. (1974) The 3' terminal sequence of *E. coli* 16S RNA: complementary to nonsense triplets and ribosome binding site. *Proc. Natl. Acad. Sci. USA* **71**, 1342–1346

Sibakov, M., Koivula, T. von Wright, A. and Palva, I. (1991) Secretion of TEM β-lactamase with signal sequences isolated from the chromosome of *Lactococcus lactis* subsp. *lactis. Appl. Environ. Microbiol.* **57**, 341–348.

Simon, D. and Chopin, A. (1988) Construction of a vector plasmid family and its use for molecular cloning in *Streptococcus lactis. Biochimie* **70**, 559–567.

Simons, G. and de Vos, W.M. (1987) Gene expression in lactic streptococci. in: *Proceedings of the Fourth European Congress on Biotechnology 1987* (O.Neyssel, R.R. van der Meer & K.Ch.M. Luyben, eds.) Volume I, Elsevier Science Publishers, Amsterdam, New York, pp. 458–460.

Simons, G., Rutten, G., Hornes, M. and de Vos, W.M. (1988) Production of bovine prochymosin by lactic acid bacteria, in: *Proceedings 2nd Netherlands Biotechnology Congress*, Breteler, H., Lelyveld, P.H., Luyben K.Ch.A.M. (eds) Netherlands Society for Biotechnology, pp. 183–187.

Simons, G., Buys, H., Hogers, R., Koehnen, E. and de Vos, W.M. (1990a) Construction of a promoter-probe vector for lactic acid bacteria using the *lacG* gene of *Lactococcus lactis. Dev. Ind. Microbiol.* **31**, 31–39.

Simons, G., van Asseldonk, M., Rutten, G., Nijhuis, M., Hornes, M. and de Vos, W.M. (1990b) Analysis of secretion signals of lactococci. In: *Proceedings of the Fifth European Congress on Biotechnology* (Christianse, C., Munck, L. and Villadsen, J. eds.) Munksgaard International Publisher, Copenhagen, pp. 290–294.

Simons, G., Rutten, G., Hornes, M. Nijhuis, M. and van Asseldonk, M. (1992) Production of prochymosin in lactococci. *Advances in Experimental Medicine and Biology*, **306**, 115–120.

Simons, G., Nijhuis, M. and de Vos, W.M. (1993) Integration and gene replacement in the *Lactococcus lactis lac* operon: Induction of a cryptic phospho-β-glucosidase in LacG-deficient strains. *J. Bacteriol.* **175**, 5168–5175.

Smith, M.D. and Clewell, D.B. (1984) Return of *Streptococcus faecalis* DNA in *Escherichia coli* via transformation of *Streptococcus sanguis* followed by conjugative mobilization. *J. Bacteriol.* **160**, 1109–1114.

Stoddard, G.W., Petzel, J.P., van Belkum, M.J., Kok, J. and McKay, L.L. (1992) Molecular analysis of the Lactococcin A gene cluster from *Lactococcus lactis* subsp. *lactis* biovar *diacetylactis* WM4. *Appl. Environ. Microbiol.* **58**, 1952–1961.

Strøman, P. (1992) Sequence of a gene (*lap*) encoding a 95.3 kDa aminopeptidase from *Lactococcus lactis* spp. *cremoris* Wg2. *Gene*, **113**, 107–112.

Summers, D.K. and Sheratt, D.J. (1984) Multimers of high copy number plasmids causes instability: ColE1 encodes a determinant essential for plasmid monomerisation and stability. *Cell* **36**, 1097–1103.

Swinfield, T.J., Oultram, J.D., Thompson, D.E., Brehm K.J. and Minton, N.P. (1990) Physical characterization of the replication region of the *Streptococcus faecalis* plasmid pAMβ1. *Gene* **87**, 79–90.

Tan, P.S.T., van Alen-Boerrigter, I.J., Poolman, B., Siezen, R.J., de Vos, W.M. and Konings, W.N. (1992) Characterization of the *Lactococcus lactis pepN* gene encoding an aminopeptidase homologous to mammalian aminopeptidase N. *FEBS Lett.* **306**, 9–16.Lett.

Tan, P.S.T., van Kessel, T.A.J.M., van de Veerdouk, F.L.M., Zuurendonk, P.F., Bruins, A.P. and Konings, W.N. (1993) Degradation and debittering of a tryptic digest from β-casein by aminopeptidase N from *Lactococcus lactis* subsp. *cremoris* Wg2. *Appl. Environ. Microbiol.* **59**, 1430–1436.

te Riele H., Michel, B. and Ehrlich, S.D. (1986) Single-stranded plasmid DNA in *Bacillus subtilis* and *Staphylococcus aureus. Proc. Natl. Acad. Sci. USA* **83**, 2541–2545.

Twigg, A.J. and Sherrat, D. (1980) Trans-complementable copy-number mutants of plasmid colE1. *Nature* **283**, 216–218.

van Alen-Boerrigter, I.J., Baankreis, R. and de Vos, W.M. (1991) Characterization and overexpression of the *Lactococcus lactis pepN* gene and localization of its product, aminopeptidase N. *Appl. Environ. Microbiol.* **57**, 2555–2561.

van Asseldonk, M., Rutten, G., Oteman, M., Siezen, R.J., de Vos, W.M. and Simons, G. (1990) Cloning of *usp45*, a gene encoding a secreted protein from *Lactococcus lactis* subsp. *lactis. Gene* **95**, 155–160.

van Asseldonk, M., de Vos, W.M. and Simons, G. (1993a) Functional analysis of the *Lactococcus lactis usp45* secretion signal in the secretion of a homologous proteinase and a heterologous α-amylase. *Mol. Gen. Genet.*, **240**, 428–434.

van Asseldonk, M., de Vos, W.M. and Simons, G. (1993b) Cloning, nucleotide sequence, and regulatory analysis of the *Lactococcus lactis dnaJ* gene. *J. Bacteriol.* **175**, 1637–1644.

van Asseldonk, M., Nijhuis, M., Doesburg, P. de Vos, W.M. and Simons, G. (1993c) Role of sequences upstream of the −35 region of the *usp45* gene in expression of a heterologous β-amylase gene in *Lactococcus lactis*, submitted for publication.

van Belkum, M.J., Hayema, B.J., Geis, A., Kok, J. and Venema, G. (1989) Cloning of two bacteriocin genes from a lactococcal bacteriocin plasmid. *Appl. Environ. Microbiol.* **55**, 1187–1191.

van Belkum, M.J., Hayema, B.J., Jeeringa, R.E., Kok, J. and Venema, G. (1991) Organization and nucleotide sequences of two lactococcal bacteriocin operons. *Appl. Environ. Microbiol.* **57**, 492–498.

van Belkum, M.J., Kok, J. and Venema, G. (1992) Cloning, sequencing, and expression in *Escherichia coli* of IcnB, a third bacteriocin determinant from the lactococcal bacteriocin plasmid pgB4–6. *Appl. Environ. Microbiol.* **58**, 572–577.

van de Guchte, M., Kok, J. and Venema, G. (1989) Construction of a lactococcal expression vector: expression of hen egg white lysozyme in *Lactococcus lactis* subsp. *lactis*. *Appl. Environ. Mircobiol.* **55**, 224–228.

van de Guchte, M., Kodde, J., van der Vossen, J.M.B.M., Kok, J. and Venema, G. (1990) Heterologous gene expression in *Lactococcus lactis* subsp. *lactis*: Synthesis, secretion, and processing of the *Bacillus subtilis* neutral protease. *Appl. Environ. Microbiol.* **56**, 2606–2611.

van de Guchte, M., Kok, J. and Venema, G. (1991) Distance-dependent translational coupling and interference in *Lactococcus lactis*. *Mol. Gen. Genet.* **227**, 65–71

van de Guchte, M., Kok, J. and Venema, G. (1992a) Gene expression in *Lactococcus lactis*. *FEMS Microbiol. Rev.* **88**, 73–92.

van de Guchte, M., van der Wal, F.J., Kok, J. and Venema, G. (1992b) Lysozyme expression in *Lactococcus lactis*. *Appl. Microbiol. Biotechnol.* **37**, 216–224.

van der Lelie, D., Bron, S., Venema, G. and Oskam, L. (1989) Similarity of minus origin of replication and flanking open reading frames of pUB110, pTB913 and pMV158. *Nucl. Acids Res.* **18**, 7283–7294.

van der Meer, J.R., Polman, J., Beerthuyzen, M.M., Siezen, R.J., Kuipers, O.P. and de Vos, W.M. (1993a) Characterization of the *Lactococcus lactis* Nisin A operon genes, *nisP*, encoding a subtilisin-like serine protease involved in the precursor processing, and *nisR*, encoding a regulatory protein involed in nisin biosynthesis. *J. Bacteriol.* **175**, 2578–2588.

van der Meer, J.R., Rollema, H.S., Siezen, R.J., Kuipers, O.P., and de Vos, W.M. (1993b) Influence of amino acid substitutions in the nisin leader peptide on biosynthesis and secretion of nisin by *Lactococcus lactis*. *J. Biol. Chem.*, submitted for publication

van der Vossen, J.M.B.M., Kok, J. and Venema, G. (1985) Construction of cloning, promoter-screening and terminator screening shuttle vectors in *Bacillus subtilis* and *Streptococcus lactis*. *Appl. Environ. Microbiol.* **50**, 540–542.

van der Vossen, J.M.B.M., van der Lelie, D. and Venema, G. (1987) Isolation and characterization of *Streptococcus cremoris* Wg2-specific promoters. *Appl. Environ. Microbiol.* **53**, 2452–2457.

van der Vossen, J.M.B.M., Kodde, J., Haandrikman, A.J., Venema, G. and Kok, J. (1992) Characterization of transcription initiation and termination signals of the proteinase genes of *Lactococcus lactis* Wg2 and enhancement of proteolysis in *L. lactis*. *Appl. Environ. Microbiol.* **58**, 3142–3149.

van Rooijen, R.J. and de Vos, W.M. (1990) Molecular cloning, characterization and nucleotide sequence of *lacR*, a gene encoding the repressor of the lactose phosphotransferase system of *Lactococcus lactis*. *J. Biol. Chem.* **265**, 18499–18503.

van Rooijen, R.J., van Schalkwijk, A. and de Vos, W.M. (1991) Molecular cloning, characterization, and nucleotide sequence of the tagatose 6-phosphate pathway gene cluster of the lactose operon of *Lactococcus lactis*. *J. Biol. Chem.* **266**, 7176–7181

van Rooijen, R.J., Gasson, M.J. and de Vos, W.M. (1992) Characterization of the promoter of the *Lactococcus lactis* lactose operon: contribution of flanking sequences and LacR repressor to its activity. *J. Bacteriol.* **174**, 2273–2280.

van Rooijen, R.J. and de Vos, W.M. (1993) Purification of the *Lactococcus lactis* LacR repressor and characterization of its DNA binding sites. R.J. van Rooijen, Ph.D. Thesis, pp. 101–119.

van Rooijen, R.J., Dechering, K.J., Wilmink, C.N.J. and de Vos, W.M. (1993a) Lysines 72, 80, 213, and aspartic acid 210 of the *Lactococcus lactis* LacR repressor are involved in response to the inducer tagatose–6-phosphate leading to induction of the *lac* operon expression. *Prot. Engin.* **6**, 201–206.

van Rooijen, R.J., Dam, W. and de Vos, W.M. (1993b) Deletion of the *Lactococcus lactis lacR* gene and its effect on the regulation of lactose operon expression. R.J. van Rooijen, Ph.D. Thesis, pp. 119–135.

van Rooijen, R.J., Dechering, K.J. and de Vos, W.M. (1993c) Characterization of the DNA binding helix of the *Lactococcus lactis* LacR repressor by site directed mutagenesii. R.J. van Rooijen, Ph.D. Thesis, pp. 151–167.

van Belkum, M.J., Kok, J. and Venema, G. (1992) Cloning, sequencing and expression in *Escherichia coli* of LCNB, a third bacteriocin determinant from the lactococcal bacteriocin plasmid p9B4–6. *Appl. Environ. Microbiol.* **58**, 572–577.

Villafane, R., Bechofer, C.H., Narayanan, C.S. and Dubnau, D. (1987) Replication control genes of plasmid pE194. *J. Bacteriol.* **169**, 4822–4829.

Viret, J.-F., Bravo, A. and Alonso, J.C. (1991) Recombination – dependent concatemeric plasmid replication. *Microbiol. Rev.* **55**, 675–683.

von Wright, A., Tynkknen, S. and Suiminen, M. (1987) Cloning of a *Streptococcus lactis* subsp. *lactis* chromosomal fragment associated with the ability to grow in milk. *Appl. Environ. Microbiol.* **53**, 1584–1588.

von Wright, A., Wesels, A., Tynkkynen, and Saarela, M. (1990) Isolation of a replication region of a large lactococcal plasmid and use in cloning of a nisin resistance determinant. *Appl. Environ. Microbiol.* **56**, 2029–2035.

von Heijne, G. and Abrahmsén, L. (1989) Species specific variation in signal peptide design. Implication for protein secretion in foreign hosts. *FEBS Letters* **244**, 439–446

von Heijne, G. (1986) A new method for predicting signal peptide cleavage sites. *Nucleic Acid Res.* **14**, 4683–4690

Vos, P., van Asseldonk, M., van Jeveren, F., Siezen, R.J., Simons, G. and de Vos, W.M. (1989a) A maturation protein is essential for production of active forms of *Lactococcus lactis* SK11 serine proteinase located in or secreted from the cell envelope. *J. Bacteriol.* **171**, 2795–2802.

Vos, P., Simons, G., Siezen, R.J. and de Vos, W.M. (1989b) Primary structure and organization of the gene for a procaryotic, cell-envelope located serine proteinase. *J. Biol. Chem.* **264**, 13579–23585.

Vosman, B. and Venema, G. (1983) Introduction of a *Streptococcus cremoris* plasmid in *Bacillus subtilis*. *J. Bacteriol.* **156**, 920–921.

Vujcic, A. and Topisirovic, L. (1993) Molecular analysis of the rolling cirle replicating plasmid pA1 of *Lactobacillus plantarum* A112. *Appl. Environ. Microbiol.* **59**, 274–280.

Wells, J.M., Wilson, P.W., Norton, P.W., Gasson, M.J. and LePage, R.W.F. (1993) *Lactococcus lactis*: high-level expression of tetanus toxin fragment C and protection against lethal challenge. *Molec. Microbiol.* **8**, 1155–1162.

Whitaker, R.D. and Batt, C.A. (1991) Characterization of the heat shock response in *Lactococcus lactis* spp. *lactis*. *Appl. Environ. Microiol.* **57**, 1408–1412.

Williams, D.M., Duvall, E.J. and Lovett, P.S. (1981) Cloning restriction fragments that promote expression of a gene in *Bacillus subtilis*. *J. Bacteriol.* **148**, 1162–1165.

Williams, A.M., Fryer, T. and Collins, M.D. (1990) *Lactococcus piscium* sp. nov. a new *Lactococcus* species from salmoid fish. *FEMS Microbiol. Lett.* **68**, 109–114.

Wirth, R., An, F.Y. and Clewell, D.B. (1986) Highly efficient protoplast transformation system for *Streptococcus faecalis* and a new *Escherichia coli-Streptococcus faecalis* shuttle vector. *J. Bacteriol.* **165**, 831–836.

Wolinowska, R., Ceglowski, P. Kok, J. and Venema, G. (1991) Isolation, sequence and expression in *Escherichia coli, Bacillus subtilis* and *Lactococcus lactis* of the DNase (streptodornase)-encoding gene from *Streptococcus equisimilius* H46A. *Gene*, **106**, 115–119.

Xu, F., Pearce, L.E. and Yu, P.-L. (1990) Molecular cloning and expression of a proteinase gene from *Lactococcus lactis* subsp. *cremoris* H2 and construction of a new lactococcal vector pFX1. *Arch. Microbiol.* **154**, 99–104.

Xu, F., Pearce, L.E. and Yu, P.-L. (1991) Construction of a family of lactococcal vectors for gene cloning and translational fusions. *FEMS Microbiol. Lett.* **77**, 55–60.

Xu, F., Pearce, L.E., Yu, P.L. (1991) Genetic analysis of a lactococcal plasmid replicon. *Mol. Gen. Genet.* **227**, 33–39.

3 Bacteriophages and bacteriophage resistance

T.R. KLAENHAMMER and G.F. FITZGERALD

3.1 Introduction

Food and dairy fermentations rely on the growth and acid producing ability of
the lactic acid bacteria. Many of these have remained as traditional fermentations, where the process is driven by the natural microflora associated with
the raw material. Increasing consistency, improved quality and processing
efficiencies have followed the development of controlled fermentations. These
rely on the activity of a starter culture which is intentionally inoculated in order
to drive the primary fermentation. However, with the increased control granted
through the repeated use of a defined starter culture comes the potential for
disruption of the fermentation by bacteriophage.

Identification of bacteriophages as a problem in the cheese and cultured dairy
products industries occurred in the 1930s. Since then the dairy industry has
devised numerous means to successfully cope with bacteriophage infection of
starter cultures (Huggins and Sandine, 1979; Klaenhammer, 1984; Sandine,
1989). Sixty years later, incidents continue to occur and bacteriophage remains
the major cause of slow acid production in dairy fermentations (Sanders, 1987).
The dairy industry is particularly susceptible to bacteriophage infection due to
the following:

(i)   phage contamination can occur since the fermentations are not sterile and
      are conducted in a non-sterile fluid medium, pasteurized milk;
(ii) processing efficiency is easily disrupted since batch culture fermentations
      are conducted under increasingly stringent manufacturing schedules;
(iii) increasing reliance on specialized strains limits the number and diversity of
      dairy starter cultures that are available;
(iv) continuous use of defined cultures provides an ever present host for
      bacteriophage attack.

Although many factors contribute to the onset of bacteriophage infections in
dairy fermentations, we should expect that more reliance on specialized cultures
and the increasing bioprocessing demands will continue to present phage
problems for the fermentation industries of the future.

Given the magnitude of the problem and its economic impact, considerable
research has been conducted on lactococcal bacteriophages and on the natural
mechanisms through which these bacteria defend themselves against phage.
Through the years this effort has identified the major species of bacteriophages
attacking lactococci and defined many of their physiological and genetic charac-

teristics. Concurrently within the past decade, the lactococci have been dissected genetically to reveal that this bacterial host has evolved multiple mechanisms of phage resistance that can naturally coexist and complement one another to prevent the infection and proliferation of bacteriophages.

Close examination of the lactococci and their bacteriophages has revealed a number of complex interactions that attests to their co-evolution through countless rounds of defense and counterdefense. This chapter examines the bacteriophages and bacteriophage resistance mechanisms that have thus far been characterized in lactococci. Where appropriate, information that has accumulated on the bacteriophages and defense systems of other lactic acid bacteria (abbreviated LAB) will also be discussed. With this knowledge base, developments in this field over the last decade have placed the dairy technologist in an exciting new position where improved starter strains can be genetically designed to resist bacteriophage attack for extended periods of time.

Bacteriophage attack occurs in many industrial fermentations, but neither the magnitude nor persistence of the problems are comparable to that encountered in dairy manufacturing (Sanders, 1987). We hope that the following information which has accumulated on the bacteriophages and bacteriophage defense systems of dairy lactic acid bacteria will also provide a working framework for the identification and solution of problems that may be encountered in any bioprocess susceptible to attack by phage.

3.2 Types and species of bacteriophages

3.2.1 Interaction with hosts

The phenomenon of phage-mediated inhibition of starter cultures used in milk fermentations was first described in New Zealand by Whitehead and Cox (1935). Since these lactococcal phages were first identified, it has subsequently become clear that other lactic acid bacteria, including *Lactobacillus* species, *Streptococcus thermophilus* and *Leuconostoc* species can also be infected when used in commercial fermentation processes. In addition, Reiter in 1949, produced evidence that lysogeny occurred in lactococci and this phenomenon has since been demonstrated in *Lactobacillus* and *Streptococcus thermophilus*, but not yet in dairy *Leuconostoc*.

The first step in the interaction between lytic phage and its host happens when the phage particle adsorbs to the cell surface. Electron microscopic studies of adsorbing lactococcal phage indicate that in general, phages attach to homologous hosts in small groups at symmetrically distributed spots usually between five and approximately 30 in number. Less commonly, some phages were found to adsorb uniformly over the whole cell surface (Budde-Niekel and Teuber, 1987). Receptors for lactococcal phages are generally cell wall located although a plasma membrane lipoprotein was implicated as the φm33 receptor in the case of *L. lactis* subsp. *lactis* ML3 (Oram, 1971). In *L. lactis* subsp. *cremoris*

EB7, the cell wall-associated components L-rhamnose, D-galactosamine and D-glucosamine, which are involved in determining serological specificity of these bacteria, were also implicated as having a role in phage adsorption (Keogh and Pettingill, 1983). Valyasevi *et al.*, (1990) have shown that a rhamnose component of the extracellular polysaccharide is likely to be the primary binding site for some phage specific for *L. lactis* subsp. *cremoris* KH while galactose, which was suggested to be located in the vicinity of the receptor, appeared to play a less direct but nonetheless essential role in adsorption. Schafer *et al.* (1991) have shown that binding of P008 and P127 phages was inhibited when the cell walls of *L. lactis* subsp. *lactis* var. *diacetylactis* F7/2 and *L. lactis* subsp. *cremoris* Wg2-1, respectively, were subjected to lysozyme, metaperiodate or acid treatment, further supporting a carbohydrate component of the peptidoglycan as phage receptor. Recently, Valyasevi *et al.* (1991) have provided experimental evidence which indicated that lactococcal phage adsorption involves an initial, reversible binding to the cell wall polysaccharide which is followed by an irreversible step whereby the phage particle interacts with a cell membrane located protein. A protein with a subunit mass of 32 kDa (and a mass of 350 kDa under non-denaturing conditions) was isolated from the cell membrane fraction of *L. lactis* subsp. *lactis* C2 which was capable of inactivating phage c2. Interestingly, this is similar in size to another phage inactivating protein isolated from membrane fraction of *L. lactis* subsp. *lactis* ML3 (Oram, 1971).

In *Lactobacillus casei*, rhamnose has also been identified as a major component of the phage receptor. Yakokura (1977) suggested that the receptor for phage J-1 involved both cell wall L-rhamnose and cytoplasmic membrane D-galactosamine of *Lb. casei* S-1. For *Lb. casei* ATCC 27092, L-rhamnosyl residues of polysaccharide, located outside the peptidoglycan layer, were reported as the main determinants of the phage PL-1 receptor site with D-glucosyl residues also possibly having some role in adsorption (Ishibashi *et al.*, 1982). While adsorption of phage PL-1 occurred normally in the absence of Ca^{++}, this cation was shown to be essential for phage DNA penetration as was ATP (Watanabe and Takesue, 1972; Watanabe *et al.*, 1979). In contrast, Sechaud *et al.* (1989) have reported that the presence of Ca^{++} was not an absolute requirement for either adsorption or subsequent lysis of *Lb. helveticus* by the closely related lysogenic and lytic phages, 0241 and 832-B1, respectively. However, Ca^{++} did result in earlier and more rapid cell lysis. Somewhat unusually the receptor for this latter phage was reported to be a protein of 45 kDa, which constitutes the S-protein layer on the surface of *Lb. helveticus* CNRZ892 and may be structurally stabilized by Ca^{++} (Callegari *et al.*, 1992). Watanabe *et al.* (1991) have examined injection of *Lb. casei* phage PL-1 DNA by electromicroscopy and have demonstrated that the process was inhibited by chloramphenicol and erythromycin which inhibit protein synthesis. Their results prompted speculation that proteins synthesized during the early stages of phage infection and prior to complete DNA transfer itself are required for injection of DNA from phage PL-1 to cells of *Lb. casei*.

Latent periods for lactococcal phages have been reported to range between 10 and 140 minutes with values between 40 and 50 minutes at 30°C being more common for most phage (Keogh, 1973). Burst sizes can vary from approximately 10 to 400 particles released per infected cell (Keogh, 1973; Klaenhammer, 1984; Moineau *et al.*, 1992b). Generally, the optimum temperature for phage replication is similar to that of the host strain (i.e. 30–32°C in the case of lactococci), although some propagate optimally at temperatures either lower or higher than those which are optimal for their hosts (Mullan *et al.*, 1981). Phage multiplication can also depend on the nutritional status of the host and on electrolyte availability, which can be active in promoting phage replication and cell lysis (Keogh, 1980; Klaenhammer, 1984).

There is also considerable variation in the latent periods and burst sizes reported for *Lactobacillus* phages with the former ranging between 40 and 75 minutes and the latter generally ranging between 80 and 300 particles per infectious centre (Sechaud *et al.*, 1988). A very low burst size of 12–14 was reported for the *Lb. plantarum* phage B2 (Nes *et al.*, 1988). Chow *et al.* (1988) have examined the growth kinetics of the *Lb. delbreuckii* subsp. *bulgaricus* phage ch2 and have shown that neither host growth nor synthesis of the enzyme beta-galactosidase was affected by phage infection until cell lysis occurred. Neve and Teuber (1991) reported that a virulent phage for *Leuconostoc mesenteroides* subsp. *cremoris* had a latent period of 28 minutes and a burst size of 41. Data describing the growth characteristics of *Streptococcus thermophilus* phages have not been reported.

Little is known regarding the intracellular development of lactococcal phages at a molecular level. Hill *et al.* (1991a) have described a method which allows intracellular phage DNA replication to be monitored and have shown that the phage DNA molecules of the cohesive ended phage 31 are synthesized in a concatameric form between 20 and 40 minutes after infection of *L. lactis* NCK203. Significantly, 60 minutes after infection, the amount of detectable phage specific DNA had decreased and this was suggested to be due to packaging and release of the phage particles from the host.

Powell *et al.* (1992) have recently performed a detailed study of the replication of the broad host range prolate-headed phage c6A. Infection of *L. lactis* C6 caused inhibition of culture growth within 10 minutes and cell lysis occurred after 25 minutes. It was shown that c6A DNA synthesis began less than 6 to 8 minutes after infection and proceeded at an approximately constant rate until cell lysis. By monitoring the fate of ^{3}H-thymidine-labelled host cell DNA, they concluded that degradation of host DNA occurred within 4 to 6 minutes after infection and the breakdown products were incorporated into DNA, but not into RNA, and these products were the precursors for new phage DNA synthesis. Their results were consistent with c6A DNA being synthesized in the form of a large replicative intermediate, similar to that indicated for phage 31 by Hill *et al.* (1991a). It was further suggested that this degradation would kill the cell in the event the infection was aborted via one of the numerous abortive resistance mechanisms defined in lactococci.

Casey (1991) has examined the production of phage-specific DNA and RNA during vegetative replication of the temperate phage Tuc2009 on its lytic host *L. lactis* subsp. *cremoris* UC526. Hybridization analysis of DNA samples isolated at 10 minute intervals from a one step growth experiment, using labelled whole phage DNA as a probe, revealed that phage-specific DNA could be detected 40 minutes after infection (the latent period of the phage was approximately 55 minutes). There was a consistent increase up to between 70 and 80 minutes and, thereafter, the concentration of DNA decreased presumably due to packaging of the DNA. This time scale for the appearance of detectable phage DNA is considerably longer than that reported by Powell *et al.* (1992) for phage c6A but the detection technique used by Casey (1991) was less sensitive and would not detect phage-specific DNA until a significant level of replication had occurred. When Hill *et al.* (1991a) monitored φ31 DNA replication using a similar approach, phage DNA could be detected 5 minutes after the infection, but significant amounts of phage 31 DNA did not accumulate until approximately 45 minutes following infection of the *L. lactis* NCK203 host.

When Tuc2009 phage-specific RNA was being monitored in infected UC526 hosts, RNA was isolated from samples which were removed from the growth cycle experiment at 2 minute intervals. Again using whole phage genome as probe, transcription products were detected 28 minutes after infection of the host, but levels were low until 48 minutes. At this point the amount of phage-specific RNA had increased significantly. It is notable that this was also the approximate time at which high levels of DNA were detected (Casey, 1991).

When various restriction fragments, which were deliberately chosen to cover the entire Tuc2009 genome, were used to probe the RNA samples, the resultant hybridization data allowed the assignment of early and late functions to the phage restriction map. Early and late expression regions were clustered at locations on the map proximal and distal to, respectively, the *pac* site at which packaging of the phage genome is initiated. The early regions do not produce a high number of transcripts, and it was speculated that genes in this part of the Tuc2009 genome may encode proteins which direct the cell to switch from cellular to phage RNA transcription. It is likely that regions that expressed a high concentration of transcripts late in the infection cycle encode proteins that are produced in high concentration and which are required late in maturation, presumably for phage morphogenesis and packaging. Interestingly, one region of Tuc2009 contained a 4.4 kb *Eco*RV fragment that did not appear to be transcribed at all suggesting that it played some regulatory or non-essential role (Casey, 1991).

Molecular studies with the *Lb. casei* phage PL-1 showed that it used unmodified host RNA polymerase to transcribe its entire genome, with about 50% of the transcript originating from each DNA strand (Stetter *et al.*, 1978). It was reported that during infection neither an additional phage-coded RNA polymerase nor a modified host polymerase could be observed. Early genes were expressed within the first 20 minutes and in that time about 36% of the heavy strand was transcribed. After 40 minutes late genes were also expressed and these were transcribed by host RNA polymerase using the light strand as a template. Thus, different regions of both strands are transcribed in opposite directions to include almost the entire length of the phage DNA.

Based on work described by Trautwetter *et al.* (1986), Sechaud *et al.* (1988) conclude that the *Lb. delbrueckii* subsp. *lactis* phage LL-H genome also contains two polycistronic units which are transcribed in opposite directions, with the early region encoding genes responsible for phage replication and the late genes encoding structural proteins and phage lysin.

Following the synthesis of phage macromolecular components, the phage particles are assembled by mechanisms which have not yet been investigated at a molecular level. However, the newly assembled phage particles are released at the end of the latent period due to the action of phage-encoded lysozyme-like lysin on the peptidoglycan layer of the host cell. Lysins from a number of lactococcal phages and for phage of some *Lactobacillus* species have been studied in detail. It has been reported that lactococcal phage lysins were activated by monovalent cations and had similar physical properties (Oram and Reiter, 1965; Mullan and Crawford, 1985). While the effective range of these lysins was restricted to lactococci and group D streptococci, they were active on strains which were not sensitive to the intact bacteriophage responsible for production of the lysin. The molecular cloning and characterization of the phage φML3 lysin will be described below (section 3.2.4.2).

The *Lactobacillus delbrueckii* subsp. *lactis* phage LL-H lysin had lytic activity on other hosts of the same species, but three out of six *Lb. helveticus* strains and four out of 10 *Lb. delbrueckii* subsp. *bulgaricus* strains were resistant, as were the *Lb. casei* and *Lb. plantarum* strains tested (Trautwetter *et al.*, 1986). Interestingly, *Bacillus subtilis* was sensitive to the lysin. The endolysin gene of the *Lb. delbrueckii* subsp. *bulgaricus* phage mvl has been cloned and sequenced (see section 3.2.4.2) and its activity on a range of hosts was examined (Boizet *et al.*, 1990). This 21 kDa protein was active on some strains of *Lb. helveticus* subsp. *bulgaricus* and *S. thermophilus* but no activity was observed on the strains of *Lactococcus lactis* examined. Based on the deduced amino acid sequence, similarities between the mvl lysin and the amino terminal portion of the muramidases of the fungus *Chalaropisis* and the *S. pneumoniae* phage Cp-1 were observed.

3.2.2 Temperate phages

While lytic bacteriophage are of major concern to those involved in the use of LAB in dairy fermentation, there is also significant interest in phages which can enter into an alternative relationship, i.e. the lysogenic cycle, with their hosts. In this case, the phage genome, after injection into the target cell, integrates, usually by recombination between two sites termed *att*P (phage attachment site) and *att*B (bacterial attachment site) on their respective genomes. Little information is available regarding the molecular or genetic basis for control and maintenance of the lysogenic relationship in LAB hosts, but it is likely that the general pattern will follow that exhibited by the very well studied lambda phage and *E. coli* (Ptashne, 1986). In the only case examined to date in the LAB, establishment of lysogeny in *Lb. gasseri* species occurs via an *att*P/*att*B-based site-specific integration process (Raya *et al.*, 1992, see below).

The study of lysogeny in LAB is very worthwhile for a number of reasons. At a practical level, lysogenic hosts could serve as a reservoir of lytic phage in dairy plants. While a genetic relationship has been established between some lytic and induced lysogenic phages in *Lactobacillus, S. thermophilus* and *Lactococcus*, it is also clear that most virulent phage are not genetically related to lysogenic phages. Although some homology between lytic and lysogenic lactococcal phages has been reported in several studies (Braun *et al.*, 1989; Coveney *et al.*, 1987; Jarvis, 1984a, 1984b; Lautier and Novel, 1987; Relano *et al.*, 1987), there are no known temperate phages in either the prolate (c2 species) or 936 species (largest) of lactococcal phages (Jarvis *et al.*, 1991). Furthermore, indicator strains for lysogenic phages are not commonly isolated, particularly in lactococci which have been the most extensively studied of the LAB. This supports the suggestion that virulent phages are typically not derived from lysogenic hosts (Jarvis, 1989). This paucity of indicator strains is not universally encountered however, as Reyrolle *et al.*, (1982) were able to identify sensitive hosts for 43% of the temperate lactococcal phages which originated from lysogenic hosts examined in their study.

With at least some lactobacilli there does appear to be a more clearcut relationship between lytic and lysogenic phages. In a series of reports, Shimizu-Kadota and her colleagues were able to show that a lytic phage capable of infecting a *Lb. casei* strain used in Yakult fermentation was derived from a lysogenic phage present in this host (Shimizu-Kadota *et al.*, 1983; Shimizu-Kadota and Tsuchida, 1984). This work will be reviewed more extensively below in section 3.2.4.2. It is notable that the 'a' homology group of *Lb. delbrueckii* phages harbours a significant number of temperate phages in addition to the lytic phages (Mata and Ritzenthaler, 1988; Lahbib-Mansais *et al.*, 1988). Sechaud *et al.*, (1989, 1990) have indicated a likely relationship between specific temperate and virulent phages of *Lb. helveticus*, although the latter report DNA homology was not assessed. Neve *et al.*, (1990) have shown that two identical temperate phages from *S. thermophilus* exhibited homology at the centre of the lytic P55 phage genome which spanned approximately 25% of the genome length. These data may suggest that lytic phages are derived from lysogenic isolates or at least that module recombination between genomes of the two types could have occurred.

The study of lysogeny is also of interest in the context of developing model systems for the regulation and control of gene expression in LAB. Although the nature of lysogenic phage repressors and how they control the cascade of reactions leading to vegetative growth has not yet been determined, elucidation of the manner in which they interact with the phage genome and maintain the lysogenic state is likely to provide valuable insight into gene control systems in these bacteria. In addition, the cloning of repressor genes may yield a family of vector plasmids whose activity can be deliberately manipulated to allow controlled gene expression in LAB. It is noteworthy that Chopin *et al.*, (1989) have generated a chromosomal integration vector harbouring a fragment of lactococcal prophage DNA which is homologous to a resident prophage genome in the host chromosome. This vector integrates via homologous recombination into the resident prophage. Raya *et al.* (1992) have recently described the construction of an integration vector for *Lactobacillus gasseri* species which harbours a DNA fragment from the temperate phage, φadh, cloned in a plasmid lacking a Gram-positive origin of replication. This recombinant plasmid contains *attP* and the phage integrase gene *intG*, and integrates by a site-specific recombination process into the *attB* site of the *Lactobacillus* chromosome (Raya *et al.*, 1992; Fremaux *et al.*, 1992). The *attB* site, with a defined core region of **5'TACACTTCTTAGGAGG 3'** was carried by all *Lb. gasseri* strains examined and site-specific integration occurred at that position, even in the presence of the residing prophage (De Antoni, Fremaux, and Klaenhammer, unpublished data). Other lysogenic *Lactobacillus* species and strains, including *Lb. casei* 393 and *Lb. delbrueckii* subsp. *bulgaricus* CNRZ668 (LT4), did not harbour the *attB* sequence nor accept the *attP* site-specific integration vector, pTRK182.

Temperate phage are also capable of mediating transduction which, in addition to transferring plasmid encoded traits, may eventually find a role in developing chromosomal maps for various members of the LAB (Fitzgerald and Gasson, 1988; Davidson *et al.*, 1990). Plasmid transduction of lactose-fermenting ability and proteolytic activity has been well documented in the lactococci (McKay and Baldwin, 1974; Fitzgerald and Gasson, 1988, Gasson, 1983). Recently, plasmid transduction has been demonstrated in a *Lactobacillus* species (Raya *et al.*, 1989; 1991). Cloning of phage fragments into a cloning vector (pGK12) improved the transduction frequencies of this plasmid by five orders of magnitude. The efficiency of this process could now facilitate the development of gene transfer and delivery systems for lactobacilli that are phage-mediated.

The majority of lactococci are considered to be lysogenic, based on their inducibility by mitomycin C or ultraviolet light and electron microscopic detection of phage particles in the lysates. Lysogeny is also common in lacto-bacilli but apparently less so in *S. thermophilus*. Carminati and Giraffa (1992) examined 45 strains of *S. thermophilus* and found only one which was inducible by mitomycin C. The temperate phage φ18 had an isometric head, long non-contractile tail and a genome of 40 kb without cohesive ends. The lysogenic status of *Leuconostoc* and *Pediococcus* is less certain, possibly because the survey type studies performed for the other LAB have not yet been done with these hosts. However, Arendt *et al.*, (1991) reported recently that lysogeny is common in *Leuconostoc oenos*. Readers interested in a more in depth discussion on lysogeny in LAB should refer to the excellent and comprehensive reviews by Davies and Gasson (1984), Davidson *et al.* (1990), Sechaud *et al.* (1988) and Jarvis (1989). In section 3.2.4.2 of this review, developments in the molecular characterization of lysogenic phages will be discussed.

3.2.3 *Classification of phages of lactic acid bacteria*

Much of the work on the characterization of phages of LAB has been directed towards developing a coherent classification scheme which would provide insight into the relationships between phages and identify the origin of phages that disturb fermentation processes. In particular, knowledge of the genetic characteristics of lactococcal phages is also proving useful in predicting the potential utility of phage resistance mechanisms that can be introduced into lactococcal starter cultures used in commercial cheese production.

Taxonomically, bacteriophage of lactococci have been examined most extensively and these studies have been based on morphology, DNA relatedness, protein composition and less commonly, serological analysis. It is not the intention in this review to examine individually the very many reports dealing with phage characterization and classification based on these criteria, and the reader is directed to a number of research and review papers relating to these topics (Keogh and Shimmin, 1974; Braun *et al.*, 1989; Coveney *et al.*, 1987; Jarvis, 1977, 1984a, 1984b, 1989, Lautier and Novel, 1987; Relano *et al.*, 1987; Klaenhammer, 1984; Lembke and Teuber, 1981; Mata and Ritzenthaler, 1988; Mata *et al.*, 1986; Prevots *et al.*, 1990; Saxelin *et al.*, 1986; Sechaud *et al.*, 1988, 1992; Lahbib-Mansais *et al.*, 1988; Neve *et al.*, 1989). However, the work has led to a definitive classification scheme for lactococcal phages, particularly the morphological and DNA homology studies, and culminated in the recent paper by Jarvis *et al.* (1991) which will be reviewed briefly.

Morphological analysis based on electron microscopic examination of negatively stained phage preparations has shown that the majority of virulent lactococcal phages and most of the temperate phages belong to the *Siphoviridae* family, displaying a B1 or B2 morphotype (Ackermann and DuBow, 1987). Those in the B1 group have heads less than 60 nm in diameter and non-contractile tails shorter than 200 nm and are commonly referred to as 'small isometric-headed' (Jarvis *et al.*, 1991). Other finer morphological structures such as collars, fibres emanating from the collar, simple to complex base plates, long fibres originating from the tail tip and cross bars on the tail have been observed and have been used to subdivide the group (Jarvis, 1989). However, these features are near the limits of resolution of the electron microscope and are, therefore, difficult to compare between different studies.

Phages of the B2 morphotype have prolate heads typically with dimensions of 60 × 40 nm and have tails considerably shorter than those of B1 group phages (i.e. approximately 100 nm, Jarvis *et al.*, 1991). Generally these phages are encountered less commonly than small isometric-headed phages and the B2 group also appears to be more homogenous than the B1 group. While prolate-headed temperate phages have occasionally been described in the literature (Teuber and Loof, 1987; Heap and Jarvis, 1980), in specific studies where large numbers of induced lysates were examined, prolate-headed phages were not observed (Teuber and Lembke, 1983; Davidson *et al.*, 1990).

Large isometric-headed phage, also of the B1 morphotype, are rarely encountered and appear to be exclusively virulent, while phages of the C2 and C3 morphotype (members of the family Podoviridae) are extremely rare and have heads of 85 × 44 and approximately 240 × 50 nm and tails of 24 and approximately 30 nm, respectively (Jarvis, 1989).

In terms of the geographical distribution of morphologically distinct lacto-coccal phages, it was reported that 85% of all virulent phages isolated in New Zealand were small isometric-headed, most of the remainder were prolate headed and the balance large isometric (Jarvis, 1977; Terzaghi, 1976). In the US and Ireland, small isometric-headed phages also appear to be far more common than any other morphotype (Jarvis and Klaenhammer, 1986; Casey 1991), while in Europe a wider variety of phages has been identified. Studies in Germany (Lembke *et al.*, 1980), France (Relano *et al.*, 1987), Finland (Saxelin *et al.*, 1986), and Canada (Moineau *et al.*, 1992b) have shown that prolate-headed phage in particular can be isolated quite commonly in addition to small isometric-headed types and occasionally rarer morphotypes. This greater diversity may be due to the fact that a larger variety of cheeses and other fermented dairy products are manufactured in mainland Europe, which rely on a greater number of starter strains and combinations. Whereas in New Zealand, the US and Ireland, in particular, Cheddar and related types predominate and these are often made with carefully selected, defined strains which have been chosen for their phage insensitivity traits in addition to the other characteristics required for cheese-making.

All the lytic phages of various *Lactobacillus* species described to date have isometric heads, although there does appear to be significant variation in tail and other minor structures with features such as cross-barred tails, fibrous collars, and tail fibres. A triple collar has been observed with specific *Lb. delbrueckii* subsp. *bulgaricus* phages (Cluzel *et al.*, 1987; Sechaud *et al.*, 1988). Prolate-headed phages have been induced from *Lb. delbrueckii* subsp. *lactis* and *Lb. salivarius* while another phage, also induced from *Lb. salivarius*, had a short tail bud (Sechaud *et al.*, 1988). *Lb. helveticus* phages (Sechaud *et al.*, 1988, 1989, 1990) and a *Lb. delbrueckii* subsp. *bulgaricus* phage described by Reinbold *et al.* (1982) are unusual in that they possess contractile tails and are therefore members of the A morphotype of Ackermann and DuBow (1987).

It appears that lytic and lysogenic phages of *S. thermophilus* are of the B1 morphotype with long non-contractile tails and isometric heads with some variation in the dimensions of these structures evident (Accolas and Spillman,

1979; Krusch *et al.*, 1987; Neve *et al.*, 1990; Reinbold *et al.*, 1982). Relatively few phages specific for the dairy *Leuconostoc* species have been described in the literature, perhaps because these hosts represent only a small component of the total population in starter cultures for specific dairy products, such as lactic butter and Dutch-type cheeses. Phage pro-2 of *Leuconostoc mesenteroides* had a B1 morphotype (Sozzi *et al.*, 1978) as did three *Leuconostoc cremoris* phages isolated in Finland from viili, although differences in head diameter and tail length between the three were apparent (Saxelin *et al.*, 1986). There is also a report (reviewed by Gasson and Davies, 1984) describing lytic and lysogenic *Leuconostoc* phages isolated in Japan which exhibited A1 and B1 morphotypes.

While morphological analysis of bacteriophages has provided much useful information about these agents and the possible relationships between them, of course this does not necessarily indicate a genetic relationship between phages. However, the wide application of hybridization analysis to phages of LAB, particularly those of the lactococci, has paid rich dividends in recent years in terms of taxonomic classification. Very many studies, which have included hundreds of lactococcal phage isolates from various geographical locations, have shown that morphologically similar phages are not necessarily related at the genome level, although the DNA of morphologically distinct phages are invariably unrelated. Due to the fact that many of the earlier studies included non-overlapping groups of phages, it was not possible to obtain a general picture of the number of genetically distinct phages present. However, this problem has been remedied and the efforts of the Lactococcal and Streptococcal Phage Study Group of the International Committee on Taxonomy of Viruses have resulted in a definitive classification scheme based primarily on DNA homology and morphology. The report, based on data collected from studies in New Zealand, Ireland, Germany, Australia and France, has defined 12 phage species with all members of a given species displaying similar morphology (although differences in minor structural features, such as the presence or absence of a collar, are allowed) and exhibiting significant homology with the type phage for that species (Jarvis *et al.*, 1991). Nine of the 12 groups are members of the family Siphoviridae (i.e. B1 or B2 morphotype) while two are Podoviridae (C2 and C3 morphotype). All the prolate-headed phages are in two major homology groups (type phages C6A and PO34); a third group contained only one isolate KSY1, which had an extremely large head. The small isometric-headed phages are divided into eight groups, but the majority of phages reside in two of these, represented by the P008 and P335 type phages. Interestingly, while one lysogenic phage (BK5-T) was placed in a group in which it was the only phage, lysogenic phage r1t was classified in the P335 group which also contains a number of lytic isolates. It was also noted in the report that phages isolated in New Zealand, Australia, Ireland, France and Germany are genetically related. Other key data relating to morphological and genomic characteristics have also been included in the report for each of the species. Although phages within a single homology group are undoubtedly genetically related, it is clear from studies in which restriction analyses of phage genomes have been compared, that there still can be significant divergence in base composition as restriction patterns generated by related phage can be quite different (Coveney *et al.*, 1987, Coveney, 1989; Jarvis and Meyer, 1986; Jarvis, 1989).

DNA homology analyses have also been used to determine the relationships between *Lactobacillus* phages. In two major studies by Mata *et al.* (1986) and Lahbib-Mansais *et al.* (1988), 19 of 22 *Lb. delbrueckii* subsp. *bulgaricus* and subsp. *lactis* phages were classified into a single group designated 'a' and interestingly, this group included four temperate phages. The other three phages constituted a second homology group, designated 'b'. The 'a' group included the well studied isolate LL-H. It is notable that a recombinant plasmid harbouring the cloned lysin gene from this phage hybridized to one or two specific DNA restriction fragments of genomes from all the other phages in the 'a' group. In a further survey, Sechaud *et al.* (1988) classified 26 phages, also specific for the *bulgaricus* and *lactis* subsp. of *Lb. delbrueckii* and again, the majority of the isolates (19) were of the 'a' homology group and seven of these were lysogenic phages. Furthermore, there were significant morphological differences between some of the lytic phages (including some with isometric heads and an unusual cross-barred tail structure). Four of the 26 phages were placed in the 'b' homology group. Two new homology groups 'c' and 'd' were created to accommodate three other phage isolates, two of which were lysogenic and had prolate heads (group c) the third being a lytic small isometric-headed phage (group d). Lahbib-Mansias *et al.* (1988) compared five *Lb. helveticus* phages that had been isolated from cheese factories over a 10-year period; all the phages constituted a single homology group which was distinct from the *Lb. delbrueckii* phage groups described above.

Generally all the phage of *S. thermophilus* examined in a number of separate studies can be classified in the three homology groups, although significant cross hybridization between phages in different sub-groups was observed (Neve *et al.*, 1989; Mercenier and Lemoine, 1989). In an earlier study, Benbadis *et al.* (1987) classified 18 *S. thermophilus* phages into two major groups based on restriction analysis of their genomic DNA.

There has been little genetic comparison of *Leuconostoc* phages. Recently, Boizet *et al.* (1992) examined nineteen *Leuconostoc mesenteroides* phages isolated from natural coffee fermentations and dairy products. DNA-DNA homology, morphology, and comparison of the major structural proteins revealed six groups, with four groups represented by a single phage. The majority of phages (12 strains) belonged to the group represented by phage φcc59a, a cohesive ended phage with a genome size of 26 kb. Genome sizes for the different phages of the six groups ranged from 24-70 kb. In the same study, Boizet *et al.* (1992) characterized eight *Leuconostoc oenos* phages into two genetic groups. Seven of the phages were isometric headed with a genome of 31 kb and contained cohesive ends.

3.2.4 Characterization of bacteriophage genomes

While genetic analysis of phages of LAB have lagged behind those of their hosts, this situation is being redressed. There is now available information relating to the sizes and gross structure of the genomes of many LAB phages. A small number of phage genes have been cloned and, in some cases, characterized at a sequence level. This progress in the analysis of phage genomes will be reviewed in detail below and the implications of this work for the future of phages will be discussed.

3.2.4.1 Genome size and gross genome structure. All phages of LAB examined thus far possess double stranded linear DNA and can have either cohesive ends or are circularly permuted with terminal redundancy. The G+C content of these phages reflects that of their hosts. Thus lactococcal phages have % G + C values ranging between approximately 37–39% (Jarvis, 1989) while values for *Lactobacillus* phages have been reported to range from approximately 37% for the *Lb. plantarum* phage B2 (Nes *et al.*, 1988) to approximately 45–48% for *Lb. casei* phages J1 and PL-1 (Sechaud *et al.*, 1988). The % G + C content for phages of other LAB have not been reported.

Based on summation of restriction fragments and, less frequently, by contour length measurements of phage DNA molecules visualized by electron microscopy, the genome lengths of the various species of lactococcal phages have been measured. Generally, the sizes range between 29 and 40 kb for small isometric, 52–134 kb for large isometric and 18–22 kb for prolate-headed phages (Jarvis *et al.*, 1991; Prevotts *et al.*, 1990). The variation in sizes between the different phage morphotypes may simply be a reflection of the physical capacity of the phage heads to accommodate different amounts of DNA.

The majority of lactococcal phages appear to have cohesive ends. Coveney (1989) has indicated that 18 of 26 small isometric and 15 of 15 prolate-headed phages had cohesive ends based on the comparison of heated and unheated restriction digests of the phage genomes. Lillehaug *et al.* (1991) have determined the sequence of the cohesive ends of the *L. lactis* subsp. *cremoris* phage φLC3 to be:

5'GTGACGGCGTGAA-3'
3'CACTGCCGCACTT-5'

While the absence of cohesive ends has been demonstrated in a number of other lactococcal phage (Coveney, 1989; Teuber and Loof, 1987), only the genomes of phages BK5-T, (Lakshmidevi *et al.*, 1988), Tuc2009 (Casey, 1991) and φ50 (Alatossava and Klaenhammer, 1991) have been categorically shown to have a circularly permuted structure with terminal redundancy. This is based on the following criteria: the presence of submolar fragments after restriction analysis; evidence for the presence of a *pac* site at which packaging of the phage genome is initiated; the absence of cohesive ends; and homology between the submolar fragment containing the terminal redundancy and the area of the genome close to the *pac* site. The amount of terminal redundancy reported for BK5-T was between 2 and 8 kb while that for φ50 was 2.2 kb and, in this latter phage, at least five rounds of packaging were found to occur from each genome concatameric intermediate.

Restriction maps have now been constructed for several lactococcal phages, both lytic and lysogenic, and as discussed above, these are generally phage specific even for phages within the same homology group. Restriction analysis of lactococcal phage DNA has indicated that there is a marked paucity of restriction sites, particularly for enzymes with 4-base recognition sequences (Coveney *et al.*, 1987; Kim and Batt, 1991a; Powell and Davidson, 1986). This is not due to methylation as passage of cloned phage DNA fragments through *E. coli* hosts, for example, does not alter the restriction pattern of the DNA. While one could speculate that phages have evolved to possess genomes with fewer recognition sites for restriction endonucleases to escape R/M system (Kruger and Bickle, 1983), it could also be argued that this could not result in the bias against sites for the large number of different enzymes which cleave less frequently than predicted.

Compared to phages of lactococci, the genome characteristics of relatively few *Lactobacillus* phages have been described, although some detailed studies of individual phages have been reported. Generally, for those which have been described, the genome sizes of phages from *Lb. delbrueckii* subsp. *lactis* and *bulgaricus* (e.g. LL-H and ch2; Trautwetter *et al.*, 1986; Chow *et al.*, 1988), *Lb. casei* (FSW, J-1 and PL-1; Shimizu-Kadota and Tsuchida, 1984; Khosaka, 1977; Stetter *et al.*, 1978) and *Lb. gasseri* (φadh; Raya *et al.*, 1989) range between 34 and 41 kb while the genome of the *Lb. plantarum* phage B2 was 73 kb (Nes *et al.*, 1988). Jimeno *et al.* (1990) have also described a group of *Lb. delbrueckii* phages with genome sizes as large as 75 kb. The *Lb. casei* phage FSW was described as having a circularly permuted genome (Shimizu-Kadota and Tsuchida, 1984) as were three *Lb. delbrueckii* subsp. *lactis* phages compared in a study by Forsman and Alatossava (1991). The other phage genomes examined at this level were described as having cohesive ends (Sechaud *et al.*, 1988). A number of *Lactobacillus* phages have now been restriction mapped (Raya *et al.*, 1989; Trautwetter *et al.*, 1986; Watanabe *et al.*, 1980) and unlike lactococcal phages, there does not appear to be a bias against restriction recognition sites for specific endonucleases (Mata and Ritzenthaler, 1988).

There are few reports describing the genomes of *S. thermophilus* phages. Neve *et al.* (1989) have reported sizes ranging from 33.8 to 44.2 kb, and an electron microscopic study of phage DNA molecules revealed the presence of circular genomes. Restriction analysis of three *S. thermophilus* phage genomes studied by Tiiro and Sarimo (1987) indicated sizes ranging from 30–40 kb. Neve *et al.* (1988) also reported that a *Leuconostoc mesenteroides* subsp. *cremoris* phage had a genome of 27.4 kb.

3.2.4.2 Genes of lactic acid bacteria bacteriophages. The characterization and molecular analysis of specific genes, regulatory elements and other sequences of lactic acid bacterial phages will undoubtedly lead to a better understanding of the following: the gross organization of phage genomes; how gene expression, particularly during intracellular vegetative replication or lysogeny is controlled; and, how virion particles are synthesized, assembled, and released. This knowledge will allow more valid comparisons with other well characterized systems such as various *E. coli* and *Bacillus* phages.

While relatively few phage genes have been isolated and characterized thus far, a review of the recent literature indicates that this situation is changing rapidly (Table 3.1). The first lactococcal phage gene cloned was the lysin gene of the prolate phage φvML3 (Shearman *et al.*, 1989). The gene was isolated following cloning of sized phage DNA fragments in *E. coli* using a lambda vector. Biological activity was detected by overlaying the lambda plaques with a lysin-sensitive *L. lactis* subsp. *cremoris* strain. While lysin was expressed in *E. coli* and *B. subtilis*, transformation of the lysin gene into *L. lactis* subsp. *lactis* MG 1363, proved difficult. When this was achieved an autolytic phenotype was created. Cultures grew normally through exponential phase but lysed in stationary phase suggesting that the lysin acted from outside the cell following some spontaneous lysis. This phenotype may have application in flavour acceleration strategies for natural cheeses (Shearman *et al.*, 1992). Sequence analysis of the DNA fragment encoding the lysin gene revealed an ORF of 561 bp which could encode a peptide of 187 amino acids with a predicted molecular weight of 21 090. This is in good agreement with the molecular weight of 24 000 for the protein detected by *in vitro* transcription and translation analysis. Significant homology between the φvML3 lysin and protein 15 of the *B. subtilis* phage PZA was observed following comparison of both the encoding genes. This latter protein has a function in DNA packaging during phage maturation of the related *B. subtilis* phage φ29. Shearman *et al.* (1989) have speculated that the φvML3 lysin may have another role in addition to lysing host cells, a concept which appears all the more likely considering the small size (23 kb) of the φvML3 genome and the consequent need to optimize its functional capacity. Recent unpublished work demonstrated the presence of a second open reading frame within the originally described gene. Evidence for the expression of a small protein has been obtained but its role in cell lysis has yet to be defined. Recently another lysin gene has been located on the genome of the isometric phage φUS3. The gene, designated *lytA*, was cloned and expressed in *E. coli* and its gene product visualized in this host. the deduced amino acid sequence of *lytA* showed homology to a muramidase from *Streptococcus pneumoniae* (Platteeuw and de Vos, 1992).

Kim and Batt (1991a, 1991b) have performed a molecular analysis of the genome of the small isometric headed *L. lactis* phage F4-1. A 7.6 kb *Hind*III/*Eco*RI fragment was cloned which encoded the major capsid protein of the phage in addition to what appears to be other minor structural proteins. *E. coli*

Table 3.1 Genes/elements cloned from phage of LAB

Gene/element	Phage	Vector	Cloning host	Sequence determined	Reference
Lactococcus					
Lysin	ΦvML3	λ gt10 pTG262	*E.coli* *B.subtilis*	+	Shearman *et al.* (1989)
p35, p43[a]	F4-1	pUC13	*E. coli* *L.lactis*	+	Kim and Batt (1991a,b)
MCP[b]	F4-1	pUC13	*E. coli*	−	Kim and Batt (1991 a,b)
ORF1365[c]	Φ7-9	pUC13	*E. coli*	+	Kim and Batt (1991c)
*Lla*I Methylase	F4-1	pBluescript pSA3	*E. coli* *L.lactis*	+	Hill *et al.* (1991c)
bpi[d]	BK5-T	pACYC194	*E. coli*	+	Lakshmidevi *et al.* (1990)
per/rep[e]	Φ50	pSA3	*E. coli* *L.lactis*	+	Hill *et al.* (1990b)
Lysin	ΦU53	pLK	*E. coli*	+	Platteeuw and de Vos (1992)
Lactobacillus					
Lysin	LL-H	pBR322	*E. coli*	+	Trautwetter *et al.* (1986)
gp[f]	LL-H	pBR325/ pBR322	*E. coli*	±	Trautwetter *et al.* (1986); Altassova *et al.* (1987)
Lysin	mv1	*	*E. coli*	+	Boizet *et al.* (1990)
ISLI[g]	ΦFSV	pTS101	*E. coli*	+	Shimizu-Kadota *et al.* (1985)
attP/integrase	Φadh	pSA34	*Lb. gasseri*	−	Raya *et al.* (1992); Fremaux *et al.* (1992)

[a]Genes encoding micro structural proteins of 35 and 43 kDa.
[b]Gene encoding major capsid protein of 35 kDa.
[c]Undefined ORF.
[d]A gene (BK-T *P*romoter *I*nhibitor) which decreased the activity of some BK5-T promoters.
[e]region of phage genome which mediates *P*hage *E*ncoded *R*esistance to some phage and contains a phage origin of replication.
[f]Genes encoding phage structural (capsid, tail) proteins.
[g]Insertion sequence of 1256 bp.
*Information not available.

and *L. lactis* clones expressing these phage-specific proteins were detected by Western blot analysis using rabbit anti-phage F4-1 antiserum. Indeed, the 7.6 kb fragment, which was located on the right arm of the genome within 9 kb of the right cohesive end, encoded at least five proteins. A 3.0 kb *Sau*3A fragment from within the 7.6 kb *Hind*III/*Eco*R1 segment of the phage F4-1 genome was sequenced and was found to contain four ORFs with coding capacities for proteins of 43, 35, 9.5 and 11 kDa. The two largest corresponded to proteins which had been previously visualized by Western blots and by *in vitro* transcription and translation analyses. An interesting observation was that ORF1 and ORF2 (encoding the two largest proteins) overlapped in the same reading frame and, in fact, ORF2 was a subset of ORF1 with the latter having an additional 75 amino acids at the amino terminus and both sharing the same termination codon. An overlap of 1 bp between ORFs 3 and 4 was also observed. The presence of overlapping genes on the phage F4-1 genome again reflects its relatively small size (33 kb) and the need to optimize its coding capacity. Each of the four ORFs was preceded by a potential ribosomal binding site and upstream of ORF 1 and 2, putative promoter sequences (which were divergent from the *E. coli* consensus) and transcriptional start sites were observed. None of the four ORFs exhibited any significant degree of homology to other reported sequences (Kim and Batt, 1991b). The initial characterization of the F4-1 phage genome, in addition to elucidating the location and, in some cases, the sequence of phage structural proteins, will provide an organizational basis for further studies. It will be interesting to determine if other lactococcal phage genomes possess similar organizational features.

Subsequently, Kim and Batt (1991c) identified a genetic element which was found to be conserved on the genomes of many of the *L. lactis* phages they surveyed. Analysis of more than 100 phages isolated from two cheese plants over a six-year period identified a subgroup of 18 phages whose genomes had different restriction patterns. Fifty percent of these isolates exhibited a common 1.6 kb fragment when restricted with *Eco*R1. When used as a probe, this fragment hybridized to more than 90% of the 18 phages. Hybridization was observed to the same 1.6 kb *Eco*R1 fragment in some phages (termed group I) and to a larger 5.3–5.8 kb *Eco*R1 fragment in others (group II). Determination of the nucleotide sequence of this fragment revealed an ORF of 1356 bp corresponding to a putative protein of 51 kDa containing 11% lysine residues. A similarly sized protein was detected when the 1.6 kb fragment was subjected to transcription/translation analysis. When the corresponding hybridizing regions of group II phages were sequenced, some differences to the group I phage sequence were observed both external to and within the ORF. A 119 bp region beginning at the the the 5' end of the group I 1.6 kb *Eco*R1 fragment was not observed in group II isolates, and a 30 bp region at the 3' of a *Hind*III site within the ORF was found only in group II phages. In a total of nine phages which possessed this conserved region, only three other minor changes were observed; two being within the ORF and none of these resulted in a change in the net charge of the protein. The regions immediately 5' to the conserved sequences in groups I and II phages are unrelated, and the authors speculate that this could be due to deletion or insertion events in these flanking sequences and that the major differences within the conserved sequence could be a consequence of the actual location of the recombination junction.

Characterization of the deduced protein sequence encoded by the 1356 bp ORF indicated that it was highly charged due to its high lysine content and that it shared homology to yeast translation initiation factor. A zinc 'finger motif' was also detected and it was noted that similar domains are found in other phage proteins involved in transcription and also in phage structural proteins. Kim and Batt (1991c) suggested that the conserved lactococcal phage protein may have a role in either regulating phage replication or specifying expression of its own genes.

An interesting application arising from the work on the conserved sequence described above was the construction of *L. lactis* strains in which the ORF was cloned in an antisense orientation (Kim and Batt, 1991d). Hosts carrying this construction exhibited a modest level of resistance to phages which harbored this sequence but were sensitive to phages which did not contain this region in their genome. In addition, neither a range of truncated genes cloned in the antisense orientation, nor the complete ORF placed in the sense orientation, conferred resistance. It was essential to turn the complete ORF in the antisense orientation to achieve any resistance, which should prompt speculation as to whether or not this is, in fact, an antisense mechanism. The phenotypic effect mediated by the antisense construct was a reduction in the EOP to 10^{-2} and an altered plaque morphology. While the data implicated an antisense RNA effect, expression of RNA from the antisense construction was not demonstrated in this study.

In the course of characterizing a number of phage isolates that had demonstrated the ability to circumvent one or both of the phage defense mechanisms mediated by the phage resistance plasmid pTR2030, Alatossava and Klaenhammer (1991) showed that a particular isolate, designated φ50, harbored a region on its genome which exhibited homology to the R/M locus on pTR2030. It was speculated that the phage may have acquired methylase sequences from the R/M region of the plasmid. Subsequent sequence analysis (Hill *et al.*, 1991c) revealed that a region 1273 bp in length was present and 100% identical in both plasmid and phage DNAs. In fact, this region encoded over 55% of the 1869 bp pTR2030-located structural gene for the methylase component of the type IIS R/M system, designated *Lla*I and included the functional amino domain and putative upstream expression signals. The phage-mediated methylase activity (*Lla*P1) was also biologically active although at a level 10-fold lower than *Lla*I. This work demonstrated that φ50 acquired a segment of host (i.e. pTR2030) DNA by *in vivo* genetic exchange which conferred a selective advantage on the phage since it allows its genome to remain methylated irrespective of the propagating host background. In a wider context, these results also indicate how lactococcal (and presumably other) phages can adapt and evolve rapidly to circumvent host-directed phage defense mechanisms or even perhaps to acquire new host ranges.

The specific mechanism by which φ50 acquired the truncated methylase gene is not clear although the authors do attach a significance to the presence of an iso-ISS*1* insertion sequence (IS*946*) upstream of the *Lla*I gene in pTR2030. This element had been previously implicated in inter- and intramolecular rearrangements involving pTR2030. Romero and Klaenhammer (1990b) and Hill *et al.* (1991c) speculated that IS*946* could have been involved in the formation of a φ50 precursor::pTR2030 intermediate which resolved to create φ50.

In addition to its capacity to counteract the pTR2030 directed R/M through the acquisition of the *lla*P1 gene, φ50 is also characterized by its ability to overcome a second phage defense mechanism encoded by pTR2030, namely the abortive infection system designated Hsp (see section 3.4.3). This phage encoded resistance (hence the designation Per) locus has been cloned from the φ50 genome (Hill *et al.*, 1990b). As well as bestowing resistance to the abortive defense encoded by pTR2030, this region when cloned *in trans* confers a resistance against a number of phages which are also resistant to Hsp. More detailed molecular analysis and sequence data revealed that the Per phenotype was associated with a region of the φ50 genome of approximately 500 bp which was rich in inverted and direct repeats and which did not encode any open reading frames. These and other character-istics, such as its negative impact on the replication of φ50 DNA when present *in trans*, and the increased replication rate of the plasmid on which *per* had been cloned following infection of the host with φ50, provided convincing evidence that the Per region was the origin of replication for φ50. Sequence analysis of the *per* locus revealed two 236 bp direct repeats with only three mismatches, separated by a 28 bp region. Within the direct repeats a number of inverted repeats capable of forming stable stem-loop structures were observed and these were typical of sequences associated with *rho*-independent termination. In addition, a 10 bp sequence was repeated four times within each of the large direct repeats and the palindromic sequence GGTACC was repeated 14 times within the 500 bp region. It was interesting that a 701 bp ORF capable of encoding a putative protein of 234 amino acids was detected immediately adjacent to the first direct repeat, however, its role, if any, in phage replication was not evaluated.

Of the temperate lactococcal phages, the genome of phage BK5-T has been studied in most detail. This genome has a unit size of 37.6 kb as measured by restriction mapping, and is circularly permuted with terminal redundancy (Lakshmidevi *et al.*, 1988). It is therefore likely that the genome is packaged by a headful mechanism, a process which is initiated from the *pac* site of a concatameric molecule consisting of several genome equivalents. The imprecise nature of packaging explains the considerable variability in the sizes of the linear phage genome molecules (39.7 to 46 kb) observed following examination by electron microscopy.

While phage BK5-T is lysogenic for *L. lactis* subsp. *cremoris* BK5, it is also capable of lysogenizing its indicator host, *L. lactis* subsp. *cremoris* H2, but only when the BK5-T phage was obtained following induction of either strains BK5 or lysogenized H2 (Lakshmidevi, 1988). When BK5-T phage is propagated lytically on strain H2 it loses the ability to lysogenize this host. This loss of lysogenic ability was linked to a deletion of between 0.6 to 2.5 kb (depending on the isolate examined) in a specific region of the genome which was separate from the phage's *attP* site. While the nature and function of this locus is not known, it has been proposed that it may be involved in mediating repressor or integrase functions.

Studies designed to investigate genetic regulation of the BK5-T genome have resulted in the isolation and sequence analysis of five phage promoters and the identification of the corresponding transcription start points (Lakshmidevi *et al.*, 1990). While the −35 and −10 regions exhibited homology to the canonical promoter sequences of *E. coli* and were functional in *Streptococcus sanguis* and *E. coli*, it is interesting that the TG dinucleotide, which precedes the −10 sequence in many lactococcal host promoters (van der Vossen *et al.*, 1987), was not observed in any of the five phage BK5-T sequences. It was present, however, in the promoter of the φvML3 lysin gene (Shearman *et al.*, 1989). By introducing cloned *Eco*R1 fragments, which spanned the entire BK5-T genome, into hosts harbouring plasmids where promoter sequences were inserted immediately upstream of a promoterless *cat* gene, it was possible to identify a region which inhibited the activity of three of the five phage promoters (based on a reduction in the levels of Cm resistance exhibited by either *E. coli* or lactococcal hosts). Further analysis revealed that this effect was mediated by the product of a 621 bp ORF which was designated *bpi* (for *B*K5-T *p*romoter *i*nhibitor). Although the mechanism by which the *bpi* gene product operates is unknown, it is likely that it acts like bacterial repressor proteins and binds to phage operators at regions exhibiting dyad symmetry. While a number of short palindromic sequences were observed in the BK5-T promoter-containing fragments, no sequence common to the three promoters, whose activity was inhibited by Bpi, was identified.

Clearly, characterization and analysis of lactococcal phage genomes has become a rapidly developing research topic and it is reasonable to expect that the accumulated data arising from the study of a number of different lactococcal phages, both lytic and lysogenic, will soon begin to yield information on phage genome organization, control and regulation of gene expression. Possibly most significant from a practical point of view will be definition of those factors which determine how a phage interacts with its host.

While studies on *Lactobacillus* phage genomes have not yet been pursued as intensively as those of lactococcal phages, it is notable that the clearest demonstration of the relationship between lysogenic and lytic phages and the mechanism through which lysogenic phage acquire virulence was provided by Shimizu-Kadota and her colleagues who undertook a detailed analysis of phages which were either lysogenic (φFSW) or lytic (φFSV) for *Lb. casei* S-1 (Shimizu-Kadota *et al.*, 1983; 1985; Shimizu-Kadota and Tsuchida, 1984). These workers were able to demonstrate that virulent phages, generically named φFSV, could be independently isolated from different Japanese factories which used *Lb. casei* S1 in a closed vat system to produce fermented milk beverages. The lytic phages were related in morphology, serology, protein composition and DNA restriction patterns to the temperate phage φFSW that lysogenized this host (Shimizu-Kadota *et al.*, 1983). Furthermore, derivatives of S-1 which were cured of φFSW were not attacked by lytic phages upon prolonged use in Yakult fermentations. Analysis of two of the lytic isolates, designated φFSV-B and φFSV-C, indicated that while their genomes were largely identical to that of φFSW, they both contained an additional 1.3 kb of host DNA which was shown subsequently to be an insertion element. IS*L1* and was the first native transposable element discovered in LAB (Shimizu-Kadota *et al.*, 1983; 1985). It was proposed that insertion of IS*L1*, which is present in the bacterial chromosome and inserted in different sites (0.7 kb apart in φFSV-B and φFSV-C), was responsible for acquisition of virulence by the φFSW prophage. This was proposed to be due to either inactivation of operator(s) or presentation of a new promoter which was not affected by the prophage repressor. Both situations would allow the constitutive expression of genes essential for lytic growth of the lysogenic phage.

Trautwetter *et al.* (1986) have localized the genes for five of the seven structural proteins and the lysin gene on the genome of the *Lb. delbrueckii* subsp. *lactis* phage LL-H. This was achieved by cloning restriction fragments in *E. coli* and detecting phage-specific products by immunoblotting techniques using rabbit anti-LL-H anitbodies. The five genes were clustered in one region, which also harboured the lysin gene. Interestingly, the authors speculated that the genes encoding the two structural proteins which they could not locate on the phage genome may actually be encoded by the host.

The location of the genes encoding the two main structural proteins and the lysin of the temperate *Lb. delbrueckii* subsp. *bulgaricus* phage mv4 has also been determined. Furthermore, the *attP* site was mapped to a 152 bp region which was sequenced and found to be A/T rich (%A+T of 75%) and to contain numerous inverted, direct and palindromic sequences (P. Ritzenthaler, personal communication). It appears that the genetic organization of the integration region is similar to that of phage lambda in that upstream of *attP* are located two ORFs, one of which encoded a protein with regional similarities to site-specific recombinases such as integrases, and the second which exhibits homology with DNA binding proteins. Raya *et al.* (1992) and Fremaux *et al.* (1992) have localized, cloned and sequenced the *attP* and the integrase gene from a 4.5 kb *Bcl*I fragment on the genome of the *Lb. gasseri* lysogenic phage adh. In addition the core attachment sequence has been defined and shown to be widely distributed among strains of the *Lb. gasseri* species. The genetic organization of the recombination functions of the temperate phage adh are clustered and similar to those described in other bacterial genera (Fremaux *et al.*, 1992).

With respect to phages of other LAB, there are not, as yet, any reports describing the genetic characterization of phage genomes. This will undoubtedly change as new information from *Lactococcus* and *Lactobacillus* phages stimulates further interest in these uncharacterized systems.

3.3 Gene directed bacteriophage resistance in lactic acid bacteria

The economic and technological problems posed by bacteriophage infection of LAB have provided a powerful motivation for the isolation and construction of strains which exhibit enhanced phage resistance properties. This is particularly true for culture systems comprising *Lactococcus* species, possibly because these are employed more intensively and on a larger scale than other LAB, and are consequently more prone to phage infection. In addition, lactococcal cultures have generally been subjected to greater scientific scrutiny and, based on this knowledge, specific starter systems consisting of very well characterized strains can be provided which will perform with increased reliability and produce cultured products of more consistent quality. These defined strain starters as they are termed, are now widely used in many parts of the world including New Zealand (Heap and Lawrence, 1988), Australia (Hull, 1983), the U.S. (Thunell *et al.*, 1981) and Ireland (Timmons *et al.*, 1988).

Clearly, it is important that strains intended for inclusion in any defined starter system should exhibit at least reasonable levels of phage resistance in addition to meeting all the other requirements of a good starter culture. Looking ahead, the ability to manipulate lactic acid bacteria at the molecular level will yield an increasing number of highly specialized and 'value-added' strains. The ability of these strains to resist bacteriophage attack will dictate their longevity in the commercial arena. If intended for continuous use, rational designs to maintain phage resistance should be an essential part of the strain development programme.

Over the course of the past decade, genetic studies on lactococci have established that these bacteria harbour a variety of plasmid coded mechanisms to defend the host against phage attack. More recently phage resistance systems have also been identified in other LAB. The mechanisms of phage resistance that have been identified include adsorption blocking, restriction/modification, and abortive infection systems. The remainder of this review will focus primarily on the categories of resistance mechanisms that have been identified in *Lactococcus* and some other LAB and will highlight how the determinants encoding these systems can be exploited to generate commercial strains with considerably improved levels of phage resistance.

3.3.1 Interference with bacteriophage adsorption

Host-mediated alterations in the ability of phage to adsorb to the lactococcal cell surface giving rise to phage-resistant mutants have frequently been reported (Limsowtin and Terzaghi, 1976; 1977; Marshall and Berridge, 1976; Hull, 1983;

King *et al.*, 1983; Huggins and Sandine, 1979; Thunell *et al.*, 1981; 1984). These bacteriophage resistant mutants (also termed bacteriophage insensitive mutants-BIMs) were generally isolated following exposure of the parent cultures to either their homologous phage or to factory-derived whey samples and in most cases were found to be unable to adsorb phage (Marshall and Berridge, 1976; Limsowtin and Terzaghi, 1977; King *et al.*, 1983). The isolation and use of resistant mutants is a logical approach to countering the phage problem in commercial practice and indeed, in specific instances resistant mutants have been employed successfully for long periods. Nevertheless, it has proved impossible to isolate resistant derivatives for some strains (Limsowtin and Terzaghi, 1977; Jarvis, 1981) and some mutants frequently grow poorly in milk, revert to phage sensitivity or acquire sensitivity to a different phage (Limsowtin and Terzaghi, 1977; Jarvis, 1981; Hull, 1983).

Phage-resistant mutants will generally be resistant to specific phage since they are likely to arise as a consequence of spontaneous point mutations on the host chromosome. Recently, a number of plasmids which encode mechanisms resulting in interference with phage adsorption have been described in lactococci. In some cases at least, these have been shown to confer resistance to more than one phage. The essential properties of these plasmids are described in Table 3.2.

Table 3.2 *Lactococcus* plasmids that interfere with phage adsorption

Plasmid	Size	Original host	Reference
pME0030	48 kb	*L.lactis* ME2	Sanders and Klaenhammer, 1983
pSK112	54 kb	*L. cremoris* SK11	de Vos *et al.*, 1984, de Vos, 1989
pCI528	46 kb	*L.lactis* UC503	Coffey *et al.*, 1991b
pKC50	80 kb	*L.lactis* 57150	Tortorello *et al.*, 1990

The genetic determinants responsible for interference with phage adsorption have not been localized, cloned or characterized.

In the prototype phage-insensitive strain *L. lactis* subsp. *lactis* ME2, plasmid pME0030 correlated with the interference of phage adsorption (Sanders and Klaenhammer, 1983). Variants of ME2 lacking the plasmid could be readily isolated following serial subculture at 30° or 42°C or treatment with acriflavine, and these adsorbed φ18 at levels of 80–90% compared with 20–40% for the parent strain. Although there was a significant increase in adsorption efficiency of phage 18, the efficiency of plating increased only by two log cycles which indicated that the pME0030 cured derivatives still retain other resistance mechanisms. It is now known that at least two additional R/M systems and two distinct abortive infection mechanisms are present in this bacterial strain. In addition to φ18, three of fourteen phages examined showed an elevated level of adsorption to the pME0030-cured isolate compared with the parental strain, demonstrating the generalized nature of the adsorption blocking mechanism involved. In a subsequent report, the mechanism mediating adsorption blocking was shown to be temperature-sensitive, in that adsorption of φ18 to ME2 pregrown at 40°C was increased two-fold compared to ME2 grown at 30°C (Sanders and Klaenhammer, 1984). However, the specific mechanism by which pME0030 mediates the adsorption blocking effect remains to be elucidated.

Plasmid pSK112, originally identified in *L. lactis* subsp. *cremoris* SK11 also mediates phage resistance via an adsorption blocking mechanism (de Vos *et al.*, 1984). Curing of the plasmid drastically increased the adsorption of phage SK11g from 1–5% to 80–90%. Linkage of the adsorption trait to pSK112 was later verified following its conjugative mobilization by pAMB1 and simultaneous transfer of the reduced phage adsorption phenotype (de Vos and Davies, 1984). Analysis of the cell walls of both SK110 and SK112 (the pSK112 cured derivative) allowed Sijtsma *et al.* (1988) to conclude that the parental strain had acquired phage resistance due to the masking of the phage receptor on the cell surface. The masking agent was resistant to treatment with agents such as SDS, Triton X-100, trypsin, pronase, alpha-amylase and HCl; and evidence suggested a galactose moiety. Treatment of the cells with a mild alkali inactivated the masking agent and allowed phage to adsorb. Further comparison of the cell surface characteristics of SK110 and SK112 showed that the latter had a hydrophobic cell surface, a relatively high negative electrophoretic mobility and was more easily resuspended following centrifugation (Sijtsma *et al.*, 1990a, 1990b). Agglutination studies with lectins also confirmed the earlier observation that SK110 produced a galactose-containing polymer which could be removed by alkali treatment which in turn, resulted in SK110 displaying properties similar to SK112. Sijtsma *et al.* (1991) finally concluded that it was in fact a galactosyl-containing lipoteichoic acid which was involved in preventing adsorption of phage to *L. lactis* subsp. *cremoris* SK110.

Another adsorption blocking plasmid, pCI528, isolated from *L. lactis* subsp. *cremoris* UC503, appears to mediate a resistance mechanism somewhat similar to that of pSK112 (Coffey *et al.*, 1991b). The presence of pCI528 in a host resulted in a decrease in cell surface hydrophobicity, caused cells to exhibit a 'fluffy' easily resuspended pellet following centrifugation and conferred complete resistance to a range of small isometric- and prolate-headed phage (Costello, 1988). All these phenotypic traits were alleviated by washing pCI528-containing cells in 0.05 M NaOH after which these hosts displayed properties similar to pCI528-cured hosts. Gas-liquid chromatographic analysis of cell wall material indicated that pCI528-containing hosts had significantly elevated levels of both galactose and rhamnose. This suggested that phage resistance was achieved as a consequence of masking of phage receptors on the cell surface by a polymer containing these sugars. The fact that some level of adsorption can be observed for some phage on pCI528-containing hosts may be explained by the observation that this agent is unevenly distributed over the cell surface thereby allowing access to a proportion of the cell receptors (Fitzgerald, unpublished data).

It is notable that the presence of a cell surface slime layer does not necessarily confer phage resistance on a host. For example, Vedamuthu and Neville (1987) have shown that production of a slime layer by *L. lactis* subsp. *lactis* var. *diacetylactis* harbouring plasmid pSRQ2202 which codes for mucoidness, does not affect adsorption of phage. In fact, the efficiency of plating of this phage is similar on hosts with or without the plasmid.

Investigations on the nature of the plasmid-encoded adsorption blocking trait associated with the 80 kb pCK50 have been performed by Tortorello *et al.* (1990). These have correlated the production of three cell surface antigens with the presence of this plasmid and have used transposon mutagenesis to show that one of the antigens may play a major role in adsorption blocking.

There are a number of features worth noting regarding plasmid-directed adsorption inhibition systems. To date, unlike the situation with other plasmid coded phage resistances, molecular cloning of specific determinants encoding adsorption blocking has not been reported. The reason for this is not clear but the information which would be provided by the detailed characterization of these genes would undoubtedly be very beneficial in elucidating the nature of the specific mechanisms involved.

De Vos (1989) has used *L. lactis* subsp. *cremoris* SK11, which harbours the adsorption blocking plasmid pSK112, to substantiate his hypothesis that the phage-carrier state can be explained by a metastable balance between two processes, namely the curing of a bacteriophage resistance plasmid resulting in phage sensitive progeny and multiplication of phage on these sensitive hosts. The phage-carrier state, which can be defined as the continuous presence of virulent bacteriophages throughout the growth of the host culture, is considered to contribute to the high phage resistance levels encountered with some starter cultures (particularly the undefined mixed strains such as the P cultures used in The Netherlands; Stadhouders and Leenders, 1984). In his study, de Vos was able to demonstrate that the phage carrier state could be maintained only with *L. lactis* subsp. *cremoris* SK110 but not with the pSK112 cured derivative, SK112. Indeed Sterkenburg *et al.* (1988) have shown that phage-sensitive variants of strain SK110, lacking pSK112 do segregate spontaneously thereby generating the type of mixed population required for the maintenance of the phage carrier state.

Also, unlike many of the plasmids mediating other types of phage resistance, adsorption blocking plasmids have not been shown to be self-transmissible. Whether this is due to the fact that comparatively few of these plasmids have yet been identified in lactococci, or whether it is an inherent property of the plasmids is unknown. It is notable however, that plasmid pCI528 which was considered originally to be non-self transmissible, can be transferred to a broad range of lactococcal hosts in agar surface matings when one specific strain is used as the donor (Fitzgerald, unpublished data). Thus, it has been possible to introduce pCI528 into commercial cultures (Kelly *et al.*, 1990) using an approach largely similar to that described by Sanders *et al.* (1986) for construction of lactococcal strains with enhanced resistance to bacteriophage (see section 3.7).

3.3.2 Restriction and modification

3.3.2.1 Distribution in lactic acid bacteria and roles in host-dependent phage replication. Following the successful phage adsorption and injection of its genome into the cell, a bacterium bearing a restriction endonuclease and modification methylase will likely recognize the invading DNA as foreign and degrade it via the action of restriction endonucleases. R/M systems are inherently leaky, however, and occasionally the phage DNA escapes restriction, incurs modification by the methylase, and is replicated to produce modified progeny that are not restricted in subsequent rounds of infection. Therefore, while R/M systems can provide a formidable barrier to infection by unmodified phages, the associated modification methylases also provide ample opportunities for the adaptation of a phage population that will no longer be contained by the action of restriction endonucleases (Klaenhammer, 1984). This potential for phage modification by R/M systems has prompted some to speculate that R/M systems are not effectual as a phage defense system. However, more recent findings have shown that R/M systems are quite effective as a defense system when combined with other complementary resistance mechanisms (Sing and Klaenhammer, 1990a, 1990b). The operation of R/M systems in lactococci can also act as a barrier to DNA transformation in protoplasts (Langella and Chopin, 1989) and to plasmid transfer via conjugation (Steenson and Klaenhammer, 1986). Langella and Chopin (1989) note that this is not always the case since plasmid DNA transfer via electroporation and conjugal transfer of pIP501 in lactococci were not retarded by the presence of R/M systems. Nevertheless, the presence and operation of native R/M systems should be evaluated if attempts to introduce and express genes in industrial strains becomes problematic.

The presence of R/M systems throughout lactococci is now well established. Since the first operational definition of R/M activity by Collins (1956), numerous studies have reported the type of host-dependent phage replication on lactococcal strains that is expected when R/M systems are operating (Pearce, 1978; Limsowtin *et al.*, 1978; Sanders and Klaenhammer, 1980; Daly and Fitzgerald, 1982, 1987; Teuber and Lembke, 1983; Boussemaer *et al.*, 1980; Chopin *et al.*, 1984; Gautier and Chopin, 1987; Higgins *et al.*, 1988; Hill *et al.*, 1989a; Josephsen and Klaenhammer, 1990; Vogensen and Josephsen, 1990; Sing and Klaenhammer, 1991; McKay *et al.*, 1989; Froseth *et al.*, 1988; Klaenhammer, 1984). Indeed, numerous lactococcal strains have been shown to contain two or more R/M systems of different specificity. Boussemaer *et al.* (1980) first alluded to the widespread distribution and multiplicity of R/M systems in lactococci after performing 23 different phage-host crosses. Genetic studies have confirmed the natural occurrence of two or three R/M systems within a single strain (Sanders and Klaenhammer, 1980; Chopin *et al.*, 1984; Gautier and Chopin, 1987; Josephsen and Govensen, 1989). The presence of multiple R/M systems of different specificity can dramatically increase the level of restriction elicited against incoming heterologous phages (Klaenhammer, 1987). In this regard, intentional stacking of plasmid-encoded R/M systems has been shown to confer a stronger level of restriction against heterologous phage attack (Sing and Klaenhammer, 1991; Josephsen and Klaenhammer, 1990). The level of restriction, however, is also dictated by a number of host and environmental factors. Paramount among these is the varied susceptibility of phages to R/M systems due to either a lack of restriction sites (Sharp, 1986; Powell and Davidson, 1986) or other antirestriction mechanisms (Kruger and Bickle, 1983). The phage species and strain used greatly influence the level of restriction elicited and determines whether or not one or more R/M systems can be detected. We have, in fact, probably underestimated the number of operational R/M systems in lactococci given the dynamics and magnitude of phage/host interactions in dairy fermentations.

Operation of R/M systems in other members of the lactic acid bacteria has recently been reported and has closely followed efforts to define and characterize bacteriophages from thermophilic dairy starters (Benbadis *et al.*, 1990; Sechaud *et al.*, 1989). Biological evidence for R/M activity in *S. thermophilus* has now been presented in two reports describing host-dependent replication of phage (Benbadis *et al.*, 1990; Mercenier *et al.*, 1987). Evidence for a plasmid linked R/M system in three *Lactobacillus helveticus* strains was recently reported by Reyes-Gavilan *et al.* (1990). The two phages used in the study exhibited classic host dependent phage replication on any one of the three host strains bearing a 34 kb plasmid correlatively linked with both restriction and modification activities in CNRZ 1094. It was interesting to note that unlike some lactococcal R/M systems (Sanders and Klaenhammer, 1980), the *Lb. helveticus* R/M activities were active at higher temperatures. The heat-induced temporary loss of R/M action in the mesophilic lactococci has been implicated as a contributor to both the loss of phage resistance and appearance of new phage in cheese plants (Klaenhammer, 1984; Pearce, 1978; Sanders and Klaenhammer, 1980). In this regard, the availability of heat tolerant R/M systems from the thermophilic dairy lactic acid bacteria could be potentially useful for genetic engineering of heat stable R/M systems in lactococci.

3.3.2.2 Restriction and modification enzymes: biochemical evidence. Biochemical investigations designed to isolate and characterize restriction and modification enzymes are relatively few in proportion to the widespread distribution of these systems among lactic acid bacteria. In the lactococci, the description of *Scr*FI, a type II restriction endonuclease, from *Streptococcus cremoris* F (currently named *L. lactis* subsp. *cremoris* UC503) provided the first biochemical evidence for a functional restriction enzyme among the R/M rich lactococci (Fitzgerald *et al.*, 1982). Restriction activity by ScrFI in strain F were initially not found (Daly and Fitzgerald, 1987), but recent studies have now confirmed a biological role for this enzyme (Fitzgerald, unpublished data). Mayo *et al.* (1991) recently reported two nucleolytic activities from *L. lactis* subsp. *lactis* NCDO 497. One was a non-specific nuclease that was located outside of the cytoplasm, whereas the second activity was that of a type II restriction endonuclease with a target sequence nearly identical to *Scr*FI (Table 3.3). The type II restriction activity from NCDO 497 was designated as *Lla*I. Two more similar restriction enzymes have now been isolated and characterized from *S. thermophilus*. Solaiman and Somkuti (1991) identified the type II restriction endonuclease *Sth*117I from one strain selected for analysis because it was recalcitrant to electroporation. Benbadis *et al.* (1991) characterized a similar type II enzyme from a strain (BSN 45) that exhibited restriction and modification activities with a series of thermophilus-specific phages. Both the *Sth*117I and *Ss*1I enzymes recognize the same restriction site and are isoschizomers of *Bst*NI. The cleavage sites for the type II restriction endonucleases characterized thus far from the lactic acid bacteria are shown in Table 3.3. This comparison illustrates that the four enzymes characterized to date are isoschizomers, recognizing a G/C rich sequence. Daly and Fitzgerald (1987) also mentioned that seven additional *Lactococcus lactis* subsp. *cremoris* strains examined from mixed strain dairy starters all contained a restriction enzyme that had an activity similar to *Scr*FI. The apparent widespread distribution of *Scr*FI activity among lactococci and now, the independent biochemical characterization of four type II restriction enzymes from the A/T rich lactococci and streptococci which recognize the identical sequence is interesting indeed. Genetic analysis of these restriction and modification systems may determine whether or not the above enzymes are identical or have common progenitors.

Table 3.3 Recognition sequence identified for Type II restriction enzymes purified from lactic acid bacteria

Type II enzyme	Recognition sequence
*Scr*FI	*
	5'-C-C-N-G-G-3'
	3'-G-G-N-G-G-5'
	*
*Sth*117I	*
	5'-C-C-W-G-G-3'
	3'-G-G-W-C-C-5'
	*
*Ss*1I	*
	5'-C-C-W-G-G-3'
	3'-G-G-W-C-C-5'
	*
*Lla*I	5'-C-C-W-G-G-3'
	5'-G-G-W-C-C-5'

N=any nucleotide; W=A or T; * = cleavage site

See section 4.2 for further details and original references

The biochemical definition and characterization of modification methylases from lactic acid bacteria remains an untouched area of investigation. Some limited progress has been made, however, on the localization and characterization of genes from lactococcal R/M systems which encode modification methylases. DNA sequence analysis of operational R/M systems has identified the presence of two distinct classes of modification enzymes in lactococci: the *Lla*I adenine methylase from the R/M system encoded by pTR2030; and a m^5-cytosine methylase encoded by the *Scr*FI R/M system from *L. lactis* subsp. *lactis* UC503. (Hill *et al.*, 1991c; Fitzgerald, unpublished data). *Lla*I has been classified as a type-IIS methylase in that its predicted protein contains two functional enzyme domains, each of which is predicted to methylate one strand of an asymmetric target sequence (Hill *et al.*, 1991c; Szybalski *et al.*, 1991).

The operation of R/M systems are practically and fundamentally significant as barriers to phage infection (Klaenhammer, 1984) as well as to 'foreign' DNA introduced via the various routes now available for gene transfer (Langella and Chopin, 1989; Solaiman and Somkuti, 1991). Given the wealth of R/M systems available, the lactic acid bacteria remain an unexploited reservoir of potentially novel restriction endonucleases and modification methylases.

3.3.2.3 Genetics of R/M systems.   Variations in the phage susceptibility of lactococcal strains due to the operation of R/M systems and destabilization of phage resistance was first recognized by Collins (1956, 1958). Genetic insta-bility in the phage resistance of lactococcal cultures continued to be observed (Pearce, 1978; Limsowtin *et al.*, 1978; Steenson and Klaenhammer, 1986), suggesting that like lactose-fermenting ability and proteolytic activity (McKay, 1982; 1983), phage-resistance may also be a plasmid-encoded trait. Evidence for the involvement of plasmid DNA in the restriction and modification activities of lactococci was initially reported in *Lactococcus cremoris* KH (Sanders and Klaenhammer, 1981). Genetic studies with the lactococci have now established that their R/M activities are commonly encoded by plasmid DNA elements, some of which are self transmissible (for reviews see Klaenhammer, 1987; Sanders, 1988; Chopin and Chopin, 1990). The common association of R/M systems with plasmids in lactococci has contributed to:

- the widespread distribution of R/M systems of different specificities;
- the natural presence of multiple R/M systems within single strains leading to additive levels of restriction;
- the genetic instability and variability in the phage resistance and susceptibility of lactococci.

A summary of plasmid-encoded R/M systems defined in lactococci is provided in Table 3.4. Evidence for chromosomal determinants exists (Sanders and Klaenhammer, 1981; Chopin *et al.*, 1984; Fitzgerald *et al.*, unpublished), but these R/M systems have not yet been well characterized. In one case chromo-somal R/M determinants may be involved with a conjugative episomal element (Ward *et al.*, 1992). Chromosomal determinants were implicated for the R/M system reported in *S. thermophilus*, whereas in *Lb. helveticus* a 34 kb plasmid was correlated with R/M action (Benbadis *et al.*, 1991; Reyes-Gavilan *et al.*, 1990).

The diversity of R/M systems and their common association with plasmid DNA elements can generate combinations that elicit higher levels of phage

Table 3.4 *Lactococcus* plasmids that encode R/M activities

R/M plasmid	Size (kb)	Phage	tra	Reference
pME100	17.5	kh,c2	–	Sanders and Klaenhammer, 1980
				Sanders and Schultz, 1990
pIL6	28	66	+	Chopin *et al.*, 1984
pIL7	31	66	–	Chopin *et al.*, 1984
pTR2030	46	31,48	+	Klaenhammer and Sanozky, 1985
				Hill *et al.*, 1989a
pLR1020	30	m12r	–	Steenson and Klaenhammer, 1986
p3085-2	15.5		–	Teuber, 1986
pTN1060	100	c2,p2	+	Higgins *et al.*, 1988
pTN20	28	c2,p2,31,	+	Higgins *et al.*, 1988
		48,50		Alatossava and Klaenhammer, 1991
pIL103	5.7	8, 66	–	Gautier and Chopin, 1987
pIL107	15	8, 66	–	Gautier and Chopin, 1987
pBF61	42	c2	–	Froseth *et al.*, 1988
pKR223	38	sk1	–	McKay *et al.*, 1989
pJW563	12	c2,p2	–	Josephsen and Vogensen, 1989
pJW565	14	c2, p2	–	Josephsen and Vogensen, 1989
pJW566	25	c2, p2	–	Josephsen and Vogensen, 1989
pFV1001	13	c2, p2	–	Josephsen and Klaenhammer, 1990
pFV1202	17	c2, p2	–	Josephsen and Klaenhammer, 1990
pTRK11	100	c2, p2	+	Sing and Klaenhammer, 1991
pTRK12	30	c2, p2	–	Sing and Klaenhammer, 1991
pTRK30	28	c2, p2	–	Sing and Klaenhammer, 1991
pTRK68	46	31	–	Hill *et al.*, 1989a
pTRK317	17	c2, p2	–	Sing and Klaenhammer, 1991
pHD131	131	c2, sk1	+	Ward *et al.*, 1992

Tra[+], self-transmissible

resistance. Lactococcal strains that are relatively impermeable to phage attack have been found to contain multiple R/M systems (Sanders and Klaenhammer, 1981; Chopin *et al.*, 1984; Gautier and Chopin, 1987; Josephsen and Vogensen, 1989). Combinations of R/M plasmids, which have been intentionally constructed by genetically stacking distinct systems, further illustrate that multiple R/M systems offer a stronger restrictive barrier to heterologous bacteriophages (Steenson and Klaenhammer, 1986; Josephsen and Klaenhammer, 1990).

R/M systems are also found in the presence of other complementary phage defense mechanisms; located either within the same strain as in the case of the *Lactococcus* strains ME2 (Sanders and Klaenhammer, 1984) and IL964 (Gautier and Chopin, 1987) or located on the same plasmid as in the cases of pTR2030 (Hill *et al.*, 1989a), pTN20 (Higgins *et al.*, 1988; Durmaz and Klaenhammer, 1992), pBF61 (Froseth *et al.*, 1988), pKR223 (McKay *et al.*, 1989), and the

131 kb conjugative plasmid described recently from DRC1 (Ward *et al.*, 1992). The combinations of R/M and abortive (Abi) systems provide complementary defenses where R/M action restricts the number of successful infections and promotes cell survival while the abortive mechanism prevents the proliferation of new or modified progeny phage by killing the cell prior to completion of the phage lytic cycle (Sing and Klaenhammer, 1990a, b; Klaenhammer, 1991). Given the effectiveness of this combination, it is curious that the two mechanisms have not been linked more often in genetic studies of phage resistant lactococci. However, the contribution of R/M action can be obscured in cases where R/M and Abi defense systems are combined (Ward *et al.*, 1992). For example, detection of R/M action by pTR2030 was possible only after using an industrial phage (i.e. isolated directly from an industrial fermentation rather than from a culture collection) that was resistant to the abortive mechanism (Hsp), but susceptible to restriction (Alatossava and Klaenhammer, 1991). Therefore, the contribution of R/M activities to the phage defense arsenal of lactococci may be substantially greater than is currently reported.

Six of the lactococcal plasmids which encode R/M systems are self-transmissible (pIL6, pTR2030, pTN20, pTR1060, pTRK11, and pHD131; see Table 4.4). Conjugal transfer frequencies can range from 10^{-7} up to 1 per input donor (Higgins *et al.*, 1988). In the transconjugants, the R/M plasmids have been found in multiple forms depending on the conjugation process and whether or not other resident plasmids are mobilized in the transfer event. For example, when the self-transmissible R/M plasmid pTN20 mobilizes a coresident Lac+ plasmid, pTR1040, the transconjugants can harbour pTN20, pTR1040, or a variety of pTN20::pTR1040 cointegrates (Higgins *et al.*, 1988). The cointegrates exhibit a combination of phenotypes (Lac$^+$ Nis^r R/M$^+$ Tra$^{+/-}$) from the component plasmids, but they may vary considerably in the frequency of conjugal transfer and level of R/M activity. Large conjugative R/M plasmids described in lactococci have been reported that also encode combinations of phenotypes including bacteriocin production (Ward *et al.*, 1992; Powell *et al.*, 1990), nisin resistance (Higgins *et al.*, 1988), lactose-fermenting ability (Higgins *et al.*, 1988), proteinase activity (Sing and Klaenhammer, 1990a), as well as other abortive phage resistance mechanisms (Froseth *et al.* 1988; McKay *et al.*, 1989; Hill *et al.*, 1989a; Romero and Klaenhammer, 1990a; Ward *et al.*, 1992). The conjugal transfer process, which can mobilize plasmids via recombinational events (Anderson and McKay, 1984), is responsible for generating a variety of lactococcal plasmids that encode many important phenotypes in addition to R/M systems.

Insertion sequences which are often associated with self-transmissible plasmids mediate the recombination events that give rise to the variety of cointegrate plasmids and phenotype combinations. In a number of cases the level of R/M activity has been altered significantly following conjugal transfer. In pIL6 transconjugants, Chopin *et al.* (1984) noted that phage 66 was still restricted, but the EOP was higher (less phage restriction) than in the donor strain bearing pIL6. The alteration in the level of restriction was attributed to the the effects, in trans, of other coresident plasmids in the original donor strain. Similar effects in other self-transmissible R/M plasmids have been observed following molecular rearrangements selected through the conjugal transfer process. For pTN20, R/M activities (EOP of 10^{-8} on phage p2) were lowered significantly after conjugal mobilization of the Lac$^+$ plasmid pTR1040 generated the cointegrant plasmid pTN1060 (EOP of 10^{-2} on phage p2; Higgins *et al.*, 1988). In contrast, IS*S1*-mediated conjugal mobilization of an R/M$^+$ recombinant plasmid, pTK6, by pRS01 resulted in a 100–1000 fold enhancement of R/M activity (Romero and Klaenhammer, 1990a). Noting these effects, it can be concluded that conjugal transfer processes and their accompanying recombinational events can cause substantial alterations in the level of R/M activity exhibited by some lactococcal plasmids (Romero and Klaenhammer, 1990a). These effects may be due to a simple change in copy number , IS-induced changes in expression or regulation signals, or other more complex interactions between R/M operons and their regulatory factors. The specific nature of these changes has not yet been investigated, but this area promises to be a fruitful topic for future research.

3.3.2.4 Molecular characterization of R/M genes. Some of the plasmids in lactococci that confer 'phage resistance', 'phage insensitivity', or 'reduced bacteriophage sensitivity' have been genetically dissected to investigate the mechanisms responsible for these phenotypes. During this process, R/M genetic determinants on the plasmids pKR223 (McKay *et al.*, 1989), pTR2030 (Hill *et al.*, 1989a, 1991c) and pME100 (Sanders and Schultz, 1990) have been localized, cloned and characterized to various degrees. The first R/M plasmid defined in lactococci, pME100 (Sanders and Klaenhammer, 1980), was later used to clone R/M functions directly in *Lactococcus lactis* (Sanders and Schultz, 1990). The vector employed (pSA34) was missing a Gram-positive origin of replication. pME100::pSA34 chimeras were constructed, positively selected in *Lactococcus*, and screened for R/M activity. The cloning strategy allowed positive selection of R/M$^+$ clones and facilitated localization of the genetic determinants. Since pSA34 could insert into various sites around pME100, R$^-$/M$^+$ and R$^+$/M$^+$ clones were generated and, thus, restriction and modification functions genetically segregated. The restriction activities were localized to a 4 kb fragment of the R/M$^+$ pSA34::pME100 clone, pME34. The complexity of R/M systems is illustrated in this study when one notes the variable phage resistance phenotypes elicited by two different *Lactococcus lactis* transformants containing pME100, as well as the various R$^+$/M$^+$ pSA34::pME100 clones.

Earlier, McKay *et al.* (1989) segregated the R/M functions of pKR223 by cloning and deletion analysis of a 19 kb fragment cloned into pGBK17. The restriction gene was localized to a 2.7 kb *Eco*RV fragment which was separate from the 1.75 kb fragment which harboured the modification gene. The R/M system was active against the small isometric headed bacteriophage sk1, but not against the prolate phage c2. pKR223 also encodes an Abi phage resistance mechanism that acts on phage c2 and is genetically distinct from the R/M functions. However, the Abi genetic determinants could not be cloned and expressed independently of the R/M functions. McKay *et al.* (1989) suggested that expression of the *abi* genes required sequences upstream of the modification genes encoded by pKR223. Although the DNA sequence of this region has not yet been reported, molecular analysis of the pKR223 phage resistance region indicates that the R/M and Abi genes may be part of a larger phage resistance operon.

The conjugative phage resistance plasmid pTR2030 encodes at least one R/M system in addition to the abortive defense mechanism, originally designated Hsp (Hill *et al.*, 1989a; Klaenhammer and Sanozky, 1985). The R/M system on pTR2030 has no effect on either c2 or sk1 phages. Therefore, its operation was not recognized until the plasmid was introduced into the *Lactococcus lactis* LMA12/NCK203 host background where industrial phages, which were sensitive to restriction by pTR2030, were used to detect the R/M system (Hill *et al.*, 1989a). A 13.6 kb phage resistance region of pTR2030 (Figure 3.1) was

Figure 3.1   Molecular organization of the phage resistance region of pTR2030.

initially cloned into pSA3 to form the recombinant plasmid pTK6, which expressed Hsp+ and R+/M+ (Hill *et al.*, 1989b). An *in vivo* deletion in pTK6 generated an R−/M− Hsp+ derivative (pTRK18) that identified the position of the R/M determinants. Subsequent molecular analysis of the entire region by subcloning, evaluation of phage resistance phenotypes, and DNA sequencing have identified the R/M region within four complete open reading frames positioned between the modification methylase *Lla*1 and the abortive infection gene *hsp* (Hill *et al.*, 1989a,b, 1991c; O'Sullivan, Zagula, and Klaenhammer, unpublished). This organization is suggestive of a phage resistance operon where a number of structural genes are located that direct R/M and Abi phenotypes. The specific functions of each putative gene product are currently being investigated and the restriction gene(s) remain to be conclusively identified within the phage resistance region from pTR2030.

The modification gene, designated *Lla*I, is the first methylase described from a biologically active R/M system in lactococci. *Lla*I occupies the first open reading frame defined within the phage resistance region on pTR2030. Molecular characterization revealed a gene of 1866 bp which could encode a 72.5 kDa protein of 622 amino acids (Hill *et al.*, 1991c). Two consensus sequences, that are found in all type II-A-methylases, are each found twice within the *Lla*I gene. Its organization reveals two domains, each potentially capable of methylating one strand of an asymetric target sequence. When the regulatory signals and amino domain of *Lla*I are cloned onto a plasmid vector, progeny phage are modified *in trans* and protected from restriction by the *Lla*I restriction system of pTR2030. *Lla*I resembles methylases that complement type IIS restriction enzymes like FokI (Szyblaski *et al.*, 1991). This class of restriction endonucleases recognizes a site 4–7 bp and cuts 1–20 bp away from the recognition sequence.

Recently, two modification methylases encoded by the chromosomally borne *Scr*FI system of *Lactococcus lactis* subsp. *cremoris* UC503 (previously strain F) have also been cloned and expressed in *E. coli* (Fitzgerald *et al.*, unpublished data). The two genes are linked on the *Lactococcus* chromosome and are not homologous. The DNA sequence for one of the methylases has been determined. It contains an ORF of 1170 bp which could encode a protein of 369 amino acids at an estimated molecular weight of 44.5 kDa. The amino acid sequence revealed motifs that are characteristic of a m^5c-methylase and shows extensive homology with the *Eco*RII and *Dcm* methylases, which recognize a subset of the *Scr*FI recognition sequence. None of the clones encoded a cognate endonuclease. Therefore, both an adenine and cytosine methylase have now been identified in lactococci by molecular characterization of R/M systems. However, in both cases, the restriction genes remain to be localized and characterized.

3.3.2.5 Anti-restriction mechanisms in bacteriophages.

Bacteriophages have evolved a variety of mechanisms to escape the action of restriction endonucleases (for a review see Kruger and Bickle, 1983). Included among these are the expression of proteins that inhibit host endonucleases, self-modification by phage-encoded methylases, DNA modification other than methylation (e.g. glycosylation and incorporation of unusual bases), and selective reduction in the number of target sequences for restriction endonucleases in the phage genome. Mechanisms by which bacteriophages of lactic acid bacteria circumvent restriction mechanisms have not been studied extensively. However, two apparent anti-restriction mechanisms have already been found in lactococcal bacteriophages.

Powell and Davidson (1986) reported that the prolate-headed bacteriophage c6A was cut infrequently by many restriction enzymes. The number of actual sites were far less than the expected frequency (based on genome length, G + C content, and the probability that a given site would be present). The data support the contention that restriction sites were selectively eliminated during the evolution of phage c6A. A number of lactococcal phages are 'hard to cut' and generate fewer than the number of expected fragments upon digestion with restriction endonucleases. Given the widespread distribution of R/M systems in these bacteria, the pressure to reduce restriction target sequences would appear to be an important selective force on the evolution and persistence of some lactococcal phages.

Lactococcal bacteriophages vary considerably in their susceptibility to R/M systems. This can be attributed to both the presence of target sequences for any given R/M system and the length of the phage genome bearing these recognition sites (Klaenhammer, 1987). In addition, two lactococcal phages have now been isolated and characterized which are not susceptible to restriction by a specific R/M system because they have incorporated the complementary bacterial methylase into their phage genome. Following the release of pTR2030 transconjugants into the dairy industry in 1985, two distinct classes of phages have now been isolated that are not restricted by the *Lla*I system encoded by pTR2030 (Alatossava and Klaenhammer, 1991; Hill *et al.*, 1991c; D'Amelio and Klaenhammer, unpublished). Phage 50 contains a circularly permuted genome of 28.9 kb, a *pac* site, and hybridizes within the B1-P335 temperate/lytic phage species defined by Jarvis *et al.* (1991). Molecular characterization of phage 50 revealed that its genome contained 1273 bp of the *Lla*I gene from pTR2030 (Hill *et al.*, 1991c; Klaenhammer, 1991). The phage sequence was identical to that from pTR2030 and it encoded the complete amino domain of *Lla*I. Since the amino domain can functionally modify phage DNA, the phage is capable of methylating itself in any propagating host and, thereby, it eludes subsequent restriction by pTR2030 (Hill *et al.*, 1991c). A second class of restriction-resistant phages has now also been isolated from industrial whey samples and characterized. These are closely related to the R/M susceptible phage (48) reported previously (Alatossava and Klaenhammer, 1991) except that they too have acquired part of the *lla*I gene from pTR2030. These are cohesive ended phages which also fall within the B1/P335 species (D'Amelio and Klaenhammer, unpublished). In our attempts to understand the development of phage anti-restriction mechanisms, it will be interesting to determine if the new *cos* phages acquired the *Lla*I methylase from pTR2030 or from phage 50. In either case, the direct acquisition of a bacterial methylase, or a functional portion thereof, is unique among the anti-restriction mechanisms that have been uncovered in bacteriophages. This type of protection against restriction is not generic, however, and those phages that harbour specific bacterial methylases have remained susceptible to restriction by R/M systems of different specificity (Alatossava and Klaenhammer, 1991).

3.3.3 *Abortive infection*

Abortive infection is a process whereby the infectious cycle is inhibited even though adsorption, DNA entry, and the initial stages of viral gene expression are normal (Duckworth *et al.*, 1981). Usually the host cells are killed, but it is unclear if this is a direct response of the resistance mechanism or occurs because the lytic cycle is aborted only after it has progressed to a point of no return. For example, production of a phage-encoded DNase by phage c6a occurs within 30 min after infection and degrades host DNA (Powell *et al.*, 1992). Phage abortion after induction of the nuclease would not rescue the infected cell. Abortive mechanisms are very effective in providing resistance to bacterial cultures since they slow phage replication and either eliminate or minimize the development and release of progeny phage (Sing and Klaenhammer, 1990a,b; Klaenhammer, 1991).

Several experimental criteria must be established in order for a phage resistance phenotype to be categorized as an abortive infection mechanism, designated Abi. First, phage DNA entry into the cell should be demonstrated; via plaque production, labelling phage DNA with radioisotopes or 4', 6'-diamidino-2-phenylindole (DAPI) and detecting injection of the genome into the cell (Sing and Klaenhammer, 1990a); or extraction of total cell DNA from infected cells and hybridization experiments with labelled phage DNA (Hill *et al.*, 1991a; Klaenhammer, 1991). Second, the contributions of R/M systems should be eliminated. Sometimes this is difficult to establish in cases where the phage does not form plaques on the strain being evaluated. In most cases, however, the contributions of R/M can be negated by using phages that are not subject to restriction or that have been modified previously via propagation through a host bearing that R/M system (Hill *et al.*, 1989a; Alatossava and Klaenhammer, 1991; Sing and Klaenhammer, 1990a). Third, cell death as a result of the infection should occur, even when no progeny phage are released. Many Abi mechanisms eliminate or severely limit plaque formation in standard plaque assays. Therefore, estimates of cell death should be combined with centre of infection assays and single step growth curves, which define the actual number of productive infections, burst size of the progeny, and the latent period for the phage. With this information the relative strength of an Abi mechanism for any given phage–host combination can be accurately assessed (Klaenhammer and Sanozky, 1985; Sing and Klaenhammer, 1990a; Laible *et al.*, 1987; Cluzel *et al.*, 1991; Coffey *et al.*, 1989; Durmaz and Klaenhammer, 1992). Among the lactic acid bacteria, host-induced abortion of phage infections have been described exclusively in lactococci. Various phenotypic effects have been described including 'phage resistance', 'heat-sensitive phage resistance' (Hsp), 'reduced bacteriophage sensitivity' (Rbs) and abortive infection (Abi). Subsequent characterization of these mechanisms reveal that these are abortive type responses and, therefore, the general usage of 'Abi' has been suggested (Coffey *et al.*, 1991b) and we recommend that this be the preferred usage hereafter.

The genetic determinants for Abi phage-resistance mechanisms characterized thus far are encoded by plasmids or conjugative elements (plasmid and episomes). A summary of plasmids and episomes that encode Abi systems is provided in Table 3.5 (adapted from E. Durmaz, M.S. Thesis, NC State University, Raleigh). Phage reactions on Abi[+] hosts and details concerning the responsible genetic determinants are also summarized to the extent possible. The common association of Abi mechanisms with self-transmissible elements and plasmids has greatly facilitated the isolation and characterization of these mechanisms. In most cases, conjugation or transformation experiments were carried out to isolate plasmids encoding phage resistance (for reviews see

Table 3.5 Abortive resistance plasmids in lactococci*

| Plasmid | | | | | Phage resistance phenotype of clone or subclones, if available | | | | | |
Name	Size	Source	Characteristics	Isolated, subcloned	EOP, plaque morphology, and temperature effect	Productive infections	Burst size**	Latent period	Sequenced	Reference
pTR2030	46.2 kb	(L) ME2 conjugation with Lac selection	R+/M+, Hsp+, Tra+	pTR2030 isolated in LM0230	†prolate φc2: 0.83 RPS †small isometric phages sk1, p2, 31: >10^{-9} †large isometric phage 949: 0.18 RPS (φp2 is not subjected to restriction by R/M on pTR2030)	†φc2: 10^{-2} φc2: 10^{-2}	†φc2: 5/40 (30°C) φp2: 13/136	†φc2: no change φc2: 15 min delay		Klaenhammer and Sanozky 1985
pTRK18	20.5 kb		Hsp+, R−/M−	spontaneous deletion derivative of pTK6, a subclone of pTR2030	φc2:0.5 RPS φc31:0.1 RPS φsk1:0.25 RPS RPS for phages 31 (NCK203 background); p2, sk1, and c2 (LM0230 background); larger plaques at 39°C than at 30°C for phages c2 and sk1 ,				yes; abiA (hsp) is 1887 bp; not homologous with pNP40	Klaenhammer and Sanozky, 1985; Hill et al., 1990a
pC1829	44 kb	(L) UC811, conjugation with Lac selection	Tra+, Abi+; DNA sequence is the same as that of pTR2030	Abi+ subcloned to pSA3 as pC1816					abiA (hsp)	Coffey, et al., 1991a
pIL611	-	(L) IL416, cloning of total cell DNA to vector pIL253	Abi+; R/M, Ads ruled out	isolated in IL1403	φhIL66: minute plaques visible after 4 days incubation	# of infective centres 0.04% of adsorbed φ's	30/150	no change	yes; abi416 is 753 bp; promoter provided by Iso-IS51 element	Cluzel et al., 1991
pTN20	28.4 kb	(L) N1, ME2 derivative, conjugation with Lac selection	Tra+, R+/M+, Abi+	Abi+ isolated on pTRK99 LM0230	pTRK99 in LM0230 background: φc2: no effect of Abi φp2: 5 x 10^{-3}; clear, mutant plaques at 10^{-4}; effective against several small isometric phages, but not against prolate phage	φp2: 10^{-2}	φp2: 7/63	no change	yes; abiC is 1026 bp; not homologous to abiA	Higgins et al., 1988; this study
pIL105	8.7 kb	(C) IL964, cotransformation with indicator plasmid pHV1301	Abi+; R/M, Ads ruled out (plasmids pIL103 and pIL107 from IL964 encode different r/M systems	isolated in IL1403	isometric φ66: lowered EOP, RPS (turbid plaques visible after 4 days incubation); clear, mutant plaques at 10^{-4} pIL105 also effective against 2 prolate and 4 other isometric phages, but not against 2 isometric phages	φ66: 3%	φ66: 50/180	similar		Gautier and Chopin, 1987

Plasmid	Size	Origin/selection	Properties	Subclone/isolate	EOP	Burst	Change	Reference
pCI750	65 kb	(C) UC653, conjugation with Lac selection	Tra+, Abi+; R/M and Ads ruled out	subcloned to pGB301 as pMM1	pMM1 in LM0230 background: total resistance to small isometric phage and partial resistance to prolate phage (RPS); no temperature effect			Baumgartner et al., 1986
pKR223	35.5 kb	(D) KR2, transformation of LM0230 with KR2 plasmid DNA, Lac selection	R+/M+ and Abi+ for φsk1; Abi+ only for φc2 (Ads ruled out); unstable in LM0230; Tra−	pKR223 isolated in LM0230 as KR223; φ^r subcloned to pGB301 as pGBK17; Abi determinant alone has not been subcloned but was localized to a 1.3 kb fragment	φc2: (for GBK17): <10% infective centers in LM0230	φc2: pin point plaques at 21 and 32°C; 0.2 RPS at 37°C; †φsk1: ~0.1–.02 RPS at all temperatures	φc2: 68/91	φc2: similar	Laible et al., 1987
pBF61	26 MDa	(L) KR5, cotransformation with pSA3; Em^r transformants screened for Lac and φ^r	R+a/M+ and Abi+ (RPS independent of R/M); Ads ruled out	isolated in BF26 (LM0230)	φc2 modified for R/M in BF26 has RPS, indicating Abi; †φc2: 10^-3, plaque sizes increased with increased temperature, but plaques smaller than on LM0230; 2–3 log decrease in sensitivity to c2, stl5, eb1; †φsk1, no plaques	φc2: 50% of normal burst	†no change	Froseth et al., 1988
not named	131 kb	(L) DRC1, conjugation with HID113, LM0230 derivative, selection for Lac, Bac^r, or φ^r	Prt+, Bac+, R/M+, Abi+, Tra(?); reduced plaque sizes indicate Abi in addition to R/M	isolated in HID600	†φc2: 10^-2 RPS; heritable mutation of phage evident in progressive increase in EOP during repeated plaquing; †φc2: 12.5% / †0.19%	†35/92 / †no burst detected after 3 hr	†10–15 min increase	Powell et al., 1990
not named	-	(L) DRC1, conjugation with HID113, selection for φ^r	Bac−, φ^r	in HID611	†φc2: no plaques at 30°C; 3 × 10^-5 RPS at 37°C; φc2: 2 × 10^-3 (30°C) 1 × 10^-2 (37°C) φsk1:3.5 × 10^-3 (30°C) 5.7 × 10^-3 (37°C)			Powell et al., 1990
pCI528	46 kb	(C) UC503	Tra−, Ads+, Abi+, but φuc2021 adsorbs normally; R/M not ruled out		††resistant to prolate and small isometric phages in lactococcal hosts, no temperature effect	φuc 2021 5-fold lower burst	φuc2021: no change	Coffey et al., 1991b

Table 3.5 *continued*

| | Plasmid | | | | Phage resistance phenotype of clone or subclones, if available | | | | | |
Name	Size	Source	Characteristics	Isolated, subcloned	EOP, plaque morphology, and temperature effect	Productive infections	Burst size**	Latent period	Sequenced	Reference
pNP40	40 MDa	(D) DRC3, conjugation with Lac selection	Nis^r, Abi$^+$, unstable at elevated temperature; R/M, Ads ruled out	isolated as Lac$^-$ D6528, LM0230 derivative	φc2: <10^{-7} (no plaques) at 21° and 32°C; 0.3 at 37°C					McKay and Baldwin, 1984
pAJ1106	106 kb (recombinant)	(D) 4942, conjugation with Lac selection	Tra+, Hsp+, Lac+; Ads ruled out, but not R/M)	isolated as T.LM0230(106), T.4193(106), and T.4862(106)	(in general less efficient at higher temperatures) φc2: no plaques at 30°C, 8×10^{-2} RPS at 37°C; also confers resistance to four other prolate phages in *L. lactis* strains; in *L. cremoris* resistance to one prolate and one small isometric phage, but sensitive to small isometric φ1483					Jarvis, 1988
pCLP51R	90 kb (recombinant)	(L) 33-4, conjugation with IOG376 (Lac, φ2 recipient), Lac selection, screen for φ^r	Lac$^+$, Hfr$^+$, Clu$^+$, Lpr$^+$; Ads ruled out, R/M not implied, possible Tra$^+$	isolated as LDP1001; also introduced to LM0230 where it is unstable without Lac selection	φLP10G (from whey sample); ~10^{-3}; not tested in LM0230 background					Dunny et al., 1988
pBU8	40 MDa	(D) BU1, conjugation with Lac selection	Conjugative, uncharacterized phage resistance	isolated in (D) Bu2-60	resistance of three transconjugants to up to 15 unidentified phages					Wetzel et al., 1986
pJS88	88 MDa	(D) 11007, conjugation with LM2336 and Lac selection	Prt$^+$, Bac^r, Tra$^+$, Rbs$^+$; Ads and R/M not ruled out	isolated in JS30 (also contains Lac+ pKB32)	φc2: 5.5×10^{-2} RPS small isometric φ712 <10^{-3}					Steele and McKay, 1989
pJS40	40 MDa (recombinant)	JS30, conjugation with Rec$^-$ MMS367	Lac$^+$, Rbs$^+$, Prt$^-$, Bac2, Tra$^+$; Ads and R/M not ruled out	isolated in JS120	φc2: 8.3×10^{-2} RPS φ712: 7.4×10^{-5} pin point plaques					Steele and McKay, 1989

pCC34	34 MDa	(C) C3by conjugation with Lac selection	Lac+, Rbs+	isolated in CC102, an LM0230 derivative	φ712: no detectable plaques φc2: reduced EOP and RPS at 20°, 30° and 37°C	Steele et al., 1989
pNP2	88 MDa	(L) WM4 by conjugation with Lac selection	Lac+, Rbs+	isolated in CS1, an LM0230 derivative	small isometric φ: no detectable plaques; prolate φ: reduced EOP and RPS at 21°, 30° and 37°C	Steele et al., 1989
pEB56	56 MDa	(C) EB7 by conjugation with Lac selection	Lac+, Rbs+	not isolated; transformant EB101 contains several plasmids	φ712 and φc2: reduced EOP and RPS	Steele et al., 1989
Tn5301	no physical evidence for plasmids	(L) ATCC11454 derivative by conjugation with (D) 18-16 derivative with selection for Suc or Nis	Suc+, Nis+, Nis+, Rbs+; Ads but not R/M ruled out	transconjugants SLA3.15 and SLA3.34 (no plasmid DNA detected)	ineffective against φc2 in SLA1.15 (an L. lactis transformant); φ18-16: no plaques on SLA3.15 or SLA3.34 at 25°, 32° or 37°C; no temperature dependence	Gonzalez and Kunka, 1985
pSRQ 2202	18.5 MDa	(C) MS	resistant φ18-16, Muc+, lac+; Ads ruled out, R/M possible	φr resulted from conjugation but was not correlated to either of the transferred plasmids		Vedamuthu and Neville, 1987

* (L) *Lactobacillus lactis* subsp. *lactis* (C) *L. lactis* subsp. *cremoris*; (D) *L. lactis* subsp. *diacetylactis*; Abi, abortive resistance; Ads, interference with adsorption; Bac+, bacteriocin producer, Bac', bacteriocin resistance; Clu+, clumping; Em', erythromycin resistance; Hfr+, high frequency transfer; Hsp, heat sensitive phage; Lac, lactose utilization; Lpr+, lytic phage resistance; Muc+, mucoid producing; Nis', nisin resistance; φ', phage resistance; Prt+, proteinase producer; Rbs+, reduced bacteriophage sensitivity; Rec, recombination deficient; R/M, restriction/modifications; RPS, reduced plaque size; sucrose utilization; Tra, conjugative.

** Burst size is represented as burst size on resistant strain/burst size on homologous, phage-sensitive host.

† φ' phenotype includes the effect of both R/M and Abi

†† φ' phenotype includes the effect of both Ads and Abi

Daly and Fitzgerald, 1987; Klaenhammer, 1987; Sanders, 1988; Sing and Klaenhammer, 1990b). For conjugation experiments, phage resistant starter strains were used as donors and phage sensitive plasmid-cured strains as recipients. Phage-resistant transconjugants were isolated by selection for other genetic markers, such as lactose-fermenting ability or nisin production and resistance, that are commonly cotransferred. Cotransformation of cryptic plasmid DNAs from 'resistant' strains with a vector encoding antibiotic-resistance (Gautier and Chopin, 1987; Froseth et al., 1988) and cloning from enzyme-digested total cell DNA (Cluzel et al., 1991) are two other approaches which have been used successfully to identify phage resistance plasmids.

Molecular analysis of phage resistance phenotypes from these plasmids has revealed that Abi mechanisms are often found in conjunction with R/M systems and it is this combination that confers the overall strength of the phage resistance phenotype (Laible et al., 1987; Froseth et al., 1988; Hill et al., 1989a; Ward et al., 1992; Coffey et al., 1991b; Sing and Klaenhammer, 1990a,b; Klaenhammer, 1991). Subcloning of Abi genetic determinants has recently allowed a more accurate assessment of the phenotypic effects of abortive resistance mechanisms. First, it is not unusual for Abi⁺ subclones to express a lower level of phage resistance than the original plasmid (Hill et al., 1989a, b; Coffey et al., 1991a) suggesting that copy number may be important or flanking DNA may regulate expression of *abi* gene systems. In two specific examples, the presence of an iso-IS*S1* element (Cluzel et al., 1991) or an upstream R/M system (McKay et al., 1989) were essential for expression of an Abi⁺ phenotype. Second, while the mechanisms by which abortive infection occur remain to be defined, it has been shown that different genes are involved and interference can occur at different points in the phage lytic cycle. The three abortive genes that have been sequenced to date (*hsp*/*abiA*, *abi*416/*abiB*, and *abiC*; Hill et al., 1990a; Coffey et al., 1991a; Cluzel et al., 1991; and Durmaz and Klaenhammer, 1991) show no significant homology or sequence similarities at the DNA or deduced protein level. Each abortive activity is confined to a single open reading frame with predicted proteins of 628, 753, and 344 amino acids, respectively. Phage DNA replication is inhibited during Hsp-induced abortive infections (Hill et al., 1990a; 1991a) whereas no effect on phage DNA replication is induced by *abiC* (Durmaz and Klaenhammer, 1991). The abortive defenses of lactococci are genotypically heterogeneous and as such they may act upon an equally wide set of phage developmental targets. The variability and range of phenotypes (Table 3.5) further suggests that different Abi mechanisms act to abort the phage lytic cycle at different points.

The following discussion will concentrate on those Abi systems and plasmids that have been characterized at the genetic level. Additional information on the various phage resistance plasmids which encode abortive defense mechanisms is available in a number of reviews published previously (Daly and Fitzgerald, 1987; Klaenhammer, 1987, 1991; Sanders, 1988; Sing and Klaenhammer, 1990b)

3.3.3.1 Abortive phage resistance plasmids

pTR2030. The first abortive gene to be sequenced was *hsp*, herein designated *abiA*, from pTR2030 (Hill *et al.*, 1990a). Plasmid pTR2030 (46.2 kb) was first detected in transconjugants generated by matings with the prototype phage insensitive strain *L. lactis* subsp. *lactis* ME2 (Klaenhammer and Sanozky, 1985). Plasmid profiles of the donor strain ME2 did not reveal the presence of monomeric pTR2030 (Hill *et al.*, 1991b; Klaenhammer and Sanosky, 1985) and hybridization experiments using pTR2030-specific probes failed to identify junction fragments; thus pTR2030 was not integrated in the chromosome. Rather, pTR2030 was present in ME2 as a high-molecular weight multimer which could not be detected by conventional plasmid analysis. Originally, the phage resistance phenotype conferred by pTR2030 was described as a modest reduction in EOP (0.83) and temperature-sensitive reduction in the burst size of the prolate phage c2 in *L. lactis* LM0230. Later, pTR2030 in both *L. lactis* and *cremoris* backgrounds was found to completely prevent plaque formation of a number of small isometric phages isolated from dairy plants (Steenson and Klaenhammer, 1986; Jarvis and Klaenhammer, 1986). Prolate phages and a large isometric-headed phage generally showed a reduction in plaque size and small reductions in the EOP. The original work conducted on pTR2030 had characterized the phage resistance as temperature dependent. At the elevated temperatures used for cheese-making (37–40°C), the phage resistance appeared to be inactivated or destabilized when LM0230 was challenged with phage c2 (Klaenhammer and Sanozky, 1985). In contrast, Sing and Klaenhammer (1986) later reported that when pTR2030 was challenged with phage during starter activity tests, acid production was not inhibited and phage proliferation was prevented. pTR2030 is generally effective against small isometric phages (Jarvis and Klaenhammer, 1986), now recognized to include most phage strains within the P335 and 336 phage species of Jarvis *et al.* (1991) (D'Amelio, Moineau, and Klaenhammer, unpublished). The effectiveness with which pTR2030 prevented phage proliferation on transconjugants in starter activity tests led to the development of conjugal strategies to introduce this plasmid into commercial starter strains that have been used in the cheese industry since 1985 (Sanders, 1988; Sanders *et al.*, 1986; Sing and Klaenhammer, 1986).

In addition to the abortive mechanism characterized initially as heat-sensitive phage resistance (Hsp[+]), pTR2030 (46.2 kb) encodes a restriction/modification system (R/M) which was not discovered for several years (Hill *et al.*, 1989a). The two different mechanisms, Abi and R/M, when combined on a large plasmid provide substantial resistance to small isometric phages. The underlying R/M phenotype remained undetected until a phage (φ48) became available which was not susceptible to the abortive mechanism encoded by pTR2030 (Alatossava and Klaenhammer, 1991). Many of the large, conjugative, phage resistance plasmids which have been termed 'abortive', are also likely to encode more than one mechanism of resistance. An estimation of the levels of death and survival among an infected population is critical to determine whether or not an abortive system is operating alone or in conjunction with an R/M system. Abortive responses typically result in death of a majority of the infected cells, whereas most infected cells with an operational R/M system will survive a first round infection (Klaenhammer, 1991; Sing and Klaenhammer, 1990a, b). The singular effect of the pTR2030 abortive system was initially evaluated with two phages (p2 and modified φ31) that are not susceptible to restriction by the R/M system encoded by pTR2030 (Sing and Klaenhammer, 1990a). In cells containing pTR2030, both phages plaqued at EOPs of $< 10^{-9}$, and cell death occurred in 70–90% of infected cells. However, both phages still formed centres of infection at frequencies of 10^{-2} indicating that only 1 in 100 infected cells released progeny phage. The phage resistance determinants were subcloned on a 13.6 kb *Bgl*II fragment of pTR2030 to pTK6 (Hsp$^+$, R$^+$/M$^+$; Hill *et al.*, 1989b), and then an *in vivo* deletion of 3.3 kb in pTK6 generated the R$^-$/M$^-$ Hsp$^+$ derivative pTRK18 (Romero and Klaenhammer, 1990b; Hill *et al.*, 1989a). The individual effect of the Abi system was again evaluated in the LM0230 and NCK203 host backgrounds. Although pTRK18 caused a reduction in plaque size and EOP for both prolate and small isometric phages (EOPs of 0.5, 0.1, and 0.25 for phages c2, φ31, and sk1, respectively), the effect was far less dramatic than for Hsp action elicited by either pTK6 or pTR2030 (Hill *et al.*, 1989b). Thus, the abortive responses of small isometric-headed phages decreased substantially upon removal of sequences upstream from the *abiA/hsp* gene.

Molecular characterization of the regions upstream from the *abiA* gene have recently identified five open reading frames that appear to be part of a R/M polycistronic operon (Figure 3.1; O'Sullivan, Zagula, and Klaenhammer, unpublished). Disruption of the four ORFs intervening between the *LlaI* and *abiA* by deletions or Tn5 insertions eliminates restriction activity and reduces the abortive response (O'Sullivan, Hill, Zagula, Klaenhammer, unpublished). Therefore, the upstream R/M region appears to have some effect on the expression of the *abiA* gene. As noted above (section 3.3.3), this molecular organization and response is similar to pKR223 where Abi expression was reported to require sequences upstream of a suspected R/M system. Noting that many conjugative phage resistance plasmids do (or are likely to) encode both Abi and R/M functions, this genetic organization and some type of co-regulation may be common in lactococci.

The Abi determinant from pTR2030 was localized by subcloning and Tn5 mutagenesis to a 3 kb region (Hill *et al.*, 1989b). DNA sequencing revealed a single 1887 base pair open reading frame, designated *hsp/abiA*, which encodes the abortive phenotype Hsp$^+$ (Hill *et al.*, 1990a). The *abiA* gene could encode a predicted protein of 628 amino acids with a molecular mass of 73.8 kDa. The *abiA* predicted protein product is hydrophilic, with a region of 20 amino acids near the amino terminus (positions 80 to 100) of marked hydrophobicity. The predicted structure of the protein is that of a cytosolic globular, soluble protein. The region upstream from *abiA* was also cloned into the promoter-screening vector pGKV210. The *cat-86* gene was constitutively expressed indicating that a functional promoter was present upstream of *abiA* and suggesting that this abortive gene product is not induced during a phage infection.

The determinant for abortive resistance from a related plasmid, pCI829 (Coffey *et al.*, 1989), was also identified, sequenced and found to be identical to *hsp/abiA* (Coffey *et al.*, 1991a, 1991b). Plasmid pCI829 was isolated by conjugation from *L. lactis* UC811 and the phage resistance determinant subcloned to pSA3 as pCI816. The two plasmids, pCI829 and pTR2030, and their subclones exhibit virtually identical physical, genetic, and phage resistance phenotypes. The *hsp/abiA* gene is, therefore, the first abortive gene cloned and sequenced in lactococci where its functional and molecular characteristics have been confirmed independently.

The mechanism of action of the *hsp/abiA* gene product is unknown, but the overall abortive mechanism does prevent or retard the replication of phage DNA (Hill *et al.*, 1991a; Klaenhammer *et al.*, 1991). Total DNA from cells containing pTR2030 was extracted at timed intervals after infection with phage $\varphi 31$, appropriately modified so as to negate action by the R/M system of pTR2030. Hybridization experiments using $\varphi 31$ DNA as a probe revealed that the injected phage DNA was neither degraded nor replicated during the course of the aborted phage infection.

Genes homologous to *abiA* have not been found through computer data-base searches of sequenced genes (GenBank and NPBRF); neither was any homology found between DNA probes internal to *abiA/hsp* and plasmid pNP40 or plasmid DNA from 17 other *L. cremoris* and *L. lactis* strains chosen from a bank of strains used by the cheese industry. Of the 17 strains screened, 11 were classified as phage insensitive. The uniqueness of *abiA/hsp* and its efficacy as an abortive mechanism may be two reasons why the starter strains containing pTR2030 have proven to be useful in the cheese and cultured dairy products industry.

pIL611. A different approach was taken by Cluzel *et al.* (1991) in order to isolate and characterize the second abortive gene described in lactococci, *abi416/abiB*. In this study, fragments of about 10 kb of enzyme-digested total cell DNA from the highly phage resistant strain *L. lactis* IL416 were ligated with the vector pIL253 and introduced to the plasmidless strain IL1403. Twenty randomly selected erythromycin resistant transformants were tested for resistance to phage bIL66 and one phage-resistant transformant, which contained a plasmid with the smallest insert of IL416 DNA (5 kb), was selected. The Abi determinant was subcloned to a 2 kb fragment, and the subclone designated pIL611. Phage bIL66 forms clear, 3 mm plaques after overnight incubation on the sensitive host strain IL140. In the presence of pIL611, phage bIL66 was able to form only minute plaques after four days of incubation. The EOP was not determined, but the number of infective centres formed was only 0.04% of adsorbed phages. Phage bIL66 adsorbed normally to cells containing pIL611, and the latent period for the phage was not affected, but burst size was reduced from 150 to 30. The 2 kb insert of pIL611 was sequenced and revealed a 753 base pair ORF, designated *abi416*, which was determined through deletion analysis to be responsible for the abortive phage resistance. Expression of *abi416* is mediated by an Iso-IS*S1* element positioned immediately upstream that provided a promoter to drive transcription of *abi416*. The presence of IS elements in both pIL611 and pTR2030 suggests that these elements may contribute to the intra- or intercellular mobility of the phage resistance genes as well as to gene expression.

pTN20. The conjugative, phage resistant plasmid, pTN20 (28.4 kb), was isolated from LM0230 after conjugation with *L. lactis* N1, an adsorption-promoting derivative of ME2 (Higgins *et al.*, 1988). The resistance conferred by pTN20 against phages p2 and c2 was determined to consist largely of R/M. However, the presence of the plasmid selected for mutant p2 phages which were restricted to lesser degrees in subsequent rounds of host-dependent phage replication. When a pTN20-modified phage was later passed through a nonrestrictive host (to remove the modification), the phage was restricted to a lesser extent (10^{-4}) than p2 phage never exposed to pTN20 (10^{-8}). Durmaz *et al.* (1991, 1992) subcloned pTN20 fragments and employed Tn5 mutagenesis to localize and characterize a third abortive gene in lactococci, designated *abiC*.

A 5.6 kb fragment subcloned from pTN20 onto pSA34 (Sanders and Schultz, 1990) generated a recombinant plasmid, pTRK99, which encoded an abortive type resistance phenotype, designated Prf. The resistance was effective against some, but not all, of the small isometric phages tested in *L. lactis* subsp. *lactis* and subsp. *cremoris* backgrounds. Prf was ineffective against the prolate phage c2 and large isometric phage 949. The EOP of phage p2 was 10^{-2} to 10^{-3} on LM0230 (pTRK99), with reduced plaque size. The burst size for phage p2 was reduced from 63 to 7, and the efficiency of formation of centres of infection was 10^{-2}. Normal-sized, clear plaques appeared at a frequency of 10^{-4} and contained-mutant p2 phage that were permanently resistant to Prf. Neither R/M nor interference with adsorption was responsible for phage inhibition by Prf. Death of the infected cells was not prevented by pTRK99, and the latent period for the phage was also unaffected. Phage DNA is replicated in cells containing pTRK99. Thus, the mechanism of action of Prf is different from that of Hsp and the phage infection is aborted at a different point in the lytic cycle.

The genetic determinant for Prf was localized through subcloning and Tn5 mutagenesis. Sequencing of a 1.3 kb fragment of pTRK99 revealed that a 1026 nucleotide ORF encodes the gene *abiC*, which is responsible for Prf. The predicted product of *abiC* is a protein of 344 amino acids. Two putative transmembrane helices were detected near the amino terminus of the predicted protein. Computer searches found no significant homology with previously sequenced genes in GenBank, and no homology was found to *abiA* or *abiB*/*abi416*.

pIL105. Gautier and Chopin (1987) identified pIL105, an abortive infection plasmid from the *L. lactis* subsp. *cremoris* strain IL964. In addition to pIL105, this strain contains two other plasmids each encoding a different R/M system. Plasmid pIL105 is 8.7 kb, and was isolated via co-transformation of IL1403 with plasmid DNA from *L. lactis* IL964 and a marker plasmid, pHV1301. pIL105 lowered the EOP of the isometric-headed phage 66, producing turbid, pin-point plaques visible only after extended incubation. Interestingly, mutant clear plaques were detected at a frequency of 10^{-4}, similar to those of pTRK99. Productive infections in IL1403 (pIL105) was reduced to 3%, and burst size was reduced from 180 to 50, but the latent period for phage 66 was unaffected. Abi from pIL105 was effective against two prolate phages and four of six other isometric-headed phages tested. This phenotype is very similar to that of pTRK99; however, no DNA homology exists between pIL105 and an internal probe for *abiC* encoding Prf (M.C. Chopin, E. Durmaz, and T.R. Klaenhammer, unpublished data).

pCI750. Baumgartner *et al.* (1986) isolated the conjugative Abi$^+$ plasmid pCI750 from the *L. lactis* subsp. *cremoris* strain UC653. Plasmid pCI750 is compatible with pCI829, which contains *abiA* (Coffey *et al.*, 1989). Either plasmid alone confers only partial resistance to phage c2, reducing the EOP 20-fold and plaque size 6-fold. However, a strain containing both plasmids, *L. lactis* ACOO3, was completely resistant to phage c2. The Abi determinant from pCI750 was subcloned, using the vector pGB301, to produce plasmid pMM1 (Steele *et al.*, 1989). In the LM0230 background, pMM1 conferred total resistance to small isometric-headed phages and partially protected against some prolate phages. Neither R/M nor adsorption interference mechanisms were involved in the phage resistance conferred by this plasmid. Steele *et al.* (1989) used the 13.9 kb fragment from pTR2030 containing Abi as a probe in DNA hybridization experiments against other Abi$^+$ strains and plasmids pTR2030, pBF61, and pGBK17 (a subclone of PKR223). Homology was found to pTR2030, but this result was erroneous because the probe used was not specific for Abi determinants due to its large size (Hill *et al.*, 1990a).

3.3.3.2 Abi plasmids encoding two resistance mechanisms. In addition to pTR2030 and pTN20 (discussed above), several other plasmids have been reported which encode an abortive resistance mechanism plus either restriction/ modification or adsorption interference. In the following cases, the abortive defense has been implied by noting reduced plaque sizes and/or lower EOPs of phage which escape the other phage resistance mechanism on the same plasmid.

pKR223. A 35.5 kb plasmid pKR223, which encodes both R/M and heat-sensitive Abi, was isolated from *L. lactis* subsp. *lactis* biovar. *diacetylactis* KR2 by Laible *et al.* (1987). The R/M system restricted the EOP of phage sk1, but had no apparent effect on prolate phage c2, which nevertheless produced pin-point plaques on cells carrying pKR223. The burst size of c2 was reduced from 91 to 68, and productive infections were reduced by over 90%, while the latent period remained unchanged. The Abi determinant was localized to a 1.3 kb fragment on pGBK17, a 28.8 kb recombinant derivative of pKR223. However, Abi+ was not expressed in pGBK17 subclones, possibly due to the absence of expression signals in the cloned fragment (McKay *et al.*, 1989). In transformation experiments with pGBK17 into LM0230, the *abi* genes of pGKB17 were inactivated by insertion of IS*981*, a recently discovered IS element found widely distributed in lactococci (Polzin and McKay, 1991).

pBF61.   Froseth *et al.* (1988) reported that the 26 MDa plasmid pBF61 isolated from *L. lactis* KR5 encodes both R/M and a heat-sensitive Abi. Adsorption resistance was eliminated as a reason for the phage resistance. Phage c2 produced smaller plaques in LM0230 in the presence of pBF61. Moreover, c2 phage, which had been modified for the restriction system encoded by pBF61, continued to produce smaller plaques in the presence of the plasmid. Thus Abi was detected and reported to elicit a reduced plaque size independent of R/M.

HID600.   An unnamed, 131 kb plasmid was identified that encodes both R/M and heat-sensitive Abi defense systems (Ward *et al.*, 1992). The plasmid was isolated in HID113, a derivative of LM0230, following conjugal matings with *L. lactis* DRC1. The transconjugant, HID600, lowered the EOP and reduced the plaque size of phages c2 and sk1. The isometric phage sk1 was completely inhibited at 30°C, but not at 37°C. Modified c2 phage plaqued at an EOP of less than one with reduced plaque size, indicating an Abi type of resistance was operating in addition to R/M. After repeated lytic growth on the transconjugant, a stepwise, heritable mutation resulted in phage populations that gradually overcame the abortive defense mechanism. A second transconjugant, HID611, had a similar, but less effective form of phage resistance (Powell *et al.*, 1990); however, no information was given about the plasmid content of HID611.

pCI528.   Coffey *et al.* (1991b) described a 46 kb plasmid, pCI528, which encodes an abortive defense combined with an adsorption blocking mechanism (See section 4.4). This plasmid, isolated from *L. cremoris* UC503, confers resistance to prolate and small isometric phages in lactococcal hosts. Abi was identified for phages which adsorbed normally to the pCI528-containing host. The two phage resistance determinants have not yet been localized or characterized at the molecular level.

3.3.3.3 Incompletely characterized phage resistance in which Abi is implied. Abi has been implied in several phage resistance-encoding plasmids isolated in conjugation experiments over the past eight years. McKay and Baldwin (1984), Jarvis (1988), and Dunny *et al.* (1988) reported plasmids pNP40, pAJ1106, and pCLP51R, respectively. All of these are large plasmids; pAJ1106 and pCLP51R are identified as recombinants. The phage resistance phenotypes conferred by these plasmids are somewhat analogous to that of pTR2030 and pCI829. In addition, uncharacterized resistance in which Abi may be implicated was reported for the following plasmids: pBU8 (Wetzel *et al.*, 1986), pJS88, pJS40, pCC34, pNP2, and pEB56 (Steele and McKay, 1989; Steele *et al.*, 1989). Gonzalez and Kunka (1985) and Vedamuthu and Neville (1987) also reported transconjugants that were resistant to phage by a probable Abi mechanism, but plasmids could not be detected. It is possible that the phage resistance determinants in these transconjugants are associated with either chromosomal insertions or high molecular weight multimers that are not detected by conventional plasmid analysis. Details of these phage resistance systems are summarized in Table 3.5.

3.3.3.4 Some unresolved questions concerning Abi. The large number of plasmids isolated so far in which Abi is implicated provide many research opportunities for the future. There are many questions which remain to be answered concerning both the varied mechanisms of action and levels of expression of Abi genes. The temperature effect on some Abi phenotypes appears at this point to be largely a host and phage-specific phenomenon that needs clarification. Although temperature effects have often been associated with abortive mechanisms, these effects may not be correlated in different strains. Accordingly, generalizations about the heat resistance or heat sensitivity of any given abortive system should be made cautiously until the defense is evaluated in a variety of host backgrounds. So far, Abi has been associated with reduced plaque size and lowered burst size for phage. However, EOPs for different abortive mechanisms and for the same mechanism cloned into different vectors vary widely. The data on pTR2030 and pGBK17 both suggest that reduction in EOP is variable and expression of the Abi response is somewhat dependent on a closely associated R/M system. In the case of *abiA*, copy number has also been shown to significantly affect expression. A single integrated copy of *abiA* elicits less resistance than when *abiA* is associated with a higher copy number replicon (Dinsmore and Klaenhammer, 1992). The expression of Abi also varies with host background and acts differently against different phage strains and species. The dramatic differences in reactions of prolate and isometric-headed phages against the various abortive phenotypes remain to be explained. In addition, the presence of an Abi mechanism has been shown to provide selection pressure for the development of Abi-resistant phages. These phages will continue to provide important clues as to the varied mechanisms by which lactococci can abort phage infections, and the creative means by which phages circumvent them. Each of the three *abi* genes characterized thus far is unique and appears to act at different points of the phage lytic cycle. Subcloning and sequencing of more abortive determinants will provide some of the answers to many currently unresolved questions. In these efforts, it is essential to first segregate *abi* genes from other phage resistance genetic determinants in order to properly define their phenotypes. Secondly, internal DNA probes or specific oligonucleotides for known *abi* genes should be made available in order to substantiate the identity of new systems which can provide new information about the abortive defense systems of lactococci.

3.4 Novel phage defense mechanisms

In addition to the native phage defense mechanisms found in lactococci described above, there are a number of additional routes whereby lactic acid bacteria could be genetically engineered to interfere with bacteriophage infection and proliferation. Creation of two unique defense systems have thus far been exploited via recombinant DNA approaches that, in effect, function to abort a phage infection. Using an antisense RNA approach (Coleman *et al.*, 1985), Kim and Batt (1991d) cloned a 1356 bp open reading frame for a 51 kDa protein from a lactococcal phage in an antisense orientation behind a functional lactococcal promoter. This construction provided a modest level of phage resistance, typical of that observed in aborted infections; EOP of 10^{-2} and reduced plaque size. Truncated versions of the ORF cloned in the antisense orientation did not interfere with phage infection and expression of antisense RNA was not demonstrated in this study. In a second study, Chung *et al.* (1992) have reported phage inhibition by antisense RNA constructions directed against the major capsid protein of phage 4–1. Again in this case, a modest level of inhibition (EOP of 0.5) was achieved using a 301 bp fragment from the 5' end of the the the gene for the major capsid protein. Expression of antisense RNA was not demonstrated, but in both this and the previous report, this seems to be the most probable mechanism based on the constructions used and level of resistance observed. The cells carrying the antisense construction pSGK1.6R were not significantly retarded in their growth and acidification capabilities relative to the parental strain. In the presence of phage, cultures carrying pSGK1.6R developed acid in milk, but at a reduced rate compared to the uninfected parent. This work suggests that antisense RNA approaches may offer an effective alternative to the native abortive-type defense systems which have evolved in lactococci. Efforts to clone and express essential phage genes in the antisense orientation should prove effective if: antisense RNA is expressed at maximum levels; the target DNA sequences are conserved among phage species and different strains within that species; and gene expression is disrupted so as to eliminate proteins that are vital to the phage developmental process. As an example of the latter, major capsid proteins are produced in excess and, thus, antisense RNA inhibition is not likely to yield significant levels of phage resistance (Chung *et al.*, 1992; Moineau *et al.*, 1992a).

Another approach was used by Hill *et al.* (1990b) to create an abortive response during an infection with a lactococcal bacteriophage. The bacterio-phage origin of replication (*ori*) of phage 50 was cloned onto the *E. coli* –

lactococcal shuttle vector, pSA3. The resulting plasmid, pTRK104, contained both a Gram positive plasmid *ori* and the phage 50 *ori*. The phage *ori* could not itself support plasmid replication in uninfected cells. However, during a phage infection of cells containing pTRK104, the phage *ori* drives plasmid replication and an explosion in plasmid copy number is detected. The presence of a phage *ori in trans* provides a false target for factors essential for replication of the phage genome and, therefore, phage development is retarded. This again is an abortive type of phage resistance, designated Per, which also provides a modest level of resistance; EOP 0.42 and small plaques. This type of resistance should go up or down with the copy number of resident plasmid encoding the phage *ori in trans*. Indeed, when the phage 50 *ori* was placed on a high copy number replicon, superinfection with phage 50 yielded no plaques (EOP $<10^{-9}$) in titration assays (O'Sullivan and Klaenhammer, unpublished). Unlike other abortive mechanisms, this type of phage resistance is not expected to act generally against different strains or species due to sequence variation among phage origins. Our current genetic technologies provide many opportunities to design traps that can interfere with intracellular phage development. In addition to the antisense RNA and Per mechanisms described above, other approaches will be forthcoming. One of the more exciting opportunities in the future may be the development and application of phage ribozymes which target conserved RNA sequence in lactococcal bacteriophages.

3.5 Genetic strategies to construct phage-insensitive strains

Both naturally occurring and recombinant plasmids are currently available which encode phage resistance mechanisms of varying scope and effectiveness. Given these genetic resources, directed strategies can now be successfully employed to introduce one or more phage resistance mechanisms into industrial strains suffering from bacteriophage attacks. The advantage of 'genetic addition strategies' (Jarvis, 1989) include the following:

- acquisition of whole gene systems is significant relative to small incremental changes incurred by host range mutations;
- many phage resistance gene systems provide broad spectrum resistance;
- defined genetic changes simplify strain selection and minimize pleotropic effects;
- creation of distinct strains that may be genetically unique relative to phage resistant strains that occur naturally.

These strategies also allow for the rational design of complementary phage resistance combinations that act at different points of the phage lytic cycle. It is now well documented that naturally occurring strains, which are recalcitrant to phage attack, harbour multiple defense systems that act at different points of the infection and lytic cycles (Klaenhammer, 1989, 1991).

Many lactococcal phage resistance plasmids are conjugative and often encode other selection markers, such as lactose-fermenting ability and nisin resistance, which can be used to monitor and score transconjugants (see Murphy *et al.*, 1988 and Tables 3.2–3.5 for other specific examples). This has been fortuitous since conjugation is a natural gene transfer mechanism which can be used in the industry to construct phage-insensitive starter cultures by addition strategies (McKay and Baldwin, 1984). The first successful application of this approach was accomplished by directing the conjugal transfer of pTR2030 (Klaenhammer and Sanozky, 1985) from a Lac⁻ donor into industrial starter cultures (Sing and Klaenhammer, 1986; Sanders *et al.*, 1986; Klaenhammer, 1991). Selection of transconjugants was based on phage resistance and retention of fast-acid producing ability (Lac⁺ Prt⁺) by the industrial recipient strain; acquisition of pTR2030 was confirmed by DNA hybridization. Jarvis *et al.* (1989) employed a similar strategy with two plasmids that were Lac⁻ deletion derivatives of pAJ1106, a conjugative plasmid which encoded both an abortive type phage resistance and lactose-fermenting ability. The Lac⁻ Hsp⁺ plasmids were transferred to a variety of Lac⁺ strains and transconjugants selected on the basis of phage resistance. The first conjugative phage resistance plasmid identified in lactococci, pNP40, encodes both a heat-sensitive phage resistance phenotype and nisin resistance (McKay and Baldwin, 1984). Harrington and Hill (1991) have recently selected pNP40 transconjugants of a Lac⁺, high-alpha acetolactate producing industrial strain, *L. lactis* 425A. The selection process was conducted in broth and required that the transconjugants exhibit resistance against two phages, nisin resistance, and lactose-fermenting ability. Low frequency conjugation events within this industrial strain were detected and the associated nisin resistance marker linked to pNP40 precluded the need to use DNA hybridization to confirm plasmid transfer. The development and use of conjugal strategies has, therefore, provided a currently acceptable means to introduce phage resistance mechanisms into lactococcal strains destined for use in dairy manufacturing. Conjugal strategies can also be used to combine complementary mechanisms of phage defense in a single strain. Recently, Coffey *et al.* (1989) introduced both pCI829 (*abiA*) and pCI751 (encoding Lac⁺ and a different abortive mechanism) into a single strain via conjugation. The combined effect of the two plasmids provided complete resistance to the small-isometric headed phage 712 and prolate-headed phage c2. Later Kelly *et al.* (1990) used the food grade conjugative strategies outlined above to introduce both pCI528 (Ads⁺) and pCI750 (Abi⁺) into Lac⁺ commercial dairy starter cultures. The industrial transconjugants showed excellent phage resistance and were subsequently introduced into the Irish dairy industry. A conjugal approach was also used to combine pTR2030 (*abiA, 11aI*) and pTN20 (*abiC*, R/M⁺) into *Lactococcus lactis* (Klaenhammer, 1989). Complete resistance to the small isometric phage p2, which is closely related to sk1 (Hill, Durmaz and Klaenhammer, unpublished data), was observed, but the prolate phage c2 still plaqued on the pTN20-pTR2030 transconjugants at an EOP of 10⁻⁴. These studies illustrate that conjugation can be employed to generate plasmid combinations that elicit higher orders of phage resistance. Rational design of those mechanisms which best complement each other will have the greatest impact on imparting phage resistance and promoting the longevity of the transconjugants in the dairy industry. Conjugation strategies do impose limits on the possible defense combinations and suitable host strains which may be amenable for directed improvements. In this regard, the availability of electroporation as an acceptable food-safe gene transfer technology would expand our potential for engineering phage-insensitive lactic acid bacteria.

Genetic technologies will provide opportunities to construct novel defense systems and express them at maximal levels. Eventually, recombinant DNA approaches will lead to the development of specialized, value added strains. The usefulness of such specialized strains will be determined by the level at which phage resistance can concommitantly be engineered into new starter culture systems (Klaenhammer, 1991). Klaenhammer and Sing (1991) have devised a rotation strategy that employs different phage resistant derivatives of a specialized parental strain. The types, combinations, and specificity of different phage defense mechanisms are varied by genetic modification of the parental strain to form a series of derivatives that are distinct from their progenitor and from each other. The different derivatives can then be used in a rotation sequence that cycles completely distinct defense systems, all within the same host background. This approach is intended to protect the parental strain from infection and minimize the potential for the emergence of new phage. Since this strategy alternates defense systems, the proliferation of any new phage capable of counteracting any one defense is minimized because the emerging phage population will confront a different set of resistance mechanisms and be 'trapped' in the next cycle of the rotation. The 'phage defense rotation strategy' is a different approach which could extend the long term usefulness of specialized strains of lactic acid bacteria (Klaenhammer, 1991).

3.6 Conclusions and perspectives

Over the past decade our knowledge of bacteriophages and their interactions with members of the lactic acid bacteria has exploded. Traditional food and dairy fermentations driven by lactic acid bacteria are coming under increasing control as both innovations in technology, starter culture microbiology and molecular biology are implemented. The evolution of the fermentation 'art' to its current level of sophistication has been accompanied by increasingly stringent tolerances for manufacturing parameters, production schedules, and expectations for starter culture performance. Accompanying the implementation of specialized strains and fermentation systems are increasing problems due to bacteriophage attack. Recent examples include the bacteriophage problems that have been encountered with *Streptococcus thermophilus* strains in yogurt fermentations, *Leuconostoc oenos* strains used to conduct the malolactate fermentation in wine, and *Pediococcus halophilus* starters developed for soy sauce fermentations. Control of these fermentations will eventually rest with control of bacteriophages that are expected to evolve once defined starters are developed and used repeatedly in these bioprocesses. Some solutions will be provided from the extensive knowledge base developed for the dairy lactococci, considering that this may be one of the most dynamic species in which to study bacteriophage-host interactions and evolutionary cycles of defense and counter-defense (Klaenhammer, 1991). In the future, the field is now postured to make significant progress in the following areas: identification of those points of the phage lytic cycle which are most susceptible to interference using natural or genetically engineered defense systems; regulation of genes and operons encoding phage defense systems in order to maximize expression in homologous as well as heterologous bacterial hosts; and definition of those factors that will maximize cooperativity between different defense mechanisms. This information, coupled with a better understanding of bacteriophage evolution and life cycles at the molecular level, will provide the foundation upon which tomorrow's specialized strains of lactic acid bacteria can best be protected from attack by lytic bacteriophages.

Acknowledgements

Paper number *FSR92-90* of the Journal Series of the Department of Food Science, Raleigh, NC 27695-7624. The use of trade names in this publication does not imply endorsement by the North Carolina Agricultural Research Service of the products named, nor criticism of similar ones not mentioned.

T.R.K. thanks Evelyn Durmaz for her written and organizational contributions to the abortive plasmids section of this text, and Sylvain Moineau, Dan O'Sullivan, and Polly Dinsmore for their suggestions and critical reading of the final manuscript. G.F.F. thanks Adrian Coffey and Patricia Garvey for their critical reading of the manuscript.

References

Accolas, J.-P. and Spillman, H. (1979) The morphology of six bacteriophages of *Streptococcus thermophilus*. *J. Appl. Bacteriol*. **47**, 135–144.

Ackermann, H.W. and DuBow, M.S. (1987) Viruses of Prokaryotes, CRC Press, Inc., Boca, Raton, FL., U.S.A.

Alatossava, T., and Klaenhammer, T.R. (1991) Molecular characterization of three small isometric-headed bacteriophages which vary in their sensitivity to the lactococcal phage resistance plasmid pTR2030. *Appl. Environ. Microbiol*. **57**, 1346–1353.

Alatossova, T., Forsman, P., Karvonen, P. and Vasala, A. (1987) Molecular biology of *Lactobacillus lactis* bacteriophage LL-H *FEMS Microbiol. Rev*. **46**, 41 (Abstract).

Anderson, D.G., and McKay, L.L. (1984) Genetic and physical characterization of recombinant plasmids associated with cell aggregation and high-frequency conjugal transfer in *Streptococcus lactis* ML3. *J. Bacteriol*. **158**, 954–962.

Arendt, E.K., Lonvaud, A. and Hammes, W.P. (1991) Lysogeny in *Leuconostoc oenos*. *J. Gen. Microbiol*. **137**, 2135–2139.

Baumgartner, A., Murphy, M., Daly, C. and Fitzgerald, G.F. (1986) Conjugative co-transfer of lactose and bacteriophage resistance plasmids from *Streptococci cremoris* UC653. *FEMS Microbiol. Lett.* **35**, 233–237.

Benbadis, L., Faelen, M., Castellino, I., Fazel, A., Mercenier, A. and Slos, P. (1987) Phages of *Streptococcus thermophilus. FEMS Microbiol. Rev.* **46**, p.43. (Abstract)

Benbadis, L., Faelen, M., Slos, P., Fazel, A., and Mercenier, A. (1990) Characterization and comparison of virulent bacteriophages of *Streptococcus thermophilus* isolated from yogurt. *Biochimie* **72**, 855–862.

Benbadis, L., Garel, J.-R., and Hartley, D.L. (1991) Purification, properties, and sequence specificity of *Ssl*I a new type II restriction endonuclease from *Streptococcus salivarius* subsp. *thermophilus. Appl. Environ. Microbiol.* **57**, 3677–3678.

Boizet, B., Lahbib-Mansais, Y., Dupont, L., Ritzenthaler, P. and Mata, M. (1990) Cloning, expression and nucleotide sequence of an endolysin gene of a *Lactobacillus bulgaricus* bacteriophage. *FEMS Microbiol. Rev.* **87**, 60 (Abstract).

Boizet, B., Mata, M., Mignot, O., Ritzenthaler, P. and Sozzi, T. (1992) Taxonomic characterization of *Leuconostoc mesenteroides* and *Leuconostoc oenos* bacteriophages. *FEMS Microbiol. Lett.* **90**, 211–216..

Boussemaer, J.P., Schrauwen, P.P., Sourrouille, J.L. and Guy, P. (1980) Multiple modification/restriction systems in lactic streptococci and their significance in defining a phage-typing system. *J. Dairy Res.* **47**, 401–409.

Braun, V., Hertwig, S., Neve, H., Geis, A. and Teuber, M. (1989) Taxonomic differentiation of bacteriophages of *Lactococcus lactis* by electron microscopy, DNA-DNA hybridization, and protein profiles. *J. Gen. Microbiol.* **135**, 2551–2560.

Budde-Niekiel, A. and Teuber, M. (1987) Electron microscopy of the adsorption of bacteriophages to lactic acid streptococci. *Milchwissenschaft* **42**, 551–554.

Callegari, M.L., Sechaud, L., Rousseau, M., Bottazzi, V. and Accolas, J.-P. (1992) The S-layer protein of *Lactobacillus helveticus* CNRZ892 contains the receptor for virulent phage 832-B1. *Appl. Environ. Microbiol.* (In press).

Carminati, D. and Giraffa, G. (1992) Evidence and characterization of temperate bacteriophage in *Streptococcus salivarius* subsp. *thermophilus* St18. *J. Gen. Microbiol.* **59**, 71–79.

Casey, C.N. (1991) Molecular and physical characterization of lysogenic and lytic lactococcal bacteriophages. PhD Thesis, National University of Ireland.

Chopin, M-C., and Chopin, A. (1990) Improvement of phage resistance in lactic acid bacteria. *Proceedings of the Sixth International Symposium on the Genetics of Industrial Microorganisms*, **Vol I**, 467–476.

Chopin, A., Chopin, M-C., Moillo-Batt, A. and Langella, P. (1984) Two plasmid-determined restriction and modification systems in *Streptococcus lactis. Plasmid* **11**, 260–263.

Chopin, M.-C., Chopin, A., Rouault, A. and Galleron, N. (1989) Insertion and amplification of foreign genes in the *Lactococcus lactis* subsp. *lactis* chromosome. *Appl. Environ. Microbiol.* **55**, 1769–1774.

Chow, J.J., Batt, C.A. and Sinskey, A.J. (1988) Characterization of *Lactobacillus bulgaricus* bacteriophage ch2. *Appl. Environ. Microbiol.* **54**, 1138–1142.

Chung, D.K., Chung, S.K., and Batt, C.A. (1992) Antisense RNA directed against the major capside protein of *Lactococcus lactis* subsp. *cremoris* bacteriophage 4–1 confers partial resistance to the host. *Appl. Microbiol. Biotechnol.* **37**, 79–83.

Cluzel, P.J., Chopin, A., Ehrlich, S.D., and Chopin, M.-C. (1991) Phage abortive infection mechanism from *Lactococcus lactis* subsp. *lactis*, expression of which is mediated by an iso-ISS*I* element. *Appl. Environ. Microbiol.* **57**, 3547–3551.

Cluzel, P.J., Vlaux, M., Rousseau, M. and Accolas, J.-P. (1987) Evidence for temperate bacteriophages in two strains of *Lactobacillus bulgaricus. J. Dairy Res.* **54**, 397–405.

Coffey, A.G., Fitzgerald, G.F., and Daly, C. (1991a) Cloning and characterization of the determinant for abortive infection of bacteriophage from lactococcal plasmid pC1829. *J. Gen. Microbiol.* **137**, 1355–1362.

Coffey, A.G., Costello, V., Daly, C. and Fitzgerald, G. (1991b) Plasmid encoded bacteriophage insensitivity in members of the genus *Lactococcus*, with special reference to pCI829. In *Genetics and Molecular Biology of Streptococci, Lactococci, and Enterococci* (Dunny, G.M., Cleary, P.P., and McKay, L.L. eds), American Society for Microbiology, Washington, DC., pp. 131–135.

Coffey, A.G., Fitzgerald, G.F. and Daly, C. (1989) Identification and characterization of a plasmid encoding abortive infection from *Lactococcus lactis* ssp. *lactis* UC811. *Neth. Milk Dairy J.* **43**, 229–244.

Coleman, J., Hirasuma, A., Inokuchi, Y. Green, P.J. and Inoye, M. (1985) A novel immune system against bacteriophage infection using complementary RNA (micRNA). *Nature* (London) **315**, 601.

Collins, E.B. (1958) Changes in the bacteriophage sensitivity of lactic streptococci. *J. Dairy Sci.* **41**, 41–48.

Collins, E.B. (1956) Host-controlled variations in bacteriophages active against lactic streptococci. *Virology* **2**, 261–271.

Costello, V. (1988) Characterization of bacteriophage-host interactions in *Streptococcus cremoris* UC503 and related lactic streptococci. PhD Thesis, The National University of Ireland.

Coveny, J.A. (1989) Characterization of lactococcal bacteriophage based on morphology, host range, DNA restriction endonuclease patterns, DNA hybridization and structural protein profiles. PhD Thesis, National University of Ireland.

Coveney, J.A., Fitzgerald, G.F., and Daly, C. (1987) Detailed characterization and comparison of four lactic streptococcal bacteriophages based on morphology, restriction mapping, DNA homology, and structural protein analysis. *Appl. Environ. Microbiol.* **53**, 1439–1447.

Daly, C. and Fitzgerald, G.F. (1982) Bacteriophage DNA restriction and the lactic streptococci. *Microbiology – 1982*, (Schlessinger, D., ed.) American Society, for Microbiology, Washington, D.C., pp. 213–216.

Daly, C. and Fitzgerald, G.F. (1987) Mechanisms of bacteriophage insensitivity in the lactic streptococci. In *Streptococcal Genetics* (Ferretti, J. and Curtiss, R. eds.). American Society for Microbiology, Washington, D.C., U.S.A. pp. 259–268.

Davidson, B.E., Powell, I.B., and Hillier, A.J. (1990) Temperate bacteriophages and lysogeny in lactic acid bacteria. *FEMS Microbiol, Rev.* **87**, 79–90.

Davies, F.L. and Gasson, M.J. (1984) Bacteriophage of dairy lactic acid bacteria. In *Advances in the Microbiology and Biochemistry of Cheese and Fermented Milk* (Davies, F.L. and Law, B.A., eds.). Elsevier Applied Science Publishers, New York, pp. 127–151.

de Vos, W.M. (1989) On the carrier state of bacteriophages in starter lactococci: an elementary explanation involving a bacteriophage resistance plasmid. *Neth. Milk Dairy J.* **43**, 221–227.

de Vos, W.M. and Davies, F.L. (1984) Plasmid DNA in lactic streptococci: bacteriophage resistance and proteinase plasmids in *Streptococcus cremoris* SK11. Third European Congress on Biotechnology. Vol III, Verlag Chemie, Weinheim, pp. 201–205.

de Vos, W.M., Underwood, H.M. and Davies, F.L. (1984) Plasmid encoded bacteriophage resistance in *Streptococcus cremoris* SK11. *FEMS Microbiol. Lett.* **23**, 175–178.

Dinsmore, P.K. and Klaenhammer, T.R. (1992) IS946-mediated integration of an abortive bacteriophage resistance gene (*hsp*) into the *L. lactis* subsp. *lactis* genome. *J. Dairy Sci.* **75**, 113 (Abstract).

Duckworth, D.H., Glenn, J. and McCorquodale, D.J. (1981) Inhibition of bacteriophage replication by extrachromosomal genetic elements. *Microbiol. Rev.* **45**, 52–71.

Dunny, G.M., Krug, D.A., Pan, C-L. and Ledford, R.A. (1988) Identification of cell wall antigens associated with a large conjugative plasmid encoding phage resistance and lactose fermentation ability in lactic streptococci. *Biochimie* **70**, 443–450.

Durmaz, E. and Klaenhammer, T.R. (1991) A fourth mechanism for bacteriophage resistance in *Lactococcus lactis* subsp. *lactis* ME2. *J. Dairy Sci.* **74**, 120 (Abstract).

Durmaz, E. and Klaenhammer, T.R. (1992) Molecular characterization of a second abortive phage resistance gene present in *L. lactis* subsp. *lactis* ME2. *J. Bacteriol.* **174**, 7463–7469.

Fitzgerald, G.F., Daly, C., Brown, L.R. and Gingeras, T.R. (1982) ScrFI: a new sequence-specific endonuclease from *Streptococcus cremoris*. *Nucleic Acids Research* **10**, 8171–8179.

Fitzgerald, G.F. and Gasson, M.J. (1988) *In vivo* gene transfer systems and transposons. *Biochimie* **70**, 489–502.

Forsman, P. and Alatossava, T. (1991) Genetic variation of *Lactobacillus delbrueckii* ssp. *lactis* phages isolated from cheese processing plants in Finland. *Appl. Environ. Microbiol.* **57**, 1805–1812.

Fremaux, C.F., De Antoni, G.L. and Klaenhammer, T.R. (1992) Genetic organization and sequence of the region encoding integrative functions from *Lactobacillus gasseri* temperate phage φadh. *Gene* **126**, 61–66.

Froseth, B.R., Harlander, S.K. and McKay, L.L. (1988) Plasmid-mediated reduced phage sensitivity in *Streptococcus lactis* KR5. *J. Dairy Sci.* **71**, 275–284.

Gasson, M.J. (1983) Genetic transfer systems in lactic acid bacteria. *Antonie van Leeuwenhoek* **49**, 275–282.

Gasson, M.J. and Davies, F.L. (1984) The genetics of dairy lactic acid bacteria. In *Advances in the Microbiology and Biochemistry of Cheese and Fermented Milk* (Davies, F.L. and Law, B.A., eds.). Elsevier Applied Science Publishers, New York, pp 99–126.

Gautier, M. and Chopin, M-C. (1987) Plasmid-determined restriction/modification systems and abortive infection in *Streptococcus cremoris. Appl. Environ. Microbiol.* **53**, 923–927.

Gonzalez, C.F. and Kunka, B.S. (1985) Transfer of sucrose-fermenting ability and nisin production phenotype among lactic streptococci. *Appl. Environ. Microbiol.* **49**, 627–633.

Harrington, A. and Hill, C. (1991) Construction of a bacteriophage-resistant derivative of *Lactococcus lactis* subsp. *lactis* 425A by using the conjugal plasmid pNP40. *Appl. Environ. Microbiol.* **57**, 3405–3409.

Heap, H.A. and Jarvis, A.W. (1980) A comparison of prolate and isometric-headed lactic streptococcal bacteriophages. *N.Z.J. Dairy Sci. Technol.* **15**, 75–81.

Heap, H.A. and Lawrence, R.C. (1988) Culture systems for the dairy industry. In *Developments in Food Microbiology* (Robinson, ed.). Vol. 4. Elsevier Applied Science, Amsterdam, pp. 149–185.

Heap, H.A. and Jarvis, A.W. (1980) A comparison of prolate- and isometric-headed lactic streptococcal bacteriophages. *N.Z.J. Dairy Sci. Technol.* **15**, 75–81.

Higgins, D.L., Sanozky-Dawes, R.B. and Klaenhammer, T.R. (1988) Restriction and modification activities from *Streptococcus lactis* ME2 are encoded by a self-transmissible plasmid, pTN20, that forms cointegrates during mobilization of lactose-fermenting ability. *J. Bacteriol.* **170**, 3435–3442.

Hill, C., Pierce, K. and Klaenhammer, T.R. (1989a) The conjugative plasmid pTR2030 encodes two bacteriophage defense mechanisms in lactococci, restriction modification (R+/M+) and abortive infection (Hsp+). *Appl. Environ. Microbiol.* **55**, 2416–2419.

Hill, C., Romero, D.A., McKenney, D.S., Finer, K.R. and Klaenhammer, T.R. (1989b) Localization, cloning and expression of genetic determinants for bacteriophage resistance (Hsp) from the conjugative plasmid pTR2030. *Appl. Environ. Microbiol.* **55**, 1684–1689.

Hill, C., Miller, L.A., and Klaenhammer, T.R. (1990a) Nucleotide sequence and distribution of the pTR2030 resistance determinant (*hsp*) which aborts bacteriophage infection in lactococci. *Appl. Environ. Microbiol.* **56**, 2255–2258.

Hill, C., Miller, L.A. and Klaenhammer, T.R. (1990b) Cloning, expression, and sequence determination of a bacteriophage fragment encoding bacteriophage resistance in *Lactococcus lactis. J. Bacteriol.* **172**, 6419–6426.

Hill, C., Massey, I.J. and Klaenhammer, T.R. (1991a) Rapid method to characterize lactococcal bacteriophage genomes. *Appl. Environ. Microbiol.* **57**, 283–288.

Hill, C., Miller, L.A. and Klaenhammer, T.R. (1991b) The bacteriophage resistance plasmid forms high-molecular weight multimers in lactococci. *Plasmid* **25**, 105–112.

Hill, C., Miller, L.A. and Klaenhammer, T.R. (1991c) *In vivo* genetic exchange of a functional domain from a type II A methylase between lactococcal plasmid pTR2030 and a virulent bacteriophage. *J. Bacteriol.* **173**, 4363–4370.

Huggins, A.R. and Sandine, W.E. (1979) Selection and characterization of phage insensitive lactic streptococci. *J. Dairy Sci.* **62**, 70–71.

Hull, R.R. (1983) Factory-derived starter cultures for the control of bacteriophage in cheese manufacture. *Aust. J. Dairy Technol.* **38**, 149–153.

Ishibashi, K., Takesue, S., Watanabe, K and Oishi, K. (1982) Use of lectins to characterise the receptor sites for bacteriophage PL–1 of *Lactobacillus casei. J. Gen. Microbiol.* **128**, 2251–2259.

Jarvis, A.W. (1977) The serological differentiation of lactic streptococcal bacteriophage. *N.Z.J. Dairy Sci. Technol.* **12**, 176–181.

Jarvis, A.W. (1981) The use of whey-derived phage-resistant starter strains in New Zealand cheese plants. *N.Z.J. Dairy Sci. Technol.* **16**, 25–31.

Jarvis, A.W. (1984a) Differentiation of lactic streptococcal phages into phage species by DNA-DNA homology. *Appl. Environ. Microbiol.* **47**, 343–349.

Jarvis, A.W. (1984b) DNA-DNA homology between lactic streptococci and their temperate and lytic phages. *Appl. Environ. Microbiol.* **47**, 1031–1038.

Jarvis, A.W. (1988) Conjugal transfer in lactic streptococci of plasmid-encoded insensitivity to prolate- and small isometric-headed bacteriophages. *Appl. Environ. Microbiol.* **54**, 777–783.

Jarvis, A.W. (1989) Bacteriophages of lactic acid bacteria. *J. Dairy Sci.* **72**, 3406–3428.

Jarvis, A.W. and Klaenhammer, T.R. (1986) Bacteriophage resistance conferred on lactic strepto-cocci by the conjugative plasmid pTR2030: effects on small isometric-, large isometric-, and prolate-headed phages. *Appl. Environ. Microbiol.* 1272–1277.

Jarvis, A.W. and Meyer, J. (1986) Electron microscopic heteroduplex study and restriction endonu-clease cleavage analysis of the DNA genomes of three lactic streptococcal bacteriophages. *Appl. Environ. Microbiol.* **51**, 566–571.

Jarvis, A.W., Fitzgerald, G.F., Mata, M., Mercenier, A., Neve, H., Powell, I.B., Ronda, C., Saxelin, M. and Teuber, M. (1991) Species and type phages of lactococcal bacteriophages. *Intervirology* **32**, 2–9.

Jarvis, A.W., Heap, H.A. and Limsowtin, G.K.Y. (1989) Resistance against industrial bacterio-phages conferred on lactococci by plasmid pAJ1106 and related plasmids. *Appl. Environ. Microbiol.* **55**, 1537–1543.

Jimeno, J., Casey, M.G., Jenni, E. and Accolas, J.-P. (1990) A new type of bacteriophage of *Lactobacillus delbrueckii*. *FEMS Microbiol. Rev.* **87**, 58 (Abstract).

Josephsen, J.J. and Klaenhammer, T.R. (1990) Stacking of three different restriction and modifi-cation systems in *Lactococcus lactis* by cotransformation. *Plasmid* **23**, 71–75.

Josephsen, J.J. and Vogensen, F.K. (1989) Identification of three different plasmid-encoded restriction/modification systems in *Streptococcus lactis* subsp. *cremoris* W56. *FEMS Microbiol. Lett.* **59**, 161–166.

Kelly, W., Dobson, J., Jorck-Ramberg, D., Fitzgerald, G. and Daly, C. (1990) Introduction of bacte-riophage resistance plasmids into commercial *Lactococcus* starter strains. *FEMS Microbiol. Rev.* **87**, P63 (Abstract).

Keogh, B.P. (1973) Adsorption, latent period and burst size of phages of some strains of lactic strep-tococci. *J. Dairy Res.* **40**, 303–309.

Keogh, B.P. (1980) Appraisal of media and methods for assay of bacteriophages of lactic strepto-cocci. *Appl. Environ. Microbiol.* **40**, 798–802.

Keogh, B.P. and Pettingill, G. (1983) Adsorption of bacteriophage lb7 on *Streptococcus cremoris* EB7. *Appl. Environ. Microbiol.* **45**, 1946–1948.

Keogh, B.P. and Shimmin, P.D. (1974) Morphology of the bacteriophages of lactic streptococci. *Appl. Microbiology* **27**, 411–415.

Khosaka, T. (1977) Physicochemical properties of virulent *Lactobacillus* phage containing DNA with cohesive ends. *J. Gen. Virol.* **37**, 209–214.

Kim, S.G. and Batt, C.A. (1991a) Molecular characterization of a *Lactococcus lactis* bacteriophage F4–1. *Food Microbiol.* **8**, 15–26.

Kim, S.G. and Batt, C.A. (1991b) Nucleotide sequence and deletion analysis of a gene coding for a structural protein of *Lactococcus lactis* bacteriophage F4–1. *Food Microbiol.* **8**, 27–36.

Kim, S.G. and Batt, C.A. (1991c) Identification of a nucleotide sequence conserved in *Lactococcus lactis* bacteriophages. *Gene* **98**, 95–100.

Kim, S.G. and Batt, C. (1991d) Antisense mRNA-mediated bacteriophage resistance in *Lactococcus lactis* subsp. *lactis*. *Appl. Environ. Microbiol.* **57**, 1109–1113.

King, W.R., Collins, E.B. and Barrett, E.L. (1983) Frequencies of bacteriophage resistant and slow acid-producing variants of *Streptococcus cremoris*. *Appl. Environ. Microbiol.* **45**, 1481–1485.

Kita, K., Kotani, H., Sugisake, H. and Takanami, M. (1989) The *Fok*1 restriction-modification system. I. Organization and nucleotide sequences of the restriction and modification genes. *J. Bio. Chem.* **264**, 5751–5756.

Klaenhammer, T.R. (1984) Interactions of bacteriophages with lactic streptococci. *Adv. Appl. Microbiol.* **30**, 1–29.

Klaenhammer, T.R. (1987) Plasmid-directed mechanisms for bacteriophage defense in lactic strep-tococci. *FEMS Microbiol. Rev.* **46**, 313–325.

Klaenhammer, T.R. (1989) Genetic characterization of multiple mechanisms of phage defense from a prototype phage-insensitive strain, *Lactococcus lactis* ME2. *J. Dairy Sci.* **72**, 3429–3442.

Klaenhammer, T.R. (1991) Development of bacteriophage-resistant strains of lactic acid bacteria. *Biochemical Soc. Transactions* **19**, 675–681.

Klaenhammer, T.R. and Sanozky, R.B. (1985) Conjugal transfer from *Streptococcus lactis* ME2 of plasmids encoding phage resistance, nisin resistance and lactose-fermenting ability: evidence for a high-frequency conjugative plasmid responsible for abortive infection of virulent bacterio-phage. *J. Gen. Microbiol.* **131**, 1531–1541.

Klaenhammer, T.R. and Sing, W.D. (1991) A novel rotation strategy using different phage defenses in a single-strain starter culture system. *J. Dairy Sci.* **74**, 120 (Abstract).

Kruger, D.H., and Bickle, T.A. (1983) Bacteriophage survival: multiple mechanisms for avoiding the deoxyribonucleic acid restriction systems of their hosts. *Microbiol. Rev.* **47**, 345–360.

Krusch, U. Neve, H., Luschei, B. and Teuber, M. (1987) Characterization of virulent phages of *Streptococcus salivarius* subsp. *thermophilus* by host specificity and electron microscopy. *Kieler Milchwirtschaftl. Forschungsberichte* **39**, 155–167.

Lahbib-Mansais, Y., Mata, M. and Ritzenthaler, P. (1988) Molecular taxonomy of *Lactobacillus* phages. *Biochimie* **70**, 429–435.

Laible, N.J., Rule, P.L., Harlander, S.K., and McKay, L.L. (1987) Identification and cloning of plasmid deoxyribonucleic acid coding for abortive phage infection from *Streptococcus lactis* ssp. *diacetylactis* KR2. *J. Dairy Sci.* **70**, 2211–2219.

Lakshmidevi, G. (1988) Molecular biology of temperate streptococcal phages. PhD Thesis, University of Melbourne, Melbourne, Australia.

Lakshmidevi, G., Davidson, B.E., and Hillier, A.J. (1988) Circular permutation of the genome of a temperate bacteriophage from *Streptococcus cremoris* BK5. *Appl. Environ. Microbiol.* **54**, 1039–1045.

Lakshmidevi, G., Davidson, B.E. and Hillier, A.J. (1990) Molecular characterization of promoters of the *Lactococcus lactis* subsp. *cremoris* temperate bacteriophage BK5-T and identification of a phage gene implicated in the regulation of promoter activity. *Appl. Environ. Microbiol.* **56**, 934–942.

Langella, P., and Chopin, A. (1989) Effect of restriction-modification systems on transfer of foreign DNA into *Lactococcus lactis* subsp. *lactis*. *FEMS Microbiol. Lett.* **59**, 301–306.

Lauster, R. (1989) Evolution of type II DNA methyltransferases. A gene duplication model. *J. Mol. Biol.* **206**, 313–321.

Lautier, M. and Novel, G. (1987) DNA-DNA hybridization in lactic streptococcal temperate and virulent phages, belonging to distinct lytic groups. *J. Ind. Microbiol.* **2**, 151–158.

Lembke, J. and Teuber, M. (1981) Serotyping of morphologically identical bacteriophages of lactic streptococci by immunoelectronmicroscopy. *Milchwissenschaft* **36**, 10–12.

Lembke, J., Krusch, U., Lompe, A. and Teuber, M. (1980) Isolation and ultrastructure of bacteriophages of group N (lactic) streptococci. *Zbl. Bakt., I. abt. Orig. C.* **1**, 79–91.

Lillehaug, D., Lindqvist, B.H. and Birkeland, N.K. (1991) Characterization of φLC3, a *Lactococcus lactis* subsp. *cremoris* temperate bacteriophage with cohesive single-stranded DNA ends. *Appl. Environ. Microbiol.* **57**, 3206–3211.

Limsowtin, G.K.Y. and Terzaghi, B.E. (1976) Phage resistant mutants: their selection and use in cheese factories. *N.Z.J. Dairy Sci. Technol.* **11**, 251–256.

Limsowtin, G.K.Y. and Terzaghi, B.E. (1977) Characterization of bacterial isolates from a phage carrying culture of *Streptococcus cremoris*. *N.Z.J. Dairy Sci. Technol.* **12**, 22–28.

Limsowtin, G.K.Y., Heap, H.A. and Lawrence, R.C. (1978) Heterogeneity among strains of lactic streptococci. *N.Z.J. Dairy Sci. Technol.* **13**, 1–8.

Marshall, R.J. and Berridge, N.J. (1976) Selection and some properties of phage-resistant starters for cheese making. *J. Dairy Res.* **43**, 449–458.

Mata, M. and Titzenthaler, P. (1988) Present state of lactic acid bacteria phage taxonomy. *Biochimie* **70**, 395–399.

Mata, M., Trautwetter, A., Luthard, G. and Ritzenthaler, P. (1986) Thirteen virulent and temperate bacteriophages of *Lactobacillus bulgaricus* and *Lactobacillus lactis* belong to a single DNA homology group. *Appl. Environ. Microbiol.* **52**, 812–818.

Mayo, B., Hardisson, C. and Brana, A.F. (1991) Nucleolytic activities in *Lactococcus lactis* subsp. *lactis* NCDO 497. *FEMS Microbiol. Lett.* **79**, 195–198.

McKay, L.L. (1982) Regulation of lactose metabolism in dairy streptococci. In *Developments in Food Microbiology*, (Davies, R., ed.), Appl. Science Publishers, London, pp. 153–182.

McKay, L.L. (1983) Functional properties of plasmids in lactic streptococci. *Antonie van Leeuwenhoek* **49**, 259–274.

McKay, L.L. and Baldwin, K.A. (1974) Simultaneous loss of proteinase and lactose-utilizing enzyme activities in *Streptococcus lactis* and reversal of loss by transduction. *Appl. Environ. Microbiol.* **28**, 342–346.

McKay, L.L. and Baldwin, K.A. (1984) Conjugative 40-megadalton plasmid in *Streptococci lactis* subsp. *diacetylactis* DRC3 is associated with resistance to nisin and bacteriophage. *Appl. Environ. Microbiol.* **47**, 68–74.

McKay, L.L., Bohanon, M.J., Polzin, K.M., Rule, P.L. and Baldwin, K.A. (1989) Localization of separate genetic loci for reduced sensitivity towards small isometric-headed bacteriophage sk1 and prolate-headed bacteriophage c2 on pGBK17 from *Lactococcus lactis* subsp. *lactis* KR2. *Appl. Environ. Microbiol.* **55**, 2702–2709.

Mercenier, A. and Lemoine, Y. (1989) Genetics of *Streptococcus thermophilus*: A Review. *J. Dairy Sci.* **72**, 3444–3454.

Mercenier, A., Robert, C., Romero, D.A., Slos, P. and Lemoine, Y. (1987) Transfection of *Streptococcus thermophilus* spheroplasts. In *Streptococcal Genetics*, (J.J. Ferretti and R. Curtiss III, eds) American Society for Microbiology, Washington, DC, USA, pp. 234–237

Moineau, S., Durmaz, E.D., Pandian, S. and Klaenhammer, T.R. (1992a) Use of monoclonal antibodies to investigate mechanisms of phage abortion in *Lactococcus lactis*. *J. Dairy Sci.* **75**, 113 (abstract).

Moineau, S., Fortier, J., Ackermann, H.W. and Pandian, S. (1992b) Characterization of lactococcal bacteriophages from Quebec cheese plants. *Can. J. Microbiol.* **38**, 875–882.

Mullan, W.M.A., Daly, C. and Fox, P. (1981) Effect of cheese making temperatures on the interactions of lactic streptococci and their phages. *J. Dairy Res.* **48**, 465–471.

Mullan, W.M.A. and Crawford, R.J.M. (1985) Partial purification and some properties of φc2 (w) lysin, a lytic enzyme produced by phage-infected cells of *Streptococcus lactis* C2. *J. Dairy Res.* **52**, 123–138.

Murphy, M.C., Steele, J.L., Daly, C. and McKay, L.L. (1988) Concomitant conjugal transfer of reduced-bacteriophage-sensitivity mechanisms with lactose-and sucrose-fermenting ability in lactic streptococci. *Appl. Environ. Microbiol.* **54**, 1951–1956.

Nes, I.F., Brendehaug, J. and von Husby, K.O. (1988) Characterization of the bacteriophage B2 of *Lactobacillus plantarum* ATCC 8014. *Biochimie* **70**, 423–427.

Neve, H., Lilischkis, R. and Teuber, M. (1988) Characterization of a virulent bacteriophage of *Leuconostoc mesenteroides* subsp. *cremoris*. *Kieler Milchwirtschaftliche Forschungsberichte* **40**, 205–212.

Neve, H., Krusch, U. and Teuber, M. (1989) Classification of virulent bacteriophages of *Streptococcus salivarius* subsp. *thermophilus* isolated from yoghurt and Swiss-type cheese. *Appl. Microbiol. Biotechnol.* **30**, 624–629.

Neve, H., Krusch, U. and Teuber, M. (1990) Virulent and temperate bacteriophages of thermophilic lactic acid streptococci. *FEMS Microbiol. Rev.* **87**, 58.

Neve, H. and Teuber, M. (1991) Basic microbiology and molecular biology of bacteriophages of lactic acid bacteria in dairies. Bulletin of the International Dairy Federation No. 263, pp. 3–15.

Ogata, S. (1980) Bacteriophage contamination in industrial processes. *Biotechnol. Bioengineering* **22**, 177–193.

Oram, J.D. (1971) Isolation and properties of a phage receptor substance from the plasma membrane of *Streptococcus lactis* ML3. *J. Gen. Virol.* **13**, 59–71.

Oram, J.D. and Reiter, B. (1965) Phage-associated lysins affecting group n and group D streptococci. *J. Gen. Microbiol.* **40**, 57–63.

Parada, J.L., La Via, M.I. and Solari, A. (1984) Isolation of *Streptococcus lactis* bacteriophages and their interaction with the host cell. *Appl. Environ. Microbiol.* **47**, 1352–1354.

Pearce, L.E. (1978) The effect of host-controlled modification on the replication rate of a lactic streptococcal bacteriophage. *N.Z.J. Dairy Sci. Technol.* **13**, 166–171.

Platteeuw, C. and de Vos, W.M. (1992) Location, characterization and expression of lytic enzyme-encoding gene, *lyt*A, of *Lactococcus lactis* bacteriophage phi US3. *Gene.* **118**, 115–120.

Polzin, K.M.,and McKay, L.L. (1991) Identification, DNA sequence, and distribution of IS*981*, a new high-copy-number insertion sequence in lactococci. *Appl. Environ. Microbiol.* **57**, 734–743.

Polzin, K.M. and Shimizu-Kadota, M. (1987) Identification of a new insertion element, similar to Gram-negative IS*26*, on the lactose plasmid of *Streptococcus lactis* ML3. *J. Bacteriol.* **169**, 5481–5488.

Powell, I.B. and Davidson, B.E. (1985) Characterization of streptococcal bacteriophage c6A. *J. Gen. Microbiol.* **66**, 2737–2741.

Powell, I.B. and Davidson, B.E. (1986) Resistance to in vitro restriction of DNA from lactic streptococcal bacteriophages c6A. *Appl. Environ. Microbiol.* **51**, 1358–1360.

Powell, I.B., Ward, A.C., Hillier, A.J. and Davidson, B.E. (1990) Simultaneous conjugal transfer in *Lactococcus* to genes involved in bacteriocin production and reduced susceptibility to bacteriophages. *FEMS Microbiol. Letts.* **72**, 209–214.

Powell, I.B., Tullock, D.L., Hillier, A.J. and Davidson, B.E. (1992) Phage DNA synthesis and host DNA degradation in the life cycle of *Lactococcus lactis* bacteriophage c6A. *J. Gen. Microbiol.* **138**, 945–950.

Ptashne, M. (1986) *A Genetic Switch: Gene Control and Phage Lambda*, Blackwell Scientific Publications, Palo Alto, CA, U.S.A.

Prevots, F., Mata, M. and Ritzenthaler, P. (1990) Taxonomic differentiation of 101 lactococcal bacteriophages and characterization of bacteriophages with unusually large genomes. *Appl. Environ. Microbiol.* **56**, 2180–2185.

Raya, R.R., Kleeman, E.G., Luchansky, J.B. and Klaenhammer, T.R. (1989) Characterization of the temperate bacteriophage φadh and plasmid transduction in *Lactobacillus acidophilus* ADH. *Appl. Environ. Microbiol.* **55**, 2206–2213.

Raya, R.R., De Antoni, G.L. Walker, D.C. and Klaenhammer, T.R. (1991) Construction of a phage φadh-mediated site-specific insertional vector, and chromosomal integration in *Lactobacillus gasseri* ADH. *J. Dairy Sci.* **74**, 122 (abstract).

Raya, R.R., Fremaux, C.F., De Antoni, G.L. and Klaenhammer, T.R, (1992) Site-specific integration of the temperate bacteriophage φadh into the *Lactobacillus gasseri* chromosome and molecular characterization of the phage (*attP*) and bacterial (*attB*) attachment sites. *J. Bacteriol.* **174**, 5584–5592.

Raya, R.R. and Klaenhammer, T.R. (1992) High frequency transduction by *Lactobacillus gasseri* bacteriophage φadh. *Appl. Environ. Microbiol.* **58**, 187–193.

Reinbold, G.W., Reddy, M.S. and Hammond, E.G. (1982) Ultrastructure of bacteriophages active against *Streptococcus thermophilus, Lactobacillus bulgaricus, Lactobacillus lactis* and *Lactobacillus helveticus. J. Food Prot.* **45**, 119–124.

Reiter, B. (1949) Lysogenic strains of lactic streptococci. *Nature* **164**, 667–668.

Relano, P., Mata, M., Bonneau, M. and Ritzenthaler, P. (1987) Molecular characterization and comparison of 38 virulent and temperate bacteriophages of *Streptococcus lactis. J. Gen. Microbiol.* **133**, 3053–3063.

Reyes-Gavilan, C.G., Limsowtin, G.K.Y., Sechaud, L., Veaux, M. and Accolas, J.P. (1990) Evidence for a plasmid-linked restriction-modification system in *Lactobacillus helveticus. Appl. Environ. Microbiol.* **56**, 3412–3419.

Reyrolle, J., Chopin, M.C., Letellier, F. and Novel, G. (1982) Lysogenic strains of lactic acid streptococci and lytic spectra of their temperate bacteriophages. *Appl. Environ. Microbiol.* **43**, 349–356.

Romero, D.A. and Klaenhammer, T.R. (1990a) Abortive phage infection and restriction/modification activities directed by pTR2030-determinants are enhanced by recombination with conjugal elements in lactococci. *J. Gen. Microbiol.* **136**, 1817–1824.

Romero, D.A. and Klaenhammer, T.R. (1990b) Characterization of Gram-positive insertion sequence IS*946*, an iso-ISS*1* element, isolated from the conjugative lactococcal plasmid pTR2030. *J. Bacteriol.* **172**, 4151–4160.

Sanders, M.E. (1987) Bacteriophages of industrial importance. In *Phage Ecology*, Goyal, S.M., Gerba, C.P., and Bitton, G. (eds), Wiley Interscience, John Wiley and Sons, NY, pp 211–244.

Sanders, M.E. (1988) Phage resistance in lactic acid bacteria. *Biochimie* **70**, 411–421.

Sanders, M.E. and Klaenhammer, T.R. (1980) Restriction and modification in group N streptococci: effect of heat on development of modified lytic bacteriophage. *Appl. Environ. Microbiol.* **40**, 500–506.

Sanders, M.E. and Klaenhammer, T.R. (1981) Evidence for plasmid linkage of restriction and modification in *Streptococcus cremoris* KH. *Appl. Environ. Microbiol.* **42**, 944–950.

Sanders, M.E. and Klaenhammer, T.R. (1983) Characterization of phage-insensitive mutants from a phage-sensitive strain of *Streptococcus lactis*: evidence for a plasmid determinant that prevents phage adsorption. *Appl. Environ. Microbiol,* **46**, 1125–1133.

Sanders, M.E. and Klaenhammer, T.R. (1984) Phage resistance in a phage-insensitive strain of *Streptococcus lactis:* temperature-dependent phage development and host-controlled phage replication. *Appl. Environ. Microbiol.* **47**, 979–985.

Sanders, M.E., Leonhard, P.J., Sing, W.E. and Klaenhammer, T.R. (1986) Conjugal strategy for construction of fast acid-producing, bacteriophage-resistant lactic streptococci for use in dairy fermentations. *Appl. Environ. Microbiol.* **52**, 1001–1007.

Sanders, M.E. and Schultz, J. (1990) Cloning of phage resistance genes from *Lactococcus lactis* ssp. *cremoris* KH. *J Dairy Sci.* **73**, 2044–2053.

Sandine, W.E. (1989) Use of bacteriophage-resistant mutants of lactococcal starters in cheese-making. *Neth. Milk Dairy J.* **43**, 211–219.

Saxelin, M.-L., Nurmiaho-Lassila, E.-L., Merilainen, V.T. and Forse, R.I. (1986) Ultrastructure and host specificity of bacteriophages of *Streptococcus cremoris, Streptococcus lactis* subsp. *diacetylactis,* and *Leuconostoc cremoris* from Finnish fermented milk viiii. *Appl. Environ. Microbiol.* **52**, 771–777.

Schafer, A., Geis, A., Neve, H. and Teuber, M. (1991) Bacteriophage receptors of *Lactococcus lactis* subsp. *diacetylactis* F7/2 and *Lactococcus lactis* subsp. *cremoris* Wg2–1. *FEMS Microbiol. Lett.* **78**, 69–74.

Sechaud, L., Cluzel, P.-J., Rousseau, M., Baumgartner, A. and Accolas, J.-P. (1988) Bacteriophages of Lactobacilli. *Biochimie* **70**, 401–410.

Sechaud, L., Callegari, M.-L., Rousseau, M., Muller, M.-C. and Accolas, J.-P. (1989) Relationship between temperate bacteriophage 0241 and virulent bacteriophage 832-B1 of *Lactobacillus helveticus Neth. Milk Dairy J.* **43**, 261–277.

Sechaud, L., Rousseau, M., Limsowtin, G.K.Y., Fayard, B., Callegari, M.-C., Quesnee, P. and Accolas, J.-P. (1992) Comparative study of 35 bacteriophages of *Lactobacillus helveticus:* morphology and host range. *Appl. Environ. Microbiol.* **58**, 1011–1018.

Sharp, P.M. (1986) Molecular evolution of bacteriophages: evidence of selection against the recognition sites of host restriction enzymes. *Mol. Biol. Evol.* **3**, 75–83.

Shearman, C., Underwood, H., Jury, K. and Gasson, M. (1989) Cloning and DNA sequence analysis of a *Lactococcus* bacteriophage lysin gene. *Mol. Gen. Genet.* **218**, 214–221.

Shearman, C.A., Hertwig S., Teuber, M., and Gasson, M.J. (1991) Characterization of the prolate-headed lactococcal bacteriophage φML3: location of the lysin gene and its DNA homology with other prolate-headed phages. *J. Gen. Microbiol.* **137**, 1285–1291.

Shearman, C.A., Jury, K. and Gasson, M.J. (1992) Autolytic *Lactococcus lactis* expressing a lactococcal bacteriophage lysin gene. *Biotechnology.* **10**, 196–199.

Shimizu-Kadota, M. and Sakurai, T. (1982) Prophage curing in *Lactobacillus casei* by isolation of a thermoinducible mutant. *Appl. Environ. Microbiol.* **43**, 1284–1287.

Shimizu-Kadota, M., Sakurai, T. and Tsuchida, N. (1983) Prophage origin of a virulent phage appearing on fermentations of *Lactobacillus casei* S–1. *Appl. Environ. Microbiol.* **45**, 669–674.

Shimizu-Kadota, M. and Tsuchida, N. (1984) Physical mapping of the virion and the prophage DNAs of a temperate *Lactobacillus* phage φFSW. *J. Gen. Microbiol.* **130**, 423–430.

Shimizu-Kadota, M., Kiwaki, M., Hirokawa, H. and Tsuchida, N. (1985) ISL*1*: a new transposable element in *Lactobacillus casei. Mol. Gen. Genet.* **200**, 193–198.

Sijtsma, L., Sterkenburg, A. and Wouters, J.T.M. (1988) Properties of the cell walls of *Lactococcus lactis* subsp. *cremoris* SK110 and SK112 and their relation to bacteriophage resistance. *Appl. Environ. Microbiol.* **54**, 2808–2811.

Sijtsma, L., Jansen, N., Hazeleger, W.C., Wouters, J.T.M. and Hellingwerf, K.J. (1990a) Cell surface characteristics of bacteriophage-resistant *Lactococcus lactis* subsp. *cremoris* SK110 and its bacteriophage sensitive variant SK112. *Appl. Environ. Microbiol,* **56**, 3230–3233.

Sijtsma, L., Wouters, J.T.M. and Hellingwerf, K.J. (1990b) Isolation and characterization of lipoteichoic acid, a cell envelope component involved in preventing phage adsorption from *Lactococcus lactis* subsp. *cremoris* SK110. *J. Bacteriol.* **172**, 7126–7130.

Sijtsma, L., Hellingwerf, K.J. and Wouters, J.T.M. (1991) Composition and phage binding capacity of cell walls isolated from *Lactococcus lactis* subsp. *cremoris* SK110 and SK112. *Neth. Milk Dairy J.* **45**, 81–95.

Simon, D., Rouault, A. and Chopin, M-C. (1985) Protoplast transformation of group N streptococci with cryptic plasmids. *FEMS Microbiol. Lett.* **26**, 239–241.

Sing, W.D. and Klaenhammer, T.R. (1986) Conjugal transfer of bacteriophage resistance determinants on pTR2030 into *Streptococcus cremoris* strains. *Appl. Environ. Microbiol.* **51**, 1264–1271.

Sing, W.D. and Klaenhammer, T.R. (1990a) Characteristics of phage abortion conferred in lactococci by the conjugal plasmid pTR2030. *J. Gen. Microbiol.* **136**, 1807–1815.

Sing, W.D. and Klaenhammer, T.R. (1990b) Plasmid-induced abortive infection in lactococci: a review. *J. Dairy Sci.* **73**, 2239–2251.

Sing, W.D. and Klaenhammer, T.R. (1991) Characterization of restriction and modification plasmids from *Lactococcus lactis* ssp. *cremoris* and their effects when combined with pTR2030. *J.Dairy Sci.* **74**, 1133–1144.

Solaiman, D.K.Y. and Somkuti, G.A. (1991) A type II restriction endonuclease of *Streptococcus thermophilus* ST117. *FEMS Microbiol. Lett* **80**, 75–80.

Sozzi, T., Poulin, J.M., Maret, R. and Pousaz, R. (1978) Isolation of a bacteriophage of *Leuconostoc mesenteroides* from dairy products. *J. Appl. Bacteriol.* **44**, 159–161.

Stadhouders, J. and Leenders, G.J.M. (1984) Spontaneously developed mixed-strain cheese starters. Their behaviour towards phages and their use in the Dutch cheese industry. *Neth. Milk Dairy J.* **38**, 157–181.

Steele, J.L., and McKay, L.L. (1989) Conjugal transfer of genetic material by *L. lactis* subsp. *lactis* 11007. *Plasmid* **22**, 32–43.

Steele, J.L., Murphy, M.C., Daly, C. and McKay, L.L. (1989) DNA-DNA homology among lactose- and sucrose-fermenting transconjugants from *Lactococcus lactis* strains exhibiting reduced bacteriophage sensitivity. *Appl. Environ. Microbiol.* **55**, 240–243.

Steenson, L.R. and Klaenhammer, T.R. (1985) *Streptococcus cremoris* M12R transconjugants carrying the conjugal plasmid pTR2030 are insensitive to attack by lytic bacteriophages. *Appl. Environ. Microbiol.* **50**, 851–858.

Steenson, L.R. and Klaenhammer, T.R. (1986) Plasmid heterogeneity in *Streptococci cremoris* M12R: effects on proteolytic activity and host-dependent phage replication. *J. Dairy Sci.* **69**, 2227–2236.

Sterkenburg, A., van Leeuwen, P. and Wouters, J. (1988) Loss of phage resistance encoded by plasmid pSK112 in chemostat cultures of *Lactococcus lactis* ssp. *cremoris* SK110. *Biochimie* **70**, 451–456.

Stetter, K.O., Priess, H. and Delius, H. (1978) *Lactobacillus casei* phage PL–1: molecular properties and first transcription studies in vivo and in vitro. *Virology* **87**, 1–12.

Szybalski, W., Kim, S.C., Hasan, N., and Podhajska, A.J. (1991) Class-IIS restriction enzymes – a review. *Gene* **100**, 13–26.

Terzaghi, B.E. (1976) Morphologies and host sensitivities of lactic streptococcal phages from cheese factories. *N.Z.J. Dairy Sci. Technol.* **11**, 155–163.

Teuber, M. (1986) Final report of the achievements of the research programme on construction of phage resistant dairy starter cultures. In, *Biomolecular Engineering in the European Community*, E. Magnien (ed.) Martinus Nijhoff, Dordecht, The Netherlands, p. 539–547.

Teuber, M. and Lembke, J. (1983) the bacteriophages of lactic acid bacteria with emphasis on genetic aspects of group N lactic streptococci. *Antonie van Leeuwenhoek* **49**, 283–295.

Teuber, M. and Loof, M. (1987) Genetic characterization of lactic streptococcal bacteriophages. In *Streptococcal Genetics* (Ferretti, J.J. and Curtiss, R., eds), 250–258. American Society for Microbiology, Washington D.C., U.S.A., pp. 250–258.

Thunell, R.K., Sandine, W.E. and Bodyfelt, F.W. (1981) Phage insensitive, multiple-strain starter approach to Cheddar cheese making. *J. Dairy Sci.* **64**, 2270–2277.

Thunell, R.K., Sandine, W.E. and Bodyfelt, F.E. (1984) Defined strains and phage insensitive mutants for commercial manufacture of cottage cheese and cultured buttermilk. *J. Dairy Sci.* **67**, 1175–1180.

Tiiro, P. and Sarimo, S.S. (1987) Characterization of the genome of some *Streptococcus thermophilus* phages by restriction endonuclease mapping. *FEMS Microbiol. Rev.* **46**, 39 (abstract).

Timmons, P., Hurley, M., Drinan, F.D., Daly, C. and Cogan, T. (1988) Development and use of a defined strain starter system for Cheddar cheese. *J. Soc. Dairy Technol.* **41**, 49–53.

Tortorello, M.L., Chang, P.-K., Ledford, R.A. and Dunny, G.M. (1990) Plasmid associated antigens associated with resistance to phage adsorption in *Lactococcus lactis*. In Abstracts of 3rd International ASM Conference on Streptococcal Genetics, Miami Beach, FL. U.S.A. A/50.

Trautwetter, A., Ritzenthaler, P., Alatossava, T. and Mata-Gilsinger, M. (1986) Physical and genetic characterization of the genome of *Lactobacillus lactis* bacteriophage LL-H. *J. Virol.* **59**, 551–555.

Valyasevi, R., Sandine, W.E. and Geller, B.L. (1990) The bacteriophage kh receptor of *Lactococcus lactis* subsp. *cremoris* KH is the rhamnose of the extracellular wall polysaccharide. *Appl. Environ. Microbiol.* **56**, 1882–1889.

Valyasevi, R., Sandine, W.E. and geller, B.L. (1991) A membrane protein is required for bacteriophage c2 infection of *Lactococcus lactis* subsp. *lactis* C2. *J. Bacteriol.* **173**, 6095–6100.

van der Vossen, J.M.B.M., van der Lelie, D. and Venema, G. (1987) Isolation and characterization of *Streptococcus cremoris* Wg2 specific promoters. *Appl. Environ. Microbiol.* **53**, 2452–2457.

Vedamuthu, E.R. and Neville, J.M. (1986) Involvement of a plasmid in production of ropiness (mucoidness) in milk cultures by *Streptococcus cremoris* MS. *Appl. Environ. Microbiol.* **51**, 677–682.

Vedamuthu, E.R. and Neville, J.M. (1987) Phage resistance in *Streptococcus lactis* ssp. *diacetylactis* transconjugant SLA3.2501 and its derivatives. *J.Dairy Sci.* **70**, 225–229.

Vogensen, F.K. and Josephson, J. (1990) Comparison of R/M systems in *Lactococcus lactis. FEMS Microbiol. Rev.* **87**, P61.

Ward, A.C., Davidson, B.E., Hillier, A.J. and Powell, I.B. (1992) Conjugally-transferable phage resistance activities from *Lactococcus lactis* DRC1. *J. Dairy Sci.* **75**, 683–691.

Watanabe, K. and Takesue, S. (1972) The requirement for calcium in infection with *Lactobacillus* phages. *J. Gen. Virol.* **17**, 19–30.

Watanabe, K., Takesue, S. and Ishibashi, K. (1979) Adenosine triphosphage content in *Lactobacillus casei* and the blender-resistant phage-cell complex-forming ability of cells on infection with PL–1 phage. *J. Gen. Virol.* **42**, 27–36.

Watanabe, K., Takesue, S. and Ishibashi, K. (1980) DNA of phage PL-1 active against *Lactobacillus casei* ATCC 27092. *Agric. Biol. Chem.* **44**, 453–455.

Watanabe, K., Shirabe, M., Nakashima, Y. and Kakita, Y. (1991) The possible involvement of protein synthesis in the injection of PL–1 phage genome into its host, *Lactobacillus casei. J. Gen. Microbiol.* **137**, 2601–2603.

Wetzel, A., Neve, H., Gels, A. and Teuber, M. (1986) Transfer of plasmid-mediated phage resistance in lactic acid streptococci. *Chem. Mikrobiol. Technol. Lebensm.* **10**, 86–89.

Whitehead, H.R. and Cox, G.A. (1935) The occurrence of bacteriophage in cultures of lactic streptococci. *N.Z.J. Dairy Sci. Technol.* **16**, 319–320.

Yokokura, T. (1977) Phage receptor material in *Lactobacillus casei. J. Gen. Microbiol.* **100**, 139–145.

4 The proteolytic system of lactic acid bacteria

J. KOK and W.M. DE VOS

4.1 Introduction

Although there are pronounced differences in proteolytic capacity between the different species of lactic acid bacteria, numerous strains are known to contain proteolytic systems that allow them to grow on protein-rich substrates such as meat, vegetables and milk. There are two characteristics that differentiate these lactic acid bacteria from many other proteolytic microorganisms. First, lactic acid bacteria are fastidious organisms with multiple amino acid auxotrophies and as a consequence their growth is critically dependent on efficient systems for the degradation of proteins and the transport of amino acids and small peptides. Second, several lactic acid bacteria contain a proteolytic system that is highly specific and results in the production of unique peptides. In retrospect, it is likely that these characteristic proteolytic properties, together with the capacity to produce lactic acid from sugars, are the primary factors that have allowed the selection of certain strains of lactic acid bacteria as starter cultures for industrial fermentations. This certainly applies to the lactic acid bacteria that initiate fermentations in milk, a medium with a high content of lactose and α_{s1}– and β-casein, the major milk proteins.

As a consequence, the proteolytic system of dairy lactic acid bacteria, notably that of *Lactococcus* and *Lactobacillus* spp., has received considerable attention during the last decades (for some recent reviews, see Thomas and Pritchard, 1987; de Vos, 1987; Laan *et al.*, 1989; Kok and Venema, 1988; de Vos *et al.*, 1989; Kok, 1990; de Vos, 1990; Kok, 1991; Visser, 1993). Most initial work has focused on the detection, isolation and biochemical characterization of (partially) purified enzymes. These studies have identified a collection of proteolytic activities in the various species of lactic acid bacteria, including proteinases and other endopeptidases, aminopeptidases, proline-specific peptidases, oligopeptidases and carboxypeptidases. In addition, several systems that are involved in the transport of amino acids and peptides have been detected.

The cascade of proteolytic reactions that occur during growth of lactococci and lactobacilli in milk is initiated by the action of well-studied extracellular endoptidases, designated proteinases, that liberate peptides of variable size from casein. It is attractive to envisage that these peptides are the natural substrates for the other peptidases that could eventually convert them into smaller peptides and amino acids. However, thus far none of these peptidases have been unequivocally implicated in casein utilization. In addition, it now appears that most, if not all, of these peptidases have an intracellular location indicating that the proteolytic breakdown occurs after transport into the starter cells.

Initial studies of the peptidases of lactic acid bacteria have been frustrated mainly because these enzymes were not purified to homogeneity and a large number of different substrates were used in analysing these enzymes. However, recent studies have eliminated those problems and have allowed a detailed analysis of the various peptidases which has strengthened the view that the peptidase complex of lactic acid bacteria consists of only a relatively small number of enzymes (Thomas and Pritchard, 1987). The availability of highly purified peptidase preparations has enabled the determination of the N-terminal amino acid sequences of the enzymes and the production of specific antibodies against the peptidases. These data and tools have allowed the genetics of the proteolytic system to be unravelled.

In what follows, we have summarized the biochemical characterization of the individual components in the proteolytic systems that have been purified to homogeneity. In addition, the genetic analysis of the proteolytic system is described in considerable detail. This genetic approach has been instrumental in confirming and extending the biochemical characterization and now allows for elucidating the physiological role of proteinases, peptidases and transport systems during growth in milk and their contribution to the quality and flavour of fermented dairy products. In addition, the availability of the genes for the key enzymes in the proteolytic degradation has opened the possibilities of engineering starter bacteria used in industrial dairy fermentations and the first examples are discussed here. As most of the work to date has been done in strains of *Lactococcus lactis*, the emphasis will inevitably be on the enzymes of those important mesophilic cheese starters.

4.2 Proteinases

It has been well established that extracellular serine proteinases are the key enzymes in the proteolytic degradation in lactococci (Thomas and Pritchard, 1987; de Vos, 1987; Kok, 1990). This is best illustrated by the initial observation that proteinase-deficient *L. lactis* cells do not grow in milk and are not suitable as starter cultures for the production of cheese (Citti *et al.*, 1965; Stadhouders *et al.*, 1988; see below). Biochemical and genetic studies have shown that the proteinases characterized so far from lactococci and lactobacilli are highly related. Their production is dependent on the simultaneous expression of the *prtP* gene encoding the proteinase, and the *prtM* gene for a maturation protein (see Figure 4.1). The pivotal role of the extracellular proteinase PrtP in the proteolytic system and its unusual biosynthesis have stimulated considerable research efforts discussed here to understand its production, processing and caseinolytic specificity.

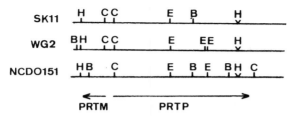

Figure 4.1 Sequence organization of sequenced proteinase genes of *L. lactis* strains SK11 (SK11) and Wg2 (WG2), and *Lactobacillus casei* NCDO 151 (NCDO151). Arrows show the direction of transcription of the two proteinase genes *prtP* and *prtM*. Restriction enzyme sites indicated are: B, *Bgl*II; C, *Cla*I; E, *Eco*RI; H, *Hind*III.

4.2.1 Biochemical characterization

The initial detection of proteolytic activities associated with the cell wall of lactococcal strains has opened the ways to purify and characterize these extra-cellular enzymes (Exterkate, 1976). A simple method involving incubation of washed cells in a calcium-free buffer has been used to isolate serine proteinases from lactic acid bacteria (Mills and Thomas, 1978; for reviews see Thomas and Pritchard, 1987, and Laan *et al.*, 1989). This procedure has allowed the purifi-cation and subsequent N-terminal sequencing of proteinases from *Lactococcus lactis* strains SK11, WG2, NCDO 763 and NCDO 1201, and a *Lactobacillus casei* strain NCDO 151, now known as *Lactobacillus paracasei* subsp. *paracasei* (Vos *et al.*, 1989; Kiwaki *et al.*, 1989; Laan and Konings, 1989; Nissen-Meyer and Sletten, 1991; Naes and Nissen-Meyer, 1992). Cell-envelope bound proteinases have also been detected in other lactobacilli but not purified in a form that allowed N-terminal sequencing. As a consequence, we limit ourselves here by referring to some recent reports describing the presence of cell wall bound-proteinases in *Lactobacillus plantarum* and *Lactobacillus delbrueckii* subsp. *bulgaricus* (El Soda *et al.*, 1986; Ezzat *et al.*, 1987; Moon *et al.*, 1988; Khalid and Marth, 1990a). The proteinase extracted from cell walls of *Lactobacillus delbrueckii* subsp. *bulgaricus* CNRZ 397 was found to consist of a 170 kDa monomer and showed partial sensitivity to inhibitors of serine and cysteine proteinases (Laloi *et al.*, 1991). In addition, a membrane-associated serine proteinase with a monomer mass of 180 kDa has recently been isolated from *Lactobacillus helveticus* NIZO L89 (Martin-Hernandez *et al.*, 1993). The latter two proteinases from thermophilic *Lactobacillus* spp. have similar biochemical properties as the intact form of the lactococcal proteinase PrtP (see below) and it is feasible that they belong to the same class of enzymes that will be discussed here in detail.

Extensive biochemical analyses of the proteinases released from *L. lactis* and *Lactobacillus paracasei* cells have shown that they have many properties in common (see Thomas and Pritchard, 1987; Laan *et al.*, 1989; Naes and Nissen-Meyer, 1992): (i) they can be classified as serine proteinases based on their sensitivity to phenylmethylsulfonyl fluoride (PMSF) or diisopropyl fluorophosphate (DFP); (ii) they have a size of approximately 110–150 kDa as estimated by sodium dodecyl sulfate–polyacrylamide gel electrophoresis (SDS–PAGE) but in the native state could be dimers or even larger multimers; (iii) their activity is dependent on calcium ions that in high concentrations are inhibitory; and (iv) they are sensitive to proteolysis resulting in the formation of a variety of different degradation products.

Genetic studies detailed below have shown that the primary translation products of the structural proteinase genes (*prtP*) have a size exceeding 200 kDa and are subject to N-terminal processing reactions leading to removal of the signal sequence during transport and the elimination of the pro-sequence (Figure 4.2). However, biochemical characterization of the proteinases and their degradation products have shown that also further N- and C-terminal degradation

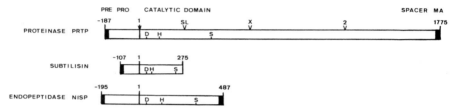

Figure 4.2 Comparison of the organization of the proteinase PrtP, subtilisin and endopeptidase NisP. For explanation see text. Positions of the catalytic residues are indicated as D (Asp), H (His) and S (Ser).

occurs (see Figure 4.2; Exterkate and de Veer, 1989; Laan and Konings, 1989; Nissen-Meyer and Sletten, 1991; Bruinenberg *et al.*, 1993a). The N-terminal degradation occurs preferentially in a surface loop at positions 205 to 216 and leads to truncated products that are inactive (site SL, Figure 4.2; Bruinenberg *et al.*, 1993a). In addition, C-terminal processing also occurs and yields degradation products that have unchanged N-terminal sequences. There are three lines of evidence that support the possibility that this further N-terminal and C-terminal processing is a consequence of autoproteolysis during the release of the proteinase from the cell envelope. First, proteinase release in *L. lactis* Wg2, monitored using sensitive immunological methods with monoclonal antibodies, could be prevented by treatment with PMSF and yielded a larger product of approximately 165 kDa (Laan and Konings, 1989). In addition, autoproteolysis of purified proteinase could be reduced by treatments that decrease the activity of the proteinase, such as the addition of calcium ions, PMSF or DFP, or reducing the temperature (Laan and Konings, 1990). Finally, it is well known that autoproteolysis of purified proteinase increases at elevated temperatures (Exterkate and de Veer, 1989; Coolbear *et al.*, 1992). Second, the formation of degradation products is not found in the active site mutant proteinases Asp30Ala and Ser433Ala from strains Wg2 or SK11, respectively, obtained by site-directed mutagenesis of the *prtP* gene (Haandrikman *et al.*, 1991; de Vos *et al.*, 1991). In addition, these mutant proteinases have a size that is expected for C-terminally unprocessed forms. Third, proteinase isolated from *L. lactis* strain H2 treated with lysozyme in the presence of calcium ions, showed a considerably larger size (180 kDa) than that isolated by the conventional release (Coolbear *et al.*, 1992). These results render alternative explanations for the processing unlikely and suggest that the proteinase is cleaved, probably in an intermolecular process, at a C-terminal site during the release. Recent experiments with C-terminally truncated proteinases have indicated that this autoproteolytic site should be located in between residues 1127 and 1272 (site 2; Bruinenberg *et al.*, 1993b). Further C-terminal degradation also occurs and another site has been located at position 623–624 (site X; Bruinenberg *et al.*, unpublished results). This model is attractive since it accommodates the results from the genetic analysis indicating the presence of a C-terminally located membrane anchor and spacer region that are likely to be involved in the interaction with the cell membrane and cell wall, respectively (see Figure 4.2 and below). The model assumes that a conformational change of the proteinase induces the autoproteolysis. This may be a direct effect of the removal of calcium ions, as suggested previously (Kok *et al.*, 1988b; Laan and Konings, 1990). However, indirect effects on the cell wall that affect the interaction with the proteinase can not be excluded. Moreover, other conformational changes also induce autoproteolysis since the C-terminal processing at site 2 is also found with various C-terminally truncated mutant proteinases that can be isolated from the calcium-containing medium without release treatment (de Vos *et al.*, 1989b; de Vos *et al.*, 1991). These results also indicate that the C-terminally located site 2 is very sensitive to autoproteolysis and is rapidly cleaved.

4.2.2 *Caseinolytic specificity and classification of proteinases*

The substrate specificity of proteinases from dairy lactic acid bacteria is an important factor in determining the nature of the degradation products generated from milk proteins. Ultimately, these products determine the growth of lactic acid bacteria in milk and contribute to the generation of flavour in fermented milk products. In addition, substrate specificity has developed in to a useful tool to classify lactococcal proteinases. Therefore, considerable attention has been focused in recent years on the substrate specificity of the *L. lactis* proteinases. Only a limited number of substrates are degraded by these proteinases and only recently synthetic chromogenic substrates have been found that are hydrolyzed (Exterkate, 1990). Therefore, most studies have been performed with the major milk protein casein as substrate.

Based on an electrophoretic analysis of the casein breakdown products, lactococcal proteinases were classified into two main groups, the PI-type and PIII-type, while also a mixed PI/PIII type specificity was found (Visser *et al.*,

1986). The PI-type, represented by the proteinase of *L. lactis* HP preferentially degrades β-casein but not or to a very limited extent α_{s1}- and κ-casein. In contrast, the PIII-type proteinase, found in the related *L. lactis* strains AM1 and SK11, degraded α_{s1}-, β- and κ-casein and differed in the breakdown of β-casein from the PI-type proteinase. The mixed PI/PIII type specificity was found in various strains including *L. lactis* strains NCDO 763 and UC317 (see Visser, 1993 for a recent review). The distinctive substrate specificities of the PI- and PIII-type proteinases were confirmed with smaller substrates, such as the chromogenic substrates and the peptide α_{s1}-CN(f1-23), that is formed from α_{s1} casein by the action of chymosin (Exterkate, 1990; Exterkate *et al.*, 1991). Recently these specificities were used to refine the classification of the lactococcal proteinases (Exterkate *et al.*, 1993). The preferential degradation of specific peptide bonds may be of practical importance since it has been found that less bitterness is generated from casein by the PIII-type proteinase than by the PI-type proteinase (Visser *et al.*, 1983).

Using representatives of PI- and PIII- type proteinases and the mixed specificity type, the peptide bonds cleaved in α_{s1} and β-casein have been determined (Visser *et al.*, 1991; Reid *et al.*, 1991a,b; Monnet *et al.*, 1992; Visser *et al.*, 1992). An important result from these degradation studies is that the lactococcal proteinases are able to liberate small peptides from caseins, eliminating the need for further extensive extracellular protein hydrolysis. Both identical and unique bonds were found to be hydrolyzed by the different classes of proteinases. A preference is found for large residues at the P4 and P1 positions and a Pro residue at the P2 position. In addition, the PIII-type proteinase appears to have a preference for a negative charge located N-terminally to the bond to be cleaved while the PI-type proteinase has a preference for a positive charge in this region. Interestingly, the deduced primary sequence of the PI-type differs from that of the PIII-type proteinases in two charge substitutions in the substrate binding region (Lys138Thr and Asn166Asp; see below). Homology modelling of the *L. lactis* proteinase based on the known three-dimensional structure of various related subtilases was used to explain these differences in specificity and suggested that electrostatic interactions in the substrate binding region can contribute to binding and hence caseinolytic specificity (Siezen, 1993; Siezen *et al.*, 1993; Exterkate *et al.*, 1993).

Various attempts have been made to classify proteinases. Polyclonal and monoclonal antibodies were used to differentiate lactococcal proteinases and provided a classification that was confirmed by the one based on caseinolytic specificity (Hugenholtz *et al.*, 1984; Laan *et al.*, 1988). The caseinolytic specificity seems a constant property of proteinases and did not depend significantly on the way the proteinase was isolated, its size and its location (Coolbear *et al.*, 1992; de Vos *et al.*, 1989a; Haandrikman *et al.*, 1991a). Therefore, classification based on substrate specificity provides a rapid and reliable way to differentiate the various lactococcal proteinases (see Exterkate *et al.*, 1993). Recently, genetic methods based on restriction length polymorphism have been proposed to be useful in the differentiation of lactococcal proteinases (Laan *et al.*, 1992). These have been extended in a recent method that is based on PCR-amplification of regions determining substrate specificity followed by sequence analysis of relevant regions (Bruinenberg and de Vos, unpublished results). This approach allowed the identification of conserved proteinase genes and analysis of sequences involved in substrate binding in various dairy lactobacilli, including *Lactobacillus helveticus* and *Lactobacillus delbrueckii* subsp. *bulgaricus*.

4.2.3 Cloning and expression of proteinase genes

Curing, transfer and hybridization studies have shown that *L. lactis* strains contain conserved genes for proteinase production that are located on plasmids ranging in size from 14 to more than 100 kb (for reviews see McKay, 1983; Kok and Venema, 1988). An extrachromosomal location is not required for proteolytic activity since a single chromosomal copy of the *prtP* and *prtM* genes has been found in *L. lactis* strain BC101 (Nissen-Meyer *et al.*, 1992). In addition, proteinase genes can be introduced into the chromosome by transduction (McKay and Baldwin, 1978; Van Alen-Boerrigter and de Vos, unpublished results; see chapter 1) and homologous recombination (Leenhouts *et al.*, 1991; see chapter 2 and below).

Several of the plasmid-encoded lactococcal proteinase genes have been cloned and expressed in *L. lactis* and some heterologous hosts (Table 4.1; for reviews see de Vos, 1986a; de Vos, 1987; Kok and Venema, 1988, and Kok, 1990). In addition, the chromosomally located proteinase genes of *Lactobacillus paracasei* have recently been cloned and expressed in *Lactobacillus plantarum* (Holck and Naes, 1992).

Most genetic attention has been focused on the proteinases from the *L. lactis* strains Wg2 and SK11, that differ in their caseinolytic specificity and resemble

Table 4.1 Proteinase genes cloned and expressed from lactic acid bacteria.

Strain	Host	Activity	Reference
L. lactis Wg2	*B. subtilis*	Yes	Kok *et al.*, 1985
	L. lactis	Yes	
L. lactis SK11	*E. coli*	No	de Vos, 1986
	L. lactis	Yes	de Vos *et al.*, 1989
L. lactis NCDO 712	*B. subtilis*	P	Gasson *et al.*, 1987
	L. lactis	P	
L. lactis NCDO 763	*L. lactis*	Yes	Kiwaki *et al.*, 1989
L. lactis H1	*L. lactis*	Yes	Yu *et al.*, 1991
L. lactis UC317	*L. lactis*	Yes	Law *et al.*, 1992
L. lactis E8	*L. lactis*	Yes	A. Haandrikman, unpublished results
Lb. paracasei NCDO 151	*L. lactis*	No	Holck and Naes, 1992
	Lb. plantarum	Yes	

Lb. designates *Lactobacillus*; P indicates partial expression.

the PI-type (Wg2) or belong to the PIII-type (SK11) (see above). From these lactococcal strains the first proteinase genes were cloned and characterized in the authors' laboratories. The strategies developed for the cloning and subsequent analysis of these proteinases genes have been followed in various later genetic studies of proteinases. In addition, the conclusions from these first analyses seem to apply to all other proteinase genes and therefore will be summarized here. In both strains proteinase plasmids have been identified, the 26 kb pWV05 in strain Wg2 and the 78 kb pSK111 in strain SK112, an isolate from the cheese starter SK11 (Otto *et al.*, 1981; de Vos and Davies, 1984). Intermediate hosts were used in cloning of the Wg2 and SK11 proteinase to compensate for the lack of efficient host–vector systems for lactococci that have been developed since (see chapter 2). Instability of proteinase genes cloned on plasmid vectors was observed in *Escherichia coli* probably to a lethal effect in this host (Kok *et al.*, 1985; de Vos, 1986b; de Vos *et al.*, 1989). This allowed the localization of the proteinase genes on plasmid pWV05 of strain Wg2 followed by their cloning in *Bacillus subtilis* (Kok *et al.*, 1985). However, when inserted in bacteriophage lambda vectors the SK11 proteinase genes were stably maintained in *E. coli* and production of an inactive proteinase could be detected with immunological methods (de Vos, 1986b). In this way the SK11 proteinase genes were located on plasmid pSK111 and subsequently subcloned in *E. coli* (de Vos, 1986b; de Vos *et al.*, 1989). Production of an active Wg2 lactococcal proteinase was observed in *Bacillus subtilis*, indicating that this Gram-positive host allows the appropriate processing reactions (Kok *et al.*, 1985). Nevertheless, subsequent manipulations in *B. subtilis* were facilitated when the proteinase gene was first inactivated by insertion of an antibiotic resistance marker (Leenhouts *et al.*, 1990). Following subcloning in *B. subtilis* or *E. coli*, the Wg2 and SK11 proteinase genes were cloned in proteinase-deficient *L. lactis* strains using broad host range vectors of the pSH71/pWV01 family of replicons (see chapter 2). Successful expression of the proteinase genes was obtained and the resulting lactococcal strains showed rapid growth in milk, caseinolytic activity and the expected phenotype on citrated milk agar, and the production of proteinase as detected by immunological and biochemical methods (Kok *et al.*, 1985; de Vos, 1986b; Kok *et al.*, 1988b; de Vos *et al.*, 1989). Surprisingly, the cloned genes specified proteinases that were found to be completely secreted (de Vos, 1986b; Haandrikman *et al.*, 1989). In later studies this was found to be due to the fact that the original clones lacked the 3' end of the proteinase genes encoding the topogenic C-terminal sequences (see Figure 4.2 and below). The functionality of those sequences was shown by the cloning of the entire SK11 and Wg2 proteinase genes that produced cell envelope-located expression products (de Vos *et al.*, 1989b; Haandrikman *et al.*, 1991a). Further studies showed that considerable regions at the C-terminus of the lactococcal proteinase could be deleted (Kok *et al.*, 1988b). However, deletions of more 500 C-terminal amino acids increasingly reduced the stability of the SK11 proteinase and hardly any proteolytic activity could be noted with strains producing a proteinase carrying a C-terminal deletion of 838 amino acids (Bruinenberg *et al.*, unpublished results). This also explains why only limited expression was obtained with the cloned proteinase gene from *L. lactis* NCDO 712 that suffered from similar C-terminal truncations (Gasson *et al.*, 1987; see Table 4.1).

4.2.4 Organization of prtM and prtP genes

The cloned lactococcal and lactobacillal proteinase genes have been located in regions that show considerable similarities in their restriction maps (Figure 4.1). At present, complete nucleotide sequences have been determined for proteinase genes from the *L. lactis* strains Wg2 (Kok *et al.*, 1988a; Haandrikman *et al.*, 1989), SK11 (Vos *et al.*, 1989a,b), NCDO 763 (Kiwaki *et al.*, 1989), and the *Lactobacillus* casei strain NCDO 151 (Holck and Naes, 1992). Interestingly, the proteinases from strains Wg2, SK11 and NCDO 763 represent the main types of caseinolytic specificities found in the lactococci, i.e. the PI-, PIII-, and the mixed type PI/PIII type specificity, respectively (see above). All sequenced regions show a similar organization and contain two divergently transcribed genes (Figure 4.1). The smaller open reading frame is the *prtM* gene for a maturation protein and the larger one is *prtP*, the structural gene for the proteinase, that will be discussed below.

The *prtP* and *prtM* genes are separated by a 0.3 kb AT-rich promoter region. In the case of the Wg2 and SK11 genes the *prtM* and *prtP* promoters overlap and transcription initiation occurs at almost oppositely located nucleotides in a region with an unusual rotational symmetry that is most pronounced in the case of the SK11 gene (Vos *et al.*, 1992; see also chapter 2). The production of proteinase appears to be dependent on the growth medium (de Vos *et al.*, 1991; Bruinenberg *et al.*, 1992; Laan *et al.*, 1993). Using transcriptional fusions and RNA studies it was recently shown that SK11 proteinase expression is controlled at the transcriptional level (Marugg, Bruinenberg and de Vos, unpublished results). Notably, the *Lactobacillus paracasei* proteinase promoter region contains a deletion of approximately 30 base pairs that removes part of the *prtM* promoter and the rotational symmetry. The consequences of this change for the regulation of proteinase production in lactobacilli have not yet been studied but the loss of the *prtM* promoter at the expected position has been suggested to cause the inability to produce a functional proteinase in *L. lactis* (Table 4.1; Holck and Naes, 1992).

The *prtP* genes are followed by sequences with the features of a rho-independent terminator, that are conserved in the *L. lactis* genes. However, recent studies have shown that the lactococcal sequence apparently is not a very strong terminator of transcription in *L. lactis* (van der Vossen *et al.*, 1992). Further downstream of these putative terminators sequences are present that resemble the sequence of *iso*-ISS*1* (see chapter 1). In strains Wg2 and NCDO 763 these *iso*-ISS*1* elements seem to be incomplete while that in SK11 is a complete copy (Haandrikman *et al.*, 1990; de Vos *et al.*, 1991). In addition, downstream of the lactococcal *prtM* genes complete copies of *iso*-ISS*1* elements were found that, in the case of strain Wg2 was flanked at one side only by part of another copy of an *iso*-ISS*1* element (Haandrikman *et al.*, 1990). Although these IS-elements have been implicated in the dissemination of the proteinase genes in *L. lactis*, it is remarkable that they are not conserved at corresponding positions near the *Lactobacillus paracasei* proteinase genes (Holck and Naes, 1992). Since the *prtM* and *prtP* genes of *L. lactis* strain SK11 are thus flanked by complete copies of the *iso*-ISS*1* elements in direct orientation, this suggests that the SK11 proteinase genes are part of a composite transposon designated Tn*5277* (de Vos *et al.*, 1991).

4.2.5 Structural characteristics of prtP

The nucleotide sequences of the *prtP* genes from *L. lactis* and *Lactobacillus paracasei* show a remarkably high degree of similarity that exceeds 95%. As a consequence, the deduced amino acid sequences of the proteinases are highly conserved and those from strains Wg2, NCDO 763 and NCDO 151 contain 1902 amino acids, while that of strain SK11 is somewhat larger due a 60-amino acids duplication at the C-terminal end of the protein (see Figure 4.2; Kok *et al.*, 1988a; Kiwaki *et al.*, 1989; Holck and Naes, 1992; Vos *et al.*, 1989a). The highest similarity is observed between the sequences of the *L. lactis* strains Wg2 and NCDO 763, that differ in only 18 amino acid residues, while the most dissimilar sequences, that of *L. lactis* SK11 and *Lactobacillus casei* NCDO 151, only differ in 77 amino acid residues. It should be noted that the exact start of the *prtP* gene has not been determined unambiguously because of the N-terminal processing of the proteinase (see below). However, several gene fusions have been made with the SK11 *prtP* gene and suggest strongly that the translational initiation is at the second ATG of the open reading frame (de Vos and Simons, 1988; de Vos *et al.*, 1989a). In addition, this initiation start is the only one conserved in the *Lactobacillus paracasei prtP* gene sequence (Holck and Naes, 1992).

The N-terminal amino acid sequences from the proteinases characterized at the DNA level are all identical and start after 188 residues from this initiator methionine of the predicted amino acid sequence. This position is taken as position 1 of the mature proteinase in the numbering used here and in many other studies (Figure 4.2; see Siezen *et al.*, 1991). Inspection of the N-terminal region of the primary translation products of the *prtP* genes indicates that this region is identical in all proteinases, has the characteristics of a sec-dependent signal sequence, and contains a consensus signal peptidase I cleavage site preceding Ala residue at position –154. A number of studies have shown that this signal peptide is sufficient to direct the secretion of heterologous proteins in *L. lactis* (see de Vos *et al.*, 1989a and chapter 2 for a detailed description). Thus these proteinases of lactic acid bacteria are synthesized as pre-pro-proteins, with a signal sequence of 33 residues and a 154-residue pro-sequence (see Figure 6.1).

The C-terminus of the lactococcal and lactobacillal proteinases appear to contain a conserved consensus membrane anchor sequence that is thought to act as a stop-transfer sequence (Vos *et al.*, 1989a; Haandrikman *et al.*, 1989). This membrane anchor is conserved in all surface proteins of Gram-positive bacteria and consists of a conserved hexapeptide with the sequence of Leu-Pro-X-Thr-Gly-Glu preceding a hydrophobic, membrane-spanning segment of about 18 residues followed by a 7-amino-acid charged tail (Fischetti *et al.*, 1990). The membrane anchor is preceded by the so called spacer domain that is thought to cross the cell wall and expose the catalytic domain of the proteinase to the outside of the lactic acid bacteria (Figure 4.2). It is possible that this spacer region also shows interaction with components of the cell wall, as is the case with other surface proteins in Gram-positive organisms, since protein engineering experiments have indicated that the membrane anchor *per se* is not sufficient to direct surface location of foreign proteins (van Alen-Boerrigter and de Vos, unpublished results).

The N-terminal part of the mature proteinases of approximately 500 residues, designated the catalytic domain, shows significant similarity to serine proteinases of the subtilisin family, recently designated subtilases (Figure 4.2; Siezen *et al.*, 1991). This includes the active site residues Asp30, His94 and Ser433, and the essential Asn196 of the oxyanion hole. Experimental evidence for the participation of the Asp and Ser residues in the catalytic activity has been provided by their conversion into Asn and Ala residues in the Wg2 and SK11 proteinases, respectively (Haandrikman *et al.*, 1991; de Vos *et al.*, 1991). In addition, the substrate binding regions identified in the subtilases are conserved and their functionality in the interaction with the substrate casein has been demonstrated in various protein engineering studies (Siezen *et al.*, 1991; Siezen *et al.*, 1993; Bruinenberg *et al.*, 1993a, b). The region of approximately 1000 residues in between the catalytic domain and the C-terminal spacer region (Figure 4.2) is characteristic for the proteinases of lactic acid bacteria and not found in other subtilases (Siezen *et al.*, 1991). The function of this region is not entirely known although the domain around residues 747 and 748 has been shown to be involved in substrate binding (Vos *et al.*, 1991).

4.2.6 *Proteinase maturation by prtM*

Upstream of and in opposite orientation to the proteinase gene *prtP*, a gene is present that is completely conserved in the three lactococcal *prt* regions and that of *Lactobacillus paracasei* (Haandrikman *et al.*, 1989; Vos *et al.*, 1989b; Kiwaki *et al.*, 1989; Holck and Naes, 1992). This gene, designated *prtM* because of its role in maturation, could encode a protein of 33 kDa. The extreme N-terminus of the primary translation product of *prtM* bears the characteristic features of a signal peptide of lipoproteins (Haandrikman *et al.*, 1989; Vos *et al.*, 1989b; see chapter 2). A consensus cleavage site of signal peptides of lipoproteins is located between Leu21-Cys22 of pre-PrtM. Immunogold labelling experiments located

PrtM in the cell envelope of lactococci, while cell fractionation studies showed that the protein was enriched in the membrane fraction (Haandrikman *et al.*, 1991b). When cells carrying *prtM* were grown in the presence of ^{3}H-labelled palmitic acid, a 32 kDa ^{3}H-labelled protein specifically precipitated with the PrtM-directed antibodies, proving that PrtM is indeed anchored to the membrane via lipid modification (Haandrikman *et al.*, 1991b).

Deletion analyses have shown that *prtM* is essential for proteolytic activity (Haandrikman *et al.*, 1989; Vos *et al.*, 1989b). Cells carrying the intact *prtP* gene but lacking *prtM* produce a proteinase that is inactive and larger than that made in the presence of *prtM*. Apparently, PrtM is involved in a proteolytic step that leads to proteinase activation. PrtM can exert its effect *in trans* (Vos *et al.*, 1989b). In the original clones, PrtM lacked four C-terminal amino acids but this did not interfere with its role in proteinase activation. Mutations destroying the active sites Asp or Ser residues of the proteinase were made by site-directed mutagenesis techniques (see above). The resulting Asp30Asn and Ser433Ala proteinases, produced in the presence of *prtM*, had a molecular mass higher than that of the active wild-type proteinase produced in the absence of *prtM* (Haandrikman *et al.*, 1991b; de Vos *et al.*, 1991). Although the exact role of PrtM in the maturation process requires further *in vivo* and *in vitro* studies, these results suggest that PrtM induces the proproteinase to adopt a conformation in which it is susceptible to autocatalytic proteolytic cleavage of the pro region at position 1 (Figure 4.2).

Recently, a gene (*prsA*) involved in a late step in protein secretion in *B. subtilis* was identified and sequenced (Kontinen and Sarvas, 1988; Kontinen *et al.*, 1991). The *prsA* gene product and PrtM share 30% identical amino acids over their entire length (292 and 299 residues, respectively). PrsA also contains the consensus lipobox of prelipoproteins. Mutations in *prsA* have a pleiotropic effect and decrease the secretion of several exoproteins in strains hyperpro-ducing α-amylase. Based on the similarity with PrtM, the authors propose that PrsA affects the conformation or folding of secreted protein(s). PrsA, in this view, would be needed to fold α-amylase into a conformation that can be released from the cytoplasmic membrane or the cell wall. Both PrtM and PrsA might, thus, represent a new class of chaperones that operate extracellularly. An interesting difference between the PrtM- and PrsA-induced effects on their respective extracellular target proteins is the observation that α-amylase secretion in the absence of PrsA is reduced, while proteinase secretion in the absence of PrtM is complete, provided that the proteinase lacks the C-terminal membrane anchor. If the anchor is present, the inactive proteinase is attached to the cell envelope and can not be released by autoproteolysis in a Ca^{2+}- free buffer (Vos *et al.*, 1989b; Haandrikman *et al.*, 1989).

4.3 Endopeptidases

In addition to the general-purpose cell envelope-located proteinase PrtP, various other endopeptidases have been found in lactococci and studied at the biochemical and genetic level. Many of these have a very narrow substrate specificity and are likely to have an intracellular location except for the endopeptidase NisP, a dedicated leader peptidase.

4.3.1 Endopeptidase NisP

Recently genetic and biochemical evidence has been provided for the presence of another cell envelope-located proteolytic enzyme, the endopeptidase NisP (see Table 4.2; van der Meer *et al.*, 1993). This enzyme is involved in the biosynthesis of nisin, an antimicrobial peptide produced by a limited number of *L. lactis* strains (see chapter 5). In *L. lactis* NIZO R5 nisin production is encoded by the conjugative transposon Tn*5276* (Rauch and de Vos, 1992). The approximately 10 kb nisin gene cluster of this strain has been characterized and found to contain the *nisABTCIPR* genes (Kuipers *et al.*, 1993; van der Meer *et al.*, 1993). The penultimate gene, *nisP*, appeared to encode a serine proteinase of the subtilase family with very similar characteristics as the proteinase PrtP (see Figure 4.2). The deduced 682-residue NisP product is probably produced as a pre-pro-enzyme and contains a C-terminal membrane anchor that is similar to that of PrtP. This is compatible with the fact that NisP is found at the outside of lactococcal cells and is probably associated with the membrane fraction. Gene deletion studies showed that in the absence of a functional *nisP* gene, a fully modified, inactive precursor of nisin is secreted to which the leader peptide is still attached. This indicates that NisP is a dedicated peptidase that is involved in proteolytic activation of nisin. Presently, the contribution of NisP in the proteolytic degradation of casein is not known. The specificity of NisP is probably very limited as deduced from modelling studies of its active site that predict a requirement for a positive charge immediately preceding the scissile bond (Siezen, 1993). In addition, NisP is dispensable for growth in milk since most starter strains are non-nisin producers and do not contain the *nisP* gene.

4.3.2 Intracellular endopeptidases

Various endopeptidases have recently been purified to homogeneity from several strains of *L. lactis* (see Table 4.2). Two different endopeptidases, LEP-I and LEP-II, were isolated from a cell-free extract of *L. lactis* H61 (Yan *et al.*, 1987a and 1987b). DEAE-Sephacel column chromatography using a linear NaCl concentration gradient resolved α_{s1}-CN(f1-23) hydrolyzing activity into two peaks. LEP-I eluted at 0.25 M NaCl and was further purified by chromato-focussing and gel filtration (Yan *et al.*, 1987a). The enzyme was purified

Table 4.2 Well-characterized lactococcal proteinases and peptidases.

Enzyme	Pure	Monomer Mw × 10³	Class	Substrate	nt seq	Leader peptide	Reference
PrtP	Y	200	Serine	casein	Y	+	(a)
NisP		54	Serine	nisin precursor	Y	+	(b)
PepA	Y	43	Metallo	Glu/Asp-pNA	Y	−	(c)
PepC	Y	50	Thiol	Leu/Lys-pNA	Y	−	(d)
PepN	Y	95	Metallo	Leu/Lys-pNA	Y	−	(e)
PepXP	Y	90	Serine	X-Pro-pNA	Y	-	(f)
PCP		25ª	Serine	PyroGlu-pNA	Y	−	(g)
dipeptidase	Y	49	Metallo	Leu-Leu			(h)
PepT	Y	52	Metallo	tripeptides	Y	−	(i)
prolidase	Y	43	Metallo	X-Pro			(j)
imino peptidase	Y	50	Metallo	Pro-X-(Y)			(k)
LEPI	Y	98	Metallo	αs1-CN (f1-23)			(l)
LEPII	Y	40	Metallo	αs1-CN (f1-23)			(m)
PepO/NOP	Y	70	Neutral	αs1-CN (f1-23)	Y	−	(n)

ª deduced from the nucleotide sequence of the gene for PCP

Y indicates purified or determined; + and − indicate present and absent, respectively

(a) Kok *et al.*, 1988; Vos *et al.*, 1989a; Kiwaki *et al.* 1989

(b) van der Meer *et al.*, 1993

(c) Exterkate and de Veer, 1987; Gasson, personal communication

(d) Neviani *et al.*, 1989; Chapot-Chartier *et al.*, 1992

(e) Tan and Konings, 1990; van Alen-Boerrigter *et al.*, 1991; Tan *et al.*, 1992; Strøman, 1992

(f) Zevaco *et al.*, 1990; Mayo *et al.*, 1991; Nardi *et al.*, 1991

(g) Baankreis, 1992; Exterkate, 1977; Haandrikman, personal communication

(h) Hwang *et al.*, 1982; van Boven *et al.*, 1988

(i) Bosman *et al.*, 1990; Mierau, personal communication

(j) Kaminogawa *et al.*, 1984

(k) Baankreis and Exterkate, 1991

(l) Yan *et al.*, 1987a

(m) Yan *et al.*, 1987b

(n) Tan *et al.*, 1991; Baankreis, 1992; Mierau *et al.*, 1993

134-fold with about 12% recovery of total activity. LEP-I has an apparent molecular weight of 98 000 as estimated by Sephacryl S-300 gel filtration. A single protein band of 98 kDa was also observed in SDS-PAGE, indicating that LEP-I is active as a monomer. Maximum activity of the enzyme is at pH 7.0 to 7.5 and at 40°C. LEP-I is not affected by inhibitors of serine and thiol proteases but is inhibited by EDTA and, 1,10-phenantroline. α_{s1}-CN(f1-23) and α_{s1}-CN(f91-100) are hydrolyzed only at the Glu-Asn bond present in both peptide fragments but larger peptides, encompassing either of these two peptide fragments, and α_{s1}-casein itself, are not hydrolyzed. A number of low molecular weight peptide hormones are also substrates for LEP-I and are cleaved at peptide bonds other than the Glu-Asn bond. Apparently, both the amino acids forming the susceptible peptide bond and the spatial conformation of the substrate determine whether it will be cleaved by LEP-I.

The α_{s1}-CN(f1-23) hydrolyzing activity in the extract of strain H61 that eluted from the DEAE-Sephacel column at 0.35 M NaCl was designated LEP-II and was further subjected to hydroxyapatite- and Sephacryl S-300 chromatography (Yan *et al.*, 1987b). LEP II has maximum activity at pH 6.0 and 37°C and is a metalloendopeptidase. It has a broader substrate specificity than LEP I and hydrolyzes various small polypeptides (up to a size of 3.5 kDa) at peptide bonds involving the amino groups of hydrophobic amino acids. LEP-II cleaves peptide bonds not attacked by LEP-I in hormone peptides such as metenkephalin, bradykinin, glucagon and oxidized insulin B chain. Milk proteins are not hydrolyzed and α_{s1}-CN(1-23) and α_{s1}-CN(91-100) are only if they are not part of a larger αs1 peptide fragment. LEP-II is considered to be a dimer of two identical subunits of 40 kDa.

Recently, an endopeptidase was partially purified from an intracellular crude extract of *L. lactis* NCDO 763. In this case, a mutant which lacked the cell envelope-bound proteinase was used to prevent contamination of the endopeptidase fraction with this enzyme (Muset *et al.*, 1989). After ion-exchange chromatography and gel filtration an extract was obtained that had high activity towards oxidized insulin B chain but still contained 3 protein bands as judged by SDS–PAGE. Attempts to further purify the enzyme were unsuccessful due to severe loss of enzyme activity. The enzyme had a molecular weight of 93 000 by gel filtration, was not inhibited by serine- or thiol enzyme inhibitors but was affected by EDTA. The enzyme only slowly degraded β-casein but readily cleaved oxidized insulin B chain at its optimum pH and temperature of 7.5 and 45°C. It has a specificity similar to other intracellular metalloproteinases, isolated from *L. lactis* and *Streptococcus thermophilus* (Yan *et al.*, 1987b; Desmazeaud, 1974; Desmazeaud and Zevaco, 1976).

4.3.3 Endopeptidase pepO.

An endopeptidase from a crude cell extract of *L. lactis* Wg2 was purified to homogeneity by a procedure involving chromatography on diethyl–aminoethane–Sephacel, phenyl–Sepharose, and hydroxylapatite, an FPLC anion-exchange step and, ultimately, a phenyl–Superose hydrophobic interaction column (Tan *et al.*, 1991). Metenkephalin (Tyr-Gly-Gly-Phe-Met) degradation was examined to follow the purification of the enzyme. During the purification procedure aminopeptidase, tripeptidase and dipeptidase were successively removed and the last purification step yielded a single protein that hydrolyzed metenkephalin between the Gly–Phe bond only. A 670-fold purification (17.4% yield) was obtained. The molecular weight was 70 000 by gel filtration and SDS–PAGE, and temperature and pH optimum 30 to 38°C and 6.0 to 6.5, respectively. Enzyme inhibition by EDTA and 1,10-phenanthroline could be restored by Co^{2+} ions only. The endopeptidase is also inhibited by PMSF, β-mercaptoethanol, and Cu^{2+} and Zn^{2+} ions. The endopeptidase of strain Wg2 has a broad substrate specificity, degrading several peptide hormones and the β-casein fragments β-CN(f184-202) and β-CN(f203-209), but not α-, β-, and κ-casein, or substrates smaller than 5 amino acid residues. The 70 kDa endopeptidase from strain Wg2 is most probably identical to the 70 kDa endopeptidase NOP purified from *L. lactis* HP (Baankreis, 1992).

Recently, the structural gene of the endopeptidase described in *L. lactis* Wg2 has been cloned and sequenced from the genomic library of *L. lactis* P8-2-47 (Mierau *et al.*, 1993). A phage lambda bank of total DNA of strain P8-2-47 was screened with polyclonal antibodies raised against the purified endopeptidase of Wg2 and several positive plaques were detected. Two of these recombinant phages expressed a protein of the expected size (70 kDa) as judged from Western hybridization using the endopeptidase antiserum. The antiserum also contained antibodies against a protein of approximately 40 kDa that copurified with the endopeptidase until the penultimate step (Tan *et al.*, 1991). All other positive phages expressed this contaminating protein. The phages expressing endopeptidase carried DNA inserts of approximately 14 kb that largely overlapped. A synthetic oligonucleotide based on the N-terminal amino acid sequence of purified endopeptidase was used to locate the structural gene for the endopeptidase, which was then subcloned in *E. coli* and sequenced. An ORF of 1881 bp (627 codons) was identified that could encode a protein of 71 527 molecular weight. The N-terminus of the deduced protein matched that of the purified endopeptidase except for the initiator methionine and the gene was termed *pepO*. The endopeptidase has similarity to mammalian neutral endopeptidase (enkephalinase; EC 3.4.24.11), but not to any bacterial enzyme. The overall similarity with human NEP is 37% while a stretch of 42 amino acids showed 72% identity. This segment in human NEP contains the motif His-Glu-X-X-Glu characteristic for zinc-dependent metalloproteinases and peptidases, as well as the putative active site of the enzyme. Assuming that PepO and NOP are the same enzymes, these protein similarities could explain the inhibition by phosphoramidon (a transition state analogue) and thiorphan (which ligates the active site zinc atom with its thiol group) of NOP (Baankreis, 1992). The lactococcal endopeptidase does not contain an N-terminally processed signal sequence, nor does it contain an obvious transmembrane sequence. The former observation agrees with the intracellular location of the enzyme (see below). No promoter was apparent immediately upstream of *pepO*. Instead, *pepO* was preceded by another ORF with the same orientation. This ORF turned out to be the gene for a binding protein of the oligopeptide permease system of *Lactococcus*, *oppA* (see below). The *pepO* gene is most likely transcribed from a promoter identified upstream of *oppA* (S. Tynkkynen, personal communication).

A mutant strain of *L. lactis* was constructed in which *pepO* was disrupted. As this strain was able to grow normally in milk with good acid production, PepO is, apparently, not essential for growth of lactococci in milk (Mierau *et al.*, 1993).

4.4 General aminopeptidases

A number of aminopeptidases have been purified to homogeneity from strains of *L. lactis* and from different lactobacilli (Geis *et al.*, 1986; Machuga and Ives, 1984; Eggimann and Bachmann, 1980; Desmazeaud and Zevaco, 1979; Neviani *et al.*, 1989; Tan and Konings, 1990; Atlan *et al.*, 1989; Baankreis, 1992; see Table 4.2). Four different aminopeptidases have been identified and characterized on the biochemical and genetic level and include aminopeptidase A, pyrrolidonyl carboxylyl peptidase, aminopeptidase N, and aminopeptidase C.

While aminopeptidase A and pyrrolidonyl carboxylyl peptidase have a relatively narrow substrate specificity, the other aminopeptidases isolated so far have been a broad substrate specificity and degrade amino acyl-*para*-nitroanilides, and certain di-, tri-, and tetrapeptides. Apparently, general aminopeptidases are common in lactic acid bacteria. All general aminopeptidases are inhibited by sulfhydryl-blocking reagents and, apart from the one isolated from *L. lactis* AM2, are metalloenzymes as they are inhibited by EDTA. The general aminopeptidases have neither carboxypeptidase nor endopeptidase activity.

4.4.1 Aminopeptidase A (PepA)

A reportedly membrane-bound aminopeptidase with narrow substrate specificity was purified from *L. lactis* strain HP (Exterkate and de Veer, 1987). It resembles mammalian aminopeptidase A in that it is highly specific for glutamyl- and aspartyl peptides and is here designated as PepA. A similar peptidase was isolated later from *L. lactis* strain 712 (Niven, 1991). The *L. lactis* PepA is a trimeric metallopeptidase of 130 000 molecular weight with identical 42 kDa subunits, although in a later study multimers of 440–520 kDa have been found (Baankreis, 1992).

Although there have been some indications for its membrane location (see also section 4.8), the N-terminal sequence of PepA has been determined and appeared to start with a Met-residue, which is unlikely for a secreted enzyme (Baankreis, 1992). Based on the determined amino acid sequences the *L. lactis pepA* gene was cloned and sequenced, which confirmed that its N-terminal end is unprocessed (M.J. Gasson, personal communication).

4.4.2 Pyrrolidonyl carboxylyl peptidase (PCP)

Another aminopeptidase of narrow specificity, pyrrolidonyl carboxylyl peptidase (PCP), that is able to release N-terminal pyroglutamyl residues, is present in lactococci and has recently been partly purified (Baankreis, 1992; Exterkate, 1977). Partly purified PCP activity was estimated at 80 kDa by HPLC gel filtration and showed a band at 40 kDa in SDS–PAGE, suggesting that the enzyme is a dimer. A second PCP-activity exhibited a relative molecular mass of

25 kDa on both denaturing and non-denaturing PAGE. On the basis of the known amino acid sequence of PCP from *B. subtilis* and *Streptococcus pneumonia*, DNA primers were designed and used to amplify the corresponding region from the *pcp* gene on the chromosome of *L. lactis*. Subsequently, the entire gene was cloned and sequenced. The *pcp* gene encodes a protein of 25 kDa that shows significant similarity to the enzymes from *B. subtilis* and *S. pneumonia* (A.J. Haandrikman, personal communication). The enzyme, as deduced from the nucleotide sequence, does not contain an N-terminal signal sequence and is, most probably, an intracellular protein.

4.4.3 Aminopeptidase N (PepN)

A lysyl-aminopeptidase activity with broad substrate specificity has been described first in *L. lactis* HP (Exterkate, 1984). In subsequent studies this general aminopeptidase activity was also found in various *L. lactis* strain AC1 and other strains and purified to homogeneity from strains Wg2 and HP (Geis *et al.*, 1985; Tan and Konings, 1990; Tan *et al.*, 1992; Baankreis, 1992). Based on its genetic characterization it was designated aminopeptidase N (van Alen-Boerrigter *et al.*, 1991). PepN of strain Wg2 was purified to homogeneity by DEAE-Sephacel and phenyl–Sepharose column chromatography, followed by gel filtration and anion-exchange HPLC using lysyl-*para*-nitroanilide (lys-*p*NA) as the substrate (Tan and Konings, 1990). In this way, the enzyme was purified approximately 30-fold with a yield of 12% over the crude extract. During SDS–PAGE, one band with molecular weight of 95 000 was detected. As the same was observed in non-denaturing PAGE, PepN is probably active as a monomer. Aminopeptidase N of Wg2 is a metalloenzyme that is irreversibly inactivated by EDTA and has a temperature optimum of 40°C and a pH optimum of 7. The enzyme has a broad substrate specificity, degrading several di-, tri-, and larger peptides by hydrolysis of the N-terminal amino acid, and has no endopeptidase or carboxypeptidase activity. It has a high activity for lys-*p*NA and leu-*p*NA.

The structural gene of lysyl-aminopeptidase, *pepN*, of *L. lactis* MG1363 was identified in a lambda library of chromosomal DNA of this strain using immunological techniques (van Alen-Boerrigter *et al.*, 1991). The *pepN* gene could be overexpressed to high levels in *E. coli* and appeared to complement a PepN-deficient mutant. The *E. coli* cells produced a 95-kDa protein that reacted with PepN-specific antibodies, indicating that the entire *pepN* gene had been cloned. A DNA fragment carrying *pepN* was cloned into pIL253 and introduced into *L. lactis* MG1363. The resulting strain showed an approximately 20-fold higher PepN-specific activity and the concomitant increase in the amount of PepN protein correlated well with the increase in the number of copies of the gene (from one copy on the chromosome to approximately 35 when cloned into pIL253; see section 2.2.1). The antiserum used to screen the phage library contained antibodies that reacted with a 35-kDa contaminant in the PepN preparation and several phages were isolated that produced this protein. It is possible that the lysyl-aminopeptidase isolated from *L. lactis* AC1, with a reported molecular size of 36 kDa, is in fact this contaminating protein and that the 95-kDa PepN was overlooked in this study (Geis *et al.*, 1985; van Alen-Boerrigter *et al.*, 1991). Nucleotide sequencing revealed that the *pepN* contained 846 codons and is transcribed from a canonical *L. lactis* promoter immediately preceding the gene and the transcription start site is at an A residue 24 nucleotides upstream of the start codon (see section 2.4.2). As the *pepN* gene is not flanked by open reading frames of any substantial length, and because the gene is followed by a putative terminator, it may constitute a monocistronic transcriptional unit. The deduced molecular weight of PepN of 95 368 is in good agreement with that determined biochemically (Tan *et al.*, 1992). Following this analysis of the *L. lactis* MG1363 *pepN* gene (van Alen-Boerrigter *et al.*, 1991; Tan *et al.*, 1992), the *L. lactis* Wg2 gene was also cloned and sequenced, and found to differ only in a few nucleotides (Stroman, 1992). The *L. lactis* PepN is homologous to aminopeptidase N of eukaryotic and bacterial origin (Tan *et al.*, 1992; Strøman, 1992). This family (EC 3.4.11.2) includes the Zn-metalloenzymes from rat, human, rabbit, pig, mouse, and *E. coli*. A highest overall homology of 27% identical residues in the C-terminal 750 residues was found with the mammalian aminopeptidase N, while the overall similarity with PepN from *E. coli* is considerably lower. Six regions with significantly higher sequence similarity were apparent from a multiple sequence alignment of PepN with the mammalian enzymes. One of these regions, a highly conserved segment comprising residues 380 to 420 of rat aminopeptidase N (amino acids 281 to 301 of PepN), shows homology with the thermolysin family of Zn-dependent neutral proteinases. In thermolysin, this region is part of the active site of the enzyme and contains an essential Zn-ion binding site. The sequence identity strongly suggests that the PepN residues His288, His292, and Glu311 are Zn-ion ligands, and that Glu289 is involved in catalysis. No signal peptide or hydrophobic (transmembrane) segments were found in the PepN sequence, which would agree with the cytosolic location of PepN (see section 4.8).

Recently, the *pepN* gene from *Lactobacillus lactis* subsp. *delbrueckii* has been cloned in *E. coli* by complementation and overexpressed in this host (R. Plapp, personal communication). The deduced amino acid sequence of this gene showed high similarity with that of the lactococcal PepN. Using a similar strategy, the *Lactobacillus helveticus* CNRZ 32 gene for a lys-β-naphthylamide (Lys-β-NA) degrading aminopeptidase has been cloned and expressed (Nowakowski *et al.*, 1993). This 97 kDa metalloenzyme APII had previously been identified in the same strain of *Lactobacillus helveticus* and is likely to represent another PepN (Khalid and Marth, 1990b).

4.4.4 Aminopeptidase C (PepC)

A general aminopeptidase with high activity towards the substrate L-histidyl-L-phenylalanine-β-naphthylamide (His-Phe-β-NA) was purified from *L. lactis*

AM2 by two FPLC ion-exchange steps (Neviani *et al.*, 1989). The enzyme was designated API in this study but based on its deduced homology later termed aminopeptidase C (PepC), as it is here. PepC was purified from strain AM2 by 136-fold with a yield of 9%. The molecular size was estimated at 300 kDa by gel filtration and at 50 kDa by SDS–PAGE, suggesting that the native aminopeptidase C is a hexamer of six identical subunits. Aminopeptidase C has optimal activity at pH 7 and 40°C, is completely or partially inhibited by sulfhydryl-group inhibitors such as iodoacetamide, *p*-chloromercuric benzoic acid, and *N*-ethylmaleimide, but is not affected by reducing agents and inhibitors of serine- and metalloproteinases. Di- and tripeptides and substrates of the type (X-)Y-β-NA are hydrolyzed provided that none of the amino acids is proline. The activity is highest with short-chain peptides. The enzyme does not have carboxypeptidase or endopeptidase activity.

The structural gene of this general aminopeptidase has been cloned from *L. lactis* AM2 by complementation of an *E. coli pepN* mutant (Chapot-Chartier *et al.*, 1992). A plasmid bank of chromosomal DNA of strain AM2 in the *E. coli* mutant was screened with an enzymatic plate assay using Leu-β-NA as the substrate. Two clones were picked up and, by immunoblotting with aminopeptidase-specific antibodies, it was shown that one expressed PepN while the other produced PepC. The DNA insert in the latter clone was 15 kb in size and specified full-size PepC. The PepC determinant was localized on a 2748 bp *Pvu*II subfragment that was sequenced completely. One ORF of 436 codons and the 3' end of a second ORF were present on the DNA fragment. The incomplete ORF did not have homology to any protein in the NBRF and Genbank databases. The N-terminal amino acid sequence of the protein inferred from the 436-codon-ORF matched with that determined of purified PepC, except that the initiation methionine was lacking in the latter. This observation proved that the structural gene for PepC had been cloned and the molecular weight of its translation product agreed well with that of the purified enzyme. The gene was termed *PepC* because of the homology of PepC (API) with cysteine proteinases such as papain, aleurain and cathepsin B and H. Two small regions encompassing the amino acids of the active site of papain are especially well conserved. The smallest of the two is 16 amino acids in length and contains the active site Cys (Cys25 of papain) as well as a conserved Gln residue (Gln19 of papain). The second region is located more to the C-terminus of the enzymes, is 35 amino acids long in PepC, and contains the fully conserved sequence Asn175-Ser176-Trp177 (numbering of papain) of which Asn and Trp have been shown to belong to the catalytic site of papain. Moreover, a His residue was found in PepC at a position equivalent to the active site His159 of papain, namely 22 amino acids upstream of the Asn-Ser-Trp stretch (in papain the distance between the two is 20 amino acids). These data are in agreement with the observation mentioned above that the enzyme is inhibited by thiol-reacting reagents. A marked difference between PepC and the other cysteine proteinases is the fact that PepC is about twice as large. Outside of the catalytic domain no homology was found, as was also observed for viral and bacterial cysteine proteinases, and calpain. Highest similarity (34% identity) was found between PepC and the partial sequence of bleomycin hydrolase, a mammalian cysteine proteinase.

4.5 Proline-specific peptidases

The degradation of the proline-rich oligopeptides liberated from casein by the action of the proteinase requires the presence of lactic acid bacteria of proline-specific peptidases, as neither the general aminopeptidases nor the broad-specificity di-and tripeptidases of these organisms are able to hydrolyze proline-containing peptides to any extent. Aminopeptidase, P, proline iminopeptidase, iminodipeptidase (prolinase), imidodipeptidase (prolidase) and X-prolyl dipeptidyl aminopeptidase have been detected in the various species of lactic acid bacteria (Bockelmann, 1987; Mou *et al.*, 1975; Kaminogawa *et al.*, 1984); Hickey *et al.*, 1983; Kaminogawa *et al.*, 1984; Casey and Meyer, 1985; Meyer and Jordi, 1987; Kiefer-Partsch *et al.*, 1989; Zevaco *et al.*, 1990; Atlan *et al.*, 1990; Khalid and Marth, 1990; Baankreis and Exterkate, 1991). Several of these enzymes have been purified to homogeneity and characterized and will be discussed here (Kaminogawa *et al.*, 1984; Meyer and Jordi, 1987; Kiefer-Partsch *et al.*, 1989; Zevaco *et al.*, 1990; Khalid and Marth, 1990, Baankreis and Exterkate, 1991; see Table 4.2).

4.5.1 Prolidases

The first report on the purification of a proline-specific peptidase involved the prolidase (proline dipeptidase or imidodipeptidase (EC 3.4.13.9)) of *L. lactis* H61 (Kaminogawa *et al.*, 1984). A combination of DEAE-cellulose, hydroxyapatite chromatography, and polyacrylamide gel electrophoresis was used to purify the enzyme from a cell-free extract. A purification of 490-fold over the cell-free extract and a yield of 30% of L-Leu-L-Pro hdyrolyzing activity was obtained in this way. The prolidase of strain H61 is highly specific for X-Pro dipeptides and does not degrade other peptides or carbobenzoxypeptides. The optimum pH of the enzyme is between 6.5 to 7.5, the optimum temperature is 40°C. The prolidase is completely inhibited by the metal chelators 1,10-phenantroline and EDTA, and by *N*-ethylmaleimide. It has a molecular weight of 43 000 as estimated by gel filtration. A prolidase was partially purified from the cytoplasmic fraction of *L. lactis* AM2 by chromatography on Sephadex G200, after which the enzyme was separated from prolinase and X-prolyl dipeptidyl aminopeptidase on a DEAE-cellulose column (Booth *et al.*, 1990). A 29-fold purification and 55% recovery were obtained after subsequent chromatography of the active Leu-Pro hydrolyzing fractions on calcium phosphate-cellulose. The pH optimum of the enzyme from strain AM2 ranged from 7.4 to 9.0 depending on the buffer system used. The metal chelators EDTA and 1,10-phenantroline, the disulfide reagent dithiotreitol and the sulfhydryl group reagent *N*-ethyl-maleimide inhibit the enzyme. X-Pro-type dipeptides (except Gly-Pro) are good substrates while of several dipeptides containing Pro as the N-terminal residue, only Pro-Val, Pro-Ala, and Pro-Pro are hydrolyzed. A number of di- and tripeptides lacking Pro are also hydrolyzed but no hydrolysis was observed with tripeptides containing a Pro residue in the central position. As yet, no report on the cloning and nucleotide sequence analysis of the structural gene of the prolidase of strain H61 or AM2 is available.

4.5.2 Proline iminopeptidase

Recently, a proline iminopeptidase from the cytoplasm of *L. lactis* HP has been purified by ammonium sulfate fractionation, DEAE ion-exchange- and hydrophobic interaction chromatography, followed by preparative PAGE or HPLC gel filtration (Baankreis and Exterkate, 1991). The enzyme is severely inhibited by EDTA and dithiotreitol, has optimum activity at pH 8.5 and 37°C and has an approximate molecular mass of 110 kDa by gel filtration. SDS–PAGE revealed a single band of approximately 50 kDa. The enzyme hydrolyzes almost all di- and tripeptides containing proline as the amino terminal residue but does not cleave X-Pro bonds or peptides longer than four residues. All lactococcal strains tested, including a plasmid-free strain, contained this enzyme activity but it was not found in strains of *Lactobacillus helveticus* and *Lactobacillus bulgaricus*.

4.5.3 X-prolyl dipeptidyl aminopeptidase (PepXP)

The enzyme that has been detected in and purified to homogeneity from the widest range of lactic acid bacterial species is, without doubt, the X-prolyl dipeptidyl aminopeptidase (X-PDAP or PepXP). The enzymes have been isolated by a variety of chromatographic methods from (crude) cell-free or intra-cellular extracts or even washed cells of various *L. lactis* strains, *Lactobacillus lactis*, *Lactobacillus bulgaricus*, *Lactobacillus acidophilus* and *Streptococcus thermophilus* (Meyer and Jordi, 1987; Kiefer-Partsch *et al.*, 1989; Zevaco *et al.*, 1990; Atlan *et al.*, 1990; Khalid and Marth, 1990; Bockelmann *et al.*, 1991; Booth *et al.*, 1990). All enzymes have rather similar properties in that they are not or only slightly inhibited by EDTA, but are severely inhibited by the serine proteinase inhibitors PMSF and DFP. The optimum pH is in the range of 6.5 to 8.0, while optimum temperatures vary from 40 to 50°C for the various enzymes. All X-PDAPs specifically degrade peptides of the structure X-Pro-Y... (X and Y can be any amino acid) but the rate of hydrolysis depends on the type of amino acids surrounding proline. Some differences exist between the enzymes with respect to the substrate which is hydrolyzed most efficiently. Slight activity with Ala-Ala-*p*NA is detected but no aminopeptidase activity is observed.

The lactococcal enzymes have been shown to split β-casomorphin (Tyr-Pro-Phe-Pro-Gly-Pro-Ile) starting from the amino terminus into the successive X-Pro dipeptides and Ile (Kiefer-Partsch *et al.*, 1989; Zevaco *et al.*, 1990). The X-PDAP from *Lactobacillus helveticus* CRNZ 32 has a subunit molecular weight of 72 000 by native and SDS–PAGE (Khalid and Marth, 1990) and the molecular weight of the partially purified enzyme of *Lactobacillus bulgaricus* was estimated at 82 000 by gel filtration (Atlan *et al.*, 1990). The enzyme from *L. lactis* AM2 has a size of 117 kDa by Sephadex G-200 gel filtration. The size under denaturing conditions was not reported (Booth *et al.*, 1990). The other enzymes are dimers of identical subunits under native conditions with molecular sizes of 160 to 200 kDa. The monomeric sizes, estimated from SDS–PAGE, range from 80 to 95 kDa (Meyer and Jordi 1987; Keifer-Patsch *et al.*, 1989; Zevaco *et al.*, 1990; Atlan *et al.*, 1990; Bockelmann *et al.*, 1991).

The structural *pepXp* genes of two lactococcal strains, P8-2-47 and NCDO 763, have been cloned and sequenced completely. The N-terminal amino acid sequence of purified PepXP of *L. lactis* P8-2-47 was determined and used to design a degenerate 17-mer oligonucleotide probe (Mayo *et al.*, 1991). A plasmid library of *Xba*I-derived fragments of chromosomal DNA of strain P8-2-47 that reacted with the probe was made in *E. coli*. The *pepXP* gene was identified by screening the *E. coli* colonies for X-PDAP activity using the chromogenic substrate Gly-Pro-β-NA. Expression of the *pepXP* gene in this heterologous host and in *B. subtilis* proved that the structural gene had been cloned. A partial *Xba*I DNA fragment of 5.3 kb carrying the gene was inserted into a pWV01-derived cloning vector (Kok *et al.*, 1984; see section 2.2) and used to complement a *pepXP* mutant of *L. lactis*. Contrary to what was published in this work (Mayo *et al.*, 1991), the PepXP specific activity was increased about 10-fold in the complemented strain (Mayo, *et al.*, 1993). This is most probably caused by a gene dosage effect, as the vector used is present in approximately 10 copies per chromosome. A 3.8-kb DNA fragment was sequenced and carried two oppositely oriented ORFs. The largest of these contained 763 coding triplets and was identified as the *pepXP* gene, as the first 25 amino acids of its translation product were the same as the N-terminus of purified X-PDAP. The ORF (ORF1) on the other DNA strand consists of 289 codons and could encode a protein of approximately 31 kDa. This putative protein is very hydrophobic and has considerable amino acid similarity to the glycerol facilitator of *E. coli*, a protein believed to be an integral membrane protein. The putative ORF1 protein has no direct role in X-PDAP activity since it could be deleted without affecting the functional expression of the *pepXP* gene *E. coli* and *L. lactis* (Mayo *et al.*, 1991, 1993). The molecular mass of PepXP deduced from its gene (88 kDa) was in good agreement with that of the monomeric form of the purified enzyme (89 kDa; Kiefer-Partsch *et al.*, 1989).

The *pepXP* gene of *L. lactis* NCDO 763 was identified by performing an enzymatic plate assay with Phe-Pro-β-NA on a plasmid library of total DNA of this strain in an isogenic X-PDAP deficient mutant (Nardi *et al.*, 1991). One colony of 48 000 transformants tested had recovered the parental phenotype. The 17.8 kb plasmid obtained not only complemented the lactococcal *pepXP* mutant but also specified X-PDAP activity in *B. subtilis*, an activity not normally present in this organism, and indicating that the entire structural gene had been cloned. The increase of the number of copies of *pepXP* of strain 763 in the X-PDAP mutant of *L. lactis* NCDO 763 did not result in a significant increase in X-PDAP specific activity, suggesting regulation of (expression of) the *pepXP* gene in this case. In a 4.0 kb sequenced region three ORFs were present and the X-PDAP gene was identified by comparison with the N-terminal amino acid sequence of the purified enzyme.

A comparison of the nucleotide sequences of *pepXP* and its flanking sequences from both lactococcal strains revealed a very high similarity, both at the DNA and amino acid level. Both enzymes consist of 763 amino acids and differ in only seven amino acids. Two of these lead to conserved amino acid substitutions, and other five are nonhomologous amino acid changes (Mayo *et al.*, 1991; Nardi *et al.*, 1991). The lactococcal PepXP aminopeptidase did not have significant amino acid sequence similarity to other proteins, not even to serine proteinases, despite the fact that the enzyme is classified as a serine proteinase (Kiefer-Partsch *et al.*, 1989; Zevaco *et al.*, 1990).

Recently, the *pepXP* gene from *Lactobacillus lactis* WS87 has been identified in an *E. coli* plasmid library by screening with Gly-Pro-β-NA (R. Plapp, personal communication). The sequence of an 870 bp *Bam*HI-*Eco*RI fragment involved in X-PDAP activity was determined and contained in ORF extending over the entire length of the fragment. The nucleotide sequence has 54% homology with that between positions 2005 to 2630 of *pepXP* of *L. lactis* P8-2-47. At the amino acid level, 45% identity is found with residues 301 to 508 of the lactococcal PepXP. In addition, a genomic library of *Lactobacillus helveticus* CNRZ 32 in *E. coli* was screened for clones expressing peptidase activity (Nowakowski *et al.*, 1993). With an enzymatic plate assay using Gly-Pro-β-NA or Lys-β-NA, colonies were identified that expressed X-PDAP, an enzyme activity previously purified from the same strain (Khalid and Marth, 1990a). So far no sequence characterization of this putative *pepXP* gene has been described.

The N-terminal amino acid sequences of X-PDAP purified from the two *Lactobacillus lactis* strains WS87 and 3043 and from *Lactobacillus acidophilus* 357 are highly similar: at least seven of the nine-amino-terminal residues are identical (Bockelmann *et al.*, 1991; R. Plapp, personal communication). The sequences also show some similarity to the N-terminus of the lactococcal PepXP; Met1, Asn4, and Val9 are present in all enzymes. These data suggest that enzymes similar to that of the lactococcal PepXP are widely disseminated in proteolytic lactic acid bacteria.

4.6 Oligo- and carboxypeptidases

4.6.1 Dipeptidases

To date, two dipeptidases of broad substrate specificity have been purified to homogeneity and characterized in detail (Hwang *et al.*, 1981, 1982; Van Boven *et al.*, 1988; see Table 4.2). A peptidase hydrolyzing Leu-Gly was purified approximately 200-fold from a cell-free extract of *L. lactis* H61 by ammonium sulfate fractionation and chromatography involving two DEAE-cellulose and two hydroxylapatite columns (Hwang *et al.*, 1981). The optimum pH of the enzyme is 8.0. Complete inhibition by EDTA and 1,10-phenantroline showed that the dipeptidase is a metalloenzyme. By gel filtration on Sephadex G-150 the native molecular weight of the enzyme was estimated to be 100 000. The peptidase has a broad specificity and is active on various dipeptides but is not active, or less so, if Gly or Pro is the N-terminal residue. Tripeptides and carbobenzoxypeptides are hardly attacked and, therefore, the enzyme was classified as a dipeptidase.

A Leu-Leu hydrolyzing dipeptidase was purified from a crude cell-free extract of *L. cremoris* Wg2 by DEAE-Sephacel chromatography followed by preparative tube-flow PAGE (Van Boven *et al.*, 1988). After SDS–PAGE, a single protein band was observed at a position corresponding with a molecular weight of 49 000. As the same molecular weight was found under native conditions, the enzyme is probably active as a monomer. The enzyme is a metallopeptidase as it is severely inhibited by EDTA and 1,10-phenantroline, has a pH optimum of 8 and is optimally active at 50°C. The enzyme is active towards various dipeptides with relatively hydrophobic and neutral N-terminal amino acids. Dipeptides containing proline, histidine, glycine or glutamate in the N-terminal position as well as larger peptides and peptides containing modified peptide bonds are not hydrolyzed. Differences in molecular weight, turnover numbers, metal dependencies and substrate specificities indicate that the Wg2 dipeptidase is distinctly different from the H61 enzyme.

Dipeptidase activities showing a broad substrate specificity have been detected in several other lactic acid bacteria (Kaminogawa *et al.*, 1984; Kolstad and Law, 1985; Rabier and Desmazeaud 1973; Desmazeaud and Zevaco, 1977; El Soda *et al.*, 1978; Sørhaug and Solberg, 1973; Abo-Elnaga and Plapp, 1987). This suggests that these types of dipeptidases may be widely distributed among lactic acid bacteria.

At present no genetic data on any of the peptidases mentioned above are available.

4.6.2 Tripeptidases (PepT)

Various authors have analysed tripeptidase activities in lactic acid bacteria (Mou *et al.*, 1975; Law, 1979; Kaminogawa *et al.*, 1984; Desmazeaud and Zevaco,

1979; Baankreis and Exterkate, 1990; Abo-Elnaga and Plapp, 1987; Bosman *et al.*, 1990). In lactococci, a tripeptidase of broad substrate specificity may be a common enzyme, while such enzymes have not been described in the other species of lactic acid bacteria. To date, there are three reports on the purification of a tripeptidase (Table 4.2). The isolation of the prolin-specific tripeptidase of *L. lactis* HP has been discussed in section 4.3.2.2 (Baankreis and Exterkate, 1991). This enzyme clearly differs from the partially purified tripeptidase (APII) from *L. lactis* CNRZ 267 and a tripeptidase purified to homogeneity from the *L. lactis* strain Wg2 (Desmazeaud and Zevaco, 1979; Bosman *et al.*, 1990) The tripeptidase of the latter strain, later designated PepT (see below), was obtained from a cell-free extract by DEAE-Sephacel and phenyl-Sepharose chromatography, followed by two gel filtration steps (Bosman *et al.*, 1990). Fractions were assayed for Leu-Gly-Gly hydrolyzing activity to follow the purification (378-fold with 19% recovery) of the tripeptidase. The temperature and pH optimum of the enzyme are 55°C and 7, respectively. The tripeptidase PepT appears to be a dimer of two identical subunits of 52 kDa with a native molecular size of 103 to 105 kDa. Of 28 tripeptides tested, all were hydrolyzed by the enzyme as long as proline was not in the second position. PepT specifically removes the N-terminal amino acid resulting in the formation of a free amino acid and a dipeptide. Di- and oligopeptides are not hydrolyzed. The tripeptidase is reversibly inactivated by EDTA. The reducing agents DTT, β-mercaptoethanol and Cu^{2+} also inhibit the enzyme while sulfhydryl group reagents stimulate the enzyme activity approximately two-fold. Like the tripeptidase PepT from strain Wg2, the enzyme APII of *L. lactis* CNRZ 267 is active on a number of different tripeptides and is reversibly inactivated by EDTA (Desmazeaud and Zevaco, 1979). However, APII differs from the Wg2 tripeptidase PepT in its molecular size, which is 75 kDa by Sephadex G-150 gel filtration.

The gene for the tripeptidase, *pepT*, of *L. lactis* Wg2 has been cloned by reverse genetics (I. Mierau, personal communication). DNA sequence analysis revealed that *pepT* encodes a protein of 413 amino acids with a calculated molecular size of 46 kDa, somewhat smaller than the 52 kDa determined biochemically. The N-terminal amino acid sequence of the deduced protein is identical to that of purified PepT, indicating that the enzyme has an intracellular location. The *pepT* gene seems to be part of an operon as no promoter terminator structures could be identified immediately upstream and downstream of the gene. Instead, *pepT* is flanked by two ORFs of which the putative proteins shown no homology to known proteins. PepT of *L. lactis* shows extensive amino acid sequence similarity to PepT of *Salmonella typhimurium*, a protein of 409 amino acids. An alignment of both tripeptidase sequences with that of carboxypeptidase G of *Pseudomonas putida* shows that the three enzymes share three regions of homology that may be part of the active site of the enzymes.

4.6.3 Carboxypeptidases

Carboxypeptidases, which attack substrates from the C-terminal end, are not present or present only in insufficient amounts to be detected in lactococci, *Streptococcus thermophilus*, *Lactobacillus helveticus* and *Lactobacillus bulgaricus* (Kaminogawa et al., 1984; Hickey et al., 1983; El Soda and Desmazeud, 1981; Cliffe and Law, 1979). Apparently, casein-derived peptides are degraded by these organisms from the N-terminus only. In contrast, in *Lactobacillus casei* and *Lactobacillus plantarum* an activity has been described on N-terminally blocked dipeptides: a carboxypeptidase which was highly specific for carbobenzoxy-glycyl-arginine was found in *Lactobacillus casei* 151 (El Soda et al., 1978a,b), although in a later study a carboxypeptidase activity with a broad substrate specificity was detected in the same strain and in *Lactobacillus plantarum* (Abo-Elnaga and Plapp, 1987). *Lactobacillus lactis* also appears to produce a carboxypeptidase hydrolyzing the substrates N-benzoyl-Gly-Arg and N-benzoyl-Gly-Phe (Eggiman and Bachmann, 1980). None of these enzymes have been purified to homogeneity and none of the genes involved in the activities have been identified.

4.7 Transport of amino acids and peptides

The last step in the utilization of casein by lactococci is the uptake of the amino acids and small peptides liberated from casein by the concerted action of the proteolytic system. Separate transport systems for amino acids, di- (and tri-), and oligopeptides are present in lactococci. The mechanisms by which these transport systems operate have received much attention during the last 5 years and, as a consequence, the knowledge of the bioenergetics of amino acid and peptide transport by lactococci has rapidly increased (for recent reviews see Driessen et al., 1987; Driessen, 1989; Konings et al., 1989). Much less is known of the genetic basis of these transport systems and in what follows a summary of the available data is given.

4.7.1 Specific amino acid carriers

Three types of amino acid transport systems are operative in lactococci: (i) pmf-driven transport, in which the uptake is primarily coupled to the proton motive force (pmf); (ii) phosphate-bond-linked transport, which is driven by the high-energy phosphate bond of ATP or an ATP-derived metabolite; and (iii) exchange transport, which is exemplified by the arginine/ornithine antiporter (Driessen 1989; Konings et al., 19889). However, presently none of these carriers has been characterized in homogeneous state and their purification is expected to be complicated by the fact that they are partly integral membrane proteins. A genetic approach could therefore be fruitful but so far no genes involved in these amino acid transport systems have been identified or characterized.

4.7.2 Di- and tripeptide transport system (DptT)

Di- and tripeptide transport by lactococci is driven by the proton motive force and takes place by a two-step process: first, the peptide is translocated across the cytoplasmic membrane via a specific peptide transport system, while in the second step the peptide is hydrolyzed by an intracellular peptidase (Smid et al., 1989a). The di-tripeptide transport system has a broad substrate specificity and a di-tripeptide-transport-deficient mutant of L. lactis ML3, selected by its resistance to the toxic dipeptide L-alanyl-β-chloro-L-alanine, was used to show that the system also transports tripeptides (Smid et al., 1989b). This mutant was used to show that lactococci posses a separate system for the uptake of tetra-, penta-, and hexapeptides (Smid, 1991). Although the di-tripeptide transport mutant had full proteinase and peptidase activity, and was unaffected in the accumulation of the larger peptides, it did not grow on media containing casein as the sole nitrogen source (Smid et al., 1989b). Apparently, one or a number of essential amino acids mainly become available as di- or tripeptides.

Recently, the L. lactis ML3 gene that is likely to encode this di-tripeptide transport system has been cloned by complementation of an E. coli proC mutant (A. Hagting and W.N. Konings, personal communication). The cloned dptT gene was sequenced and appeared to encode an 462-residue polypeptide with 12 putative membrane-spanning α-helices.

4.7.3 Oligopeptide transport system (Opp)

It has been known for a number of years that certain strains of lactococci, although containing a full proteinase and peptidase complement, cannot grow in milk to the cell densities reached by a wild-type strain. It was reported that L. lactis MG1614 did not grow in milk even when the proteinase plasmid pVS9 of strain SSL135 was present (von Wright et al., 1987; Tynkkynen et al., 1989). Later experiments showed that the L. lactis strain used is probably a mutant of MG1614, since the original MG1614 strain, a plasmid-free streptomycin- and rifampicin resistant derivative of L. lactis NCDO 712, shows wild-type growth in milk when equipped with a lactose-proteinase plasmid (Gasson and de Vos, unpublished observations). This was confirmed by the original authors (S. Tynkkynen, personal communication) and studies discussed below showed that the strain they used is actually a MG1614-derivative deficient in an oligopeptide transport system and is designated here MG1614-opp.

Von Wright et al. (1987) cloned a DNA fragment derived from the chromosome of L. lactis strain SSL135 that, when introduced on plasmid pVS8 into strain MG1614-opp, restored the ability of this strain to grow in autoclaved milk but not in pasteurized milk (Tynkkynen et al., 1989). Strain MG1614-opp did grow to normal cell densities in both types of milk when pVS8 and pVS9, the proteinase plasmid of strain SSL135, were jointly present (Tynkkynen and von Wright, 1988; Tynkkynen *et al.*, 1989). The presence of pVS8 and MG1614-opp enabled the strain to utilize tryptic peptides of up to 3600 Da, purified by FPLC from tryptone, and to grow in a synthetic medium supplemented with peptides of this size. In contrast, strain MG1614-opp itself was only able to grow to some extent on tryptic peptides if they were smaller than 2100 Da (Tynkkynen *et al.*, 1989). The 9.4 kb chromosomal DNA fragment in pVS8 that complemented the growth deficiency of MG1614-opp did not hybridize with a DNA probe carrying most of the *prtP/prtM* gene region of *L. lactis* NCDO 712 (von Wright *et al.*, 1987). The fragment specified four proteins of 69, 42, 38 and 36 kDa as was shown by *in vitro* transcription and translation experiments on pVS8 DNA (Tynkynnen *et al.*, 1989). All proteins appeared to be essential for lactic acid production in autoclaved milk, although deletion or truncation of the 69 kDa protein resulted in a 'leaky' phenotype: strains carrying a plasmid specifying a mutated 69 kDa protein produced acid at a lower rate than a strain containing pVS8. However, this acid production was significantly higher than that of strains containing plasmids lacking the genes for one or all of

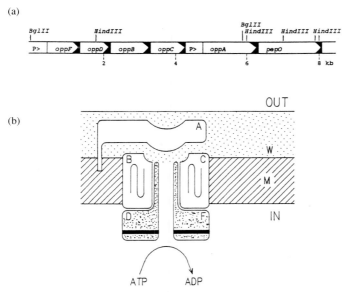

Figure 4.3 (a) Genetic organization of the *opp* operon of *L. lactis*. The direction of transcription of the genes is indicated by the arrowheads. The size of the DNA (in kb) is shown below the bar. P>, putative promoter and its direction. (b) Tentative model of the *L. lactis* oligopeptide permease system Opp, based on the model proposed for Opp of *S. typhimurium* (Ames, 1986; Higgins *et al.*, 1990). A, B, C, D, and F are OppA, OppB, OppC, OppD and OppF, respectively. OppA is shown attached to the membrane (M) via lipid modification of its Cys23 residue. The black bars in OppD and OppF represent the conserved putative ATP binding sites. Possible energization of the uptake system through ATP hydrolysis (by either OppD or OppF, or both) is also shown. W, cell wall.

the other proteins. Tynkkynen *et al.* (1983) have now sequenced over 8 kb of DNA of the chromosomal insert in pVS8 and identified five complete ORFs (see Figure 4.3). The sizes of the deduced products of these ORFs agree well with the sizes of the five proteins obtained in the *in vitro* transcription/translation experiment. Comparison of the five inferred proteins with the SwissProt bank revealed significant similarity (25 to more than 40% identity) to the five proteins encoded by the *opp* operons of *B. subtilis* and *Salmonella typhimurium* (Hiles *et al.*, 1987; Perego *et al.*, 1991). The five *S. typhimurium* proteins, OppA, OppB, OppC, OppD, and OppF, form a periplasmic binding protein-dependent transport system that mediates the uptake of peptides containing up to five amino acid residues and is involved in the recycling of cell wall peptides during growth (Goodell and Higgins, 1987). Opp is a member of a large family of both prokaryotic and eukaryotic import and export systems, collectively known as the ABC transporters (Higgins *et al.*, 1987; Hyde *et al.*, 1990). In Gram-negative bacteria, a periplasmic substrate-binding protein is the initial receptor of the molecule to be transported. In the oligopeptide transport system of *S. typhimurium* this function is executed by OppA. The peptides are delivered by OppA to a complex of four membrane-associated proteins (Higgins *et al.*, 1990; Ames, 1986; see Figure 4.3). Two of these, OppB and OppC, are highly hydrophobic membrane-spanning proteins, while OppD and OppF are ATP-binding proteins associated with the cytoplasmic side of the inner membrane. The latter two proteins couple ATP hydrolysis to the transport process (Higgins, 1990). Recently, it has been shown that oligopeptide transport in *L. lactis* occurs in the absence of a proton motive force and that it is most likely driven by ATP or a related energy-rich, phosphorylated intermediate (Kunji *et al.*, 1993). In the examples of binding protein-dependent transport systems in Gram-positive bacteria, a binding protein, anchored to the cytoplasmic membrane via lipid modification, seems to be the equivalent of the Gram-negative periplasmic binding protein (Gilson *et al.*, 1988; Perego *et al.*, 1991). In line with these results, the lactococcal OppA has a typical N-terminal lipobox (see chapter 2; and Figure 4.3).

Immediately downstream of *oppA*, the last gene in the lactococcal *opp* gene cluster, the gene encoding the 70 kDa endopeptidase (PepO), is located (see above and Figure 6.3). Two integration mutants were constructed, one defective in OppA and the other lacking PepO. Growth experiments using a chemically defined medium with oligopeptides showed that the transport system, but not the endopeptidase, is essential for growth of *L. lactis* on peptides longer than three residues. Growth experiments in pasteurized milk revealed that transport of oligopeptides forms an essential part of the proteolytic system of lactococci. Apparently, even in the presence of a full proteinase and peptidase complement, and a functional di- and tripeptidase uptake system, one or a number of essential amino acids are only available in oligopeptides.

4.8 Cellular localization of proteolytic enzymes

While the extracellular surface location of the general purpose proteinase PrtP and the dedicated endopeptidase NisP is evident from biochemical experiments and genetic data, the location of other peptidases has been the subject of debate and various studies. Attempts to determine the cellular location of the various peptidases have been frustrated by the fact that in cell fractionation studies cross-contamination of the fractions was inevitable. In many of the early studies contamination of the presumed extracellular fraction with intracellular enzymes was not properly addressed and appropriate controls were lacking with bonafide cytoplasmic enzymes such as aldolase, lactate dehydrogenase or (phospho-) β-galactosidase. This is especially important in the light of the presence of highly active intracellular peptidases. Contamination of fractions with membranes could not be assessed due to the absence of a proper marker enzyme. Recently, the issue of peptidase location in the cell has received renewed attention due to the fact that antibodies against highly purified peptidases and nucleotide sequence data of a number of peptidase genes became available.

Polyclonal antibodies were used in localization studies of the two aminopeptidases PepC and PepN, the X-prolyl dipeptidyl aminopeptidase PepXP, the tripeptidase PepT and the endopeptidase PepO (Tan *et al.*, 1992). The antibodies were peptidase-specific, did not cross react with other proteins and were used in immunoblotting and immunogold labelling experiments. In these studies, anti-PrtM antibodies were used as a control to detect PrtM, a lipoprotein outwardly exposed on the cytoplasmic membrane (Haandrikman *et al.*, 1991; see section 4.2.6). Protoplasts were treated with proteinase K, sonicated after inactivation of proteinase K with PMSF and analysed in Western hybridization with the various antibody preparations. As expected, PrtM was lost after this treatment. All five peptidases were protected from proteinase K digestion, indicating that the enzymes were, at least partially, located intracellularly. Only very small amounts of PepC and no PepN could be detected in cytoplasmic membrane vesicles of *L. lactis* and these peptidases are, therefore, true intracellular enzymes. An intracellular location of PepN was also deduced from immunogold labelling studies by van Alen-Boerrigter *et al.* (1991). This was confirmed by the immunogold labelling studies of Tan *et al.* (1992) that also showed that PepC is located intracellularly. Therefore, it is concluded that both aminopeptidases PepC and PepN are intracellular enzymes, which is supported by the observation that, apart from removal of the initiator methionine, no further processing takes place at the N-terminus of these enzymes (Chapot-Chartier *et al.*, 1992; van Alen-Boerrigter *et al.*, 1991).

The endopeptidase PepO, PepT, and PepXP were clearly present in the vesicle preparation, suggesting some interaction of the enzymes with the membrane. Immunogold labelling showed that although cytoplasmic labelling was evident with the antibodies against PepO, PepT, and PepXP, significant peripheral labelling was also observed (Tan *et al.*, 1992). From a comparison of the deduced amino acid sequences of PepXP, PepT and PepO with the N-termini of the purified enzymes it is clear that, apart from removal of the initiator methionine in PepO, none of them are N-terminally processed. Although this does not exclude secretion by a sec-independent secretion pathway (Pugsley, 1988; see chapter 2), it supports the proposed intracellular location. The peripheral location might explain why under certain conditions the enzyme can be specifically washed from the producing cells (Meyer and Jordi, 1987).

A similar conclusion should be reached for the location of the aminopeptidase PepA. Cell fractionation studies and immunogold labelling studies with PepA antibodies suggested a membrane-associated location (Baaukreis, 1992). However, the partial nucleotide sequence of the gene for aminopeptidase A, *pepA*, shows that also in this case the deduced N-terminus of the peptidase and that of the purified enzyme are the same (M.J. Gasson, personal communication).

The general picture that emerges from all these biochemical and genetic data is that, except for the proteinase PrtP and the endopeptidase NisP, all lactococcal peptidases purified to date are intracellular enzymes. It is very likely that *L. lactis* does not secrete peptidases into the medium and that the general purpose proteinase PrtP is the only extracellular enzyme participating in the degradation of casein into peptide fragments of ingestible size. As described above, it has been found that the proteinase PrtP may produce small peptides from casein that could be transported into the growing cells. However, it should be stressed that this does not preclude the participation of intracellular peptidases in growth in milk or cheese ripening. Limited lysis of cells could account for the release of some peptidases in those systems. In this context it is of interest to note that some of the intracellular peptidases, such as PepN and PepXP, have very high specific activities.

4.9 Engineering of the proteolytic system

The availability of mutants defective in the production of proteinases or peptidases is of crucial importance for the unravelling of the complexity of the proteolytic system and assessing its impact in industrial processes. The genetic analysis of the proteolytic system has greatly facilitated the verification of spontaneous or induced mutations and moreover allowed for the construction of mutants by replacement recombination (see above and chapter 2). Furthermore, the availability of genes for the individual components of the proteolytic system allows it to engineer its production level, biochemical characteristics and control. Recent results in this area and some future perspectives are discussed here.

4.9.1 Construction and use of mutants

By screening for the inability to hydrolyze chromogenic substrates in a direct enzymatic assay on bacterial colonies a number of peptidase mutants were identified among nitrosoguanidine-treated *Lactobacillus delbrueckii* subsp. *bulgaricus* cells (Atlan *et al.*, 1989). In lysyl-aminopeptidase (APII) deficient mutants the arginyl- and leucyl-aminopeptidase activities are reduced considerably and the remaining activity was attributed to the presence of three other aminopeptidases. Mutants in the X-PDAP of the same strain of *Lactobacillus delbrueckii* subsp. *bulgaricus* have also been isolated (Atlan *et al.*, 1990). Both types of mutants are proteinase-positive and have wild-type amino acid requirements and sugar fermentation patterns. Both peptidases were shown to be non-essential for growth in milk. Although the mutants have wild-type doubling times in rich (MRS) medium they produce a significantly lower biomass due to a shorter exponential growth phase. Loss of X-PDAP activity results in an enhancement of proteolytic activity while mutants lacking APII display an increase in X-PDAP activity when cells are grown in milk. These data are suggestive of a common regulatory mechanism governing the biosynthesis of the three enzymes in this strain of *Lactobacillus delbrueckii* subsp. *bulgaricus* (Atlan *et al.*, 1989, 1990).

Both random chemical mutagenesis and plasmid integration strategies have been used to make mutations in a number of lactococcal peptidase genes. These peptidase mutants have been employed in a number of different ways. An X-PDAP-negative mutant of *L. lactis* NCDO 763 was instrumental in the cloning of the *pepXP* gene (see above). Defined mutants, obtained by replacement recombination, have been used to assess the importance of X-PDAP, PepN and PepO for growth in milk and these studies have shown that all three are dispensable (Mierau *et al.*, 1993; Mayo *et al.*, 1993; van Alen-Boerrigter and de Vos, unpublished results). This approach also allows for the combination of mutations in a single strain. Recent results indicate that *L. lactis* cells deficient for PepN and PepXP show reduced growth rates in milk, indicating their physiological role (Mierau, personal communication).

Two chemically induced mutants, a *pepN* strain and a strain mutated in *pepXP*, have been used in cheese making trials to examine the role of the peptidases in cheese flavour development (Baankreis, 1992). Gouda-type cheeses were made with 0.25% starter cultures of varying composition. All starters contained 10% proteolytically active cells of *L. lactis* HP, a 'bitter'-producing strain, and 90% Prt⁻ cells. The Prt⁻ fraction consisted of HP cells and increasing concentrations of either the *pepN* or the *pepXP* mutant. The organoleptic quality as well as amino acid and salt soluble nitrogen were examined. The extent of peptide hydrolysis was estimated by reversed-phase HPLC.

Cheeses manufactured with increasing concentrations of the PepXP-deficient strain did not accumulate bitterness but exhibited decreasing organoleptic quality. In contrast, raising the amount of *pepN* mutants in the starter culture led to a pronounced increase in bitterness. With high concentrations of *pepN* strains this was already noticeable after 2 weeks of ripening. The increase of bitter was ascribed to the accumulation of low molecular weight bitter peptides that are normally degraded by PepN of wild-type starter strains (Baankreis, 1992). This debittering role of PepN in cheese is consistent with the activity of PepN in debittering trypsin-digested casein (Tan, *et al.,* 1993).

4.9.2 Engineering of the proteinase PrtP

The principal cell envelope-associated proteinase PrtP has been engineered in various ways, aimed at overproduction, altered caseinolytic specificity and stability.

Proteinase overproduction has been achieved by putting the two proteinase genes *prtP* and *prtM* of *L. lactis* Wg2 in an operon structure and providing them with an *L. lactis* chromosomal promoter that was at least five-fold stronger than either of the original *prt* promoters (Van der Vossen *et al.,* 1992). An increase in proteolytic activity of five-fold was reported. Chromosomal stabilization and amplification of the *prt* genes of strain Wg2 has also been achieved in *L. lactis*. The proteolytic activity of a strain carrying approximately eight copies of the proteinase genes on the chromosome was eleven times higher than that of a strain carrying only two copies of the *prt* genes (Leenhouts *et al.,* 1991). In the case of the cloned *prt* genes from *L. lactis* subsp. *lactis* strain UC317, a three-fold overproduction of proteinase was observed by SDS–PAGE when a mutant of strain UC317, lacking its original proteinase plasmid but carrying the cloned *prt* genes, was compared with the wild-type strain UC317 (Law *et al.,* 1992). This resulted in an increase in the rate of casein breakdown by the recombinant strain. In both studies (Leenhouts *et al.,* 1991; Law *et al.,* 1992), no or only minor effects of the increased proteinase production on growth rate and rate of acid formation were observed. Overproduction of the SK11 proteinase has been accomplished by cloning the SK11 *prt* genes, with the *prtP* terminator, on a plasmid with an elevated copy number in *L. lactis* (Bruinenberg *et al.,* 1992). Proteinase overproduction appeared to be host dependent and resulted in three-fold more proteinase in derivatives of *L. lactis* NCDO 712, which showed an increased growth and acidification rate in milk independent of the location of the proteinase. The latter results seem to confirm previous physiological experiments indicating that casein hydrolysis proceeds too slowly for lactococci to reach maximum specific growth rates in milk (Hugenholtz *et al.,* 1987).

In a first approach to engineer the caseinolytic specificity of the proteinase PrtP, hybrids were constructed between the Wg2 proteinase (PI-type specificity) and that of SK11 (PIII-type specificity) (Vos *et al.,* 1991). This resulted in the identification of an additional domain involved in caseinolytic specificity located around amino acid residues 747 and 748 (see Figure 4.2). Furthermore, proteinases with novel specificities were obtained. In subsequent studies protein engineering was used to alter the catalytic properties of the SK11 proteinase by the introduction of specific mutations in the *prtP* gene (Siezen *et al.*, 1993; Bruinenberg *et al.*, 1993a, 1993b). The selection of these mutations, which affected single amino acid residues or complete domains, was based on differences in primary structure of proteinases with different caseinolytic specificity and modelling of the catalytic domain of the SK11 proteinase using known three-dimensional structures of members of the subtilase family (Siezen *et al.*, 1991; Siezen, 1993). Mutants of the SK11 proteinase were thus obtained that showed new substrate specificities and increased stability towards autoproteolysis (Siezen *et al.*, 1993). In addition, a third domain was identified in the SK11 proteinase that affects substrate binding and hence autoproteolysis (Bruinenberg *et al.*, 1993a).

4.9.3 Future perspectives

A variety of lactococcal strains have now been generated with changed proteolytic properties. Based on the rapid accumulation of genes for important components of the proteolytic system, it can be expected that many other, possibly improved, strains will follow. A recent example is the construction of food-grade peptidase-overproducing strains (van Alen-Boerrigter *et al.*, 1991 and chapter 2). However, it is evident that this panoply of new strains will not be limited to lactococci but will also include other lactic acid bacteria. In addition, heterologous genes from food-grade bacteria also have perspectives to be used in altering the proteolytic properties of lactic acid bacteria. This is illustrated by the cloning and expression in *L. lactis* of the *B. subtilis* neutral proteinase, an enzyme that had shown potential in accelerating cheese ripening (van der Guchte *et al.*, 1991).

In spite of the wealth of information on the proteolytic system discussed here it is difficult to predict the effect of these changes on the products made with these lactococci. This can partly be ascribed to our limited knowledge of what determines exactly the desired technological and organoleptical properties of fermented dairy products. Therefore, careful studies are required to evaluate the properties of the products made using lactic acid bacteria with altered proteolytic properties. Currently, these studies have been initiated and, irrespective of the possible selection of improved starter strains, they will allow for a better understanding of the contribution of the proteolytic system of lactic acid bacteria in the manufacture of fermented dairy products.

Acknowledgements

We are grateful to Gerard Venema, Alfred Haandrikman and Roland Siezen for valuable discussions and are indebted to our colleagues for communicating their results prior to publication.

References

Abo-Elnaga, I.G. and Plapp, R. (1987) Peptidases of *Lactobacillus casei* and *L. plantarum*. *J. Basic Microbiol*. **27**: 123–130.

Ames, G.F.L. (1986) Bacterial periplasmic transport systems: structure, mechanism, and evolution. *Annu. Rev. Biochem*. **55**: 397–425.

Atlan, D., Laloi, P. and Portalier, R. (1989) Isolation and characterization of aminopeptidase-deficient *Lactobacillus bulgaricus* mutants. *Appl. Environ. Microbiol*. **55**: 1717–1723.

Atlan, D., Laloi, P. and Portalier, R. (1990) X-prolyl dipeptidyl aminopeptidase of *Lactobacillus delbrueckii* subsp. *bulgaricus*: characterization of the enzyme and isolation of deficient mutants. *Appl. Environ. Microbiol*. **56**: 2174–2179.

Baankreis, R. (1992) The role of lactococcal peptidases in cheese ripening, Thesis, University of Amsterdam, Amsterdam, The Netherlands.

Baankreis, R. and Exterkate, F.A. (1991) Characterization of a peptidase from *Lactococcus lactis* ssp. *cremoris* HP that hydrolyses di- and tripeptides containing proline or hydrophobic residues as the aminoterminal amino acid. *System. Appl. Microbiol*. **14**: 317–323.

Bockelmann, W. (1987) Das proteolytische system von milchsäurebakterien. PhD thesis, University of Kiel, Kiel, Germany.

Bockelmann, W., Fobker, M. and Teuber, M. (1991) Purification and characterization of the X-prolyl-dipeptidyl-aminopeptidase from *Lactobacillus delbrueckii* subsp. *bulgaricus* and *Lactobacillus acidophilus*. *Int. Dairy J*. **1**: 51–66

Booth, M., Ni Fhaolain, I., Jennings, P.V. and O'Cuinn, G. (1990a) Purification and characterization of a post-proline dipeptidyl aminopeptidase from *Streptococcus cremoris* AM2. *J. Dairy Res*. **57**: 89–99.

Booth, M., Jennings, V., Ni Fhaolain, I. and O'Cuinn (1990b) Prolidase activity of *Lactococcus lactis* subsp. *cremoris* AM2: partial purification and characterization. *J. Dairy Res*. **57**: 245–254.

Bosman, B.W., Tan, P.S.T. and Konings, W.N. (1990) Purification and characterization of a tripeptidase from *Lactococcus lactis* subsp. *cremoris* Wg2. *Appl. Environ. Microbiol*. **56**: 1839–1843.

Bruinenberg, P.G., Vos, P. and de Vos, W.M. (1992) Proteinase overproduction in *Lactococcus lactis* strains: regulation and effect on growth and acidification in milk. *Appl. Environ. Microbiol*. **58**: 78–84.

Bruinenberg, P.G., de Vos, W.M. and Siezen, R.J. (1993a) Engineering of a surface loop of *Lactococcus lactis* SK11 cell-envelope proteinase affects activity, specificity and autoprocessing. submitted for publication.

Bruinenberg, P.G., Vos, P., Exterkate, F.A., Alting, A.C., de Vos, W.M. and Siezen, R.J. (1993b) Engineering stability and specificity of the *Lactococcus lactis* SK11 proteinase. In: *Stability and Stabilization of Enzymes* (W. van der Tweel, ed.), pp. 231–238.

Casey, M.G. and Meyer, J. (1985) Presence of X-prolyl-dipeptidyl-peptidase in lactic acid bacteria. *J. Dairy Sci*. **68**: 3212–3215.

Chapot-Chartier, M.-P., Nardi, M., Chopin, M.-C., Chopin, A. and Gripon, C. (1992) Cloning and sequencing of *pepC*, a cysteine aminopeptidase gene from *Lactococcus lactis* subsp. *cremoris* AM2. *Appl. Environ. Microbiol*. **59**: 330–333.

Citti, J.E., Sandiine W.E. and Elliker, P.R. (1965) Comparison of slow and fast acid-producing *Streptococcus lactis*. *J. Dairy Sci*. **48**: 1253–1258.

Cliffe, A.J. and Law, B.A. (1979) An electrophoretic study of peptidases in starter streptococci and in cheddar cheese. *J. Appl. Bacteriol*. **47**: 65–73.

Coolbear, T., Reid, J.R. and Pritchard, G.G. (1992) Stability and specificity of the cell wall-associated proteinase from *Lactococcus lactis* subsp. *cremoris* H2 released by tretament with lysozyme in the presence of calcium ions. *Appl. Environ. Microbiol.* **58**: 3263–3270.

De Vos, W.M. (1986a) Gene cloning in lactic streptococci. *Neth. Milk Dairy J.* **40**: 141–154.

De Vos, W.M. (1986b) Genetic improvement of starter streptococci by the cloning and expression of the gene coding for a non-bitter proteinase. In: *Biomolecular Engineering in the European Community: Achievements of the Research Programme(1982–1986): Final Report* (E. Magnien, ed.), Martinus Nijhoff Publishers, Dordrecht, pp. 465–472.

De Vos, W.M. (1987) Gene cloning and expression in lactic streptococci. *FEMS Microbiol. Rev.* **46**: 281–295.

De Vos, W.M. and Davies, F.L. (1984) Plasmid DNA in lactic streptococci. Bacteriophage resistance and proteinase plasmids in *Streptococcus cremoris* SK11. In: *Third European Congress on Biotechnology, Vol. III*, Verlag Chemie, Basel, pp. 202–206.

De Vos, W.M., Vos, P., Simons, G and David, S. (1989a) Gene organization and expression in mesophilic lactic acid bacteria. *J. Dairy Sci.* **72**: 3398–3405.

De Vos, W.M., Vos, P., De Haard, H. and Boerrigter, I. (1989b) Cloning and expression of the *Lactococcus lactis* subsp. *cremoris* SK11 gene encoding an extracellular serine proteinase. *Gene* **85**: 169–176.

De Vos, W.M., Boerrigter, I., Vos, P., Bruinenberg, P. and Siezen, R.J. (1991) Production, processing, and engineering of the *Lactococcus lactis* SK11 proteinase. In: *Genetics and Molecular Biology of Streptococci, Lactococci and Enterococci* (Dunny, G.M., Cleary, P.P. and McKay, L.L., eds.) American Society for Microbiology, Washington D.C., pp. 115–119.

Desmazeaud, M.J. (1974) General properties and specificity of an intracellular neutral endopeptidase from *Streptococcus thermophilus*. *Biochimie* **56**: 1173–1181.

Desmazeaud, M.J. and Zevaco, C. (1976) General properties and substrate specificity of an intracellular neutral protease from *Streptococcus diacetylactis*. *Ann. Biol. Anim. Bioch. Biophys.* **16**: 851–868.

Desmazeaud, M.J. and Zevaco, C. (1977) General properties and substrate specificity of an intracellular soluble dipeptidase from *Streptococcus diacetylactis*. *Ann. Biol. Anim. Bioch. Biophys.* **17**: 723–736.

Desmazeaud, M.J. and Zevaco, C. (1979) Isolation and general properties of two intracellular amino peptidases of *Streptococcus diacetylactis*. Milchwisenschaft **34**: 606–610.

Driessen, A.J.M. (1989) Secondary transport of amino acids by membrane vesicles derived from lactic acid bacteria. *Antonie van Leeuwenhoek* **56**: 139–160.

Driessen, A.J.M., Hellingwerf, K.J. and Konings, W.N. (1987) Membrane systems in which foreign protein pumps are incorporated. *Micro. Sci.* **4**: 173–180.

Eggimann, B. and Bachmann, H. (1980) Purification and partial characterization of an aminopeptidase from *Lactobacillus lactis*. *Appl. Environ. Microbiol.* **40**: 876–882.

El Soda, M. and Desmazeaud, M.J. (1981) General properties of a new aryl-peptidyl amidase in *Lactobacillus casei*. *Agric. Biol. Chem.* **45**: 1693–1700.

El Soda, M., Desmazeaud, M.J., Le Bars, D. and Zevaco (1986) Cell-wall-associated proteinases in *Lactobacillus casei* and *Lactobacillus plantarum*. *J. Food Protect.* **49**: 361–365.

El Soda, M., Desmazeaud, M. J. and Bergère, J.-L. (1978a) Peptide hydrolases of *Lactobacillus casei*: isolation and general properties of various peptidase activities. *J. Dairy Res.* **48**: 445–455.

El Soda, M., Bergère, J.-L. and Desmazeaud, M.J. (1978b) Detection and localization of peptide hydrolases in *Lactobacillus casei*. *J. Dairy Res.* **45**: 519–524.

Exterkate, F.A. (1976) Comparison of strains of *Streptococcus cremoris* for proteolytic activities associated with the cell wall. *Neth. Milk Dairy J.* **30**: 95–105.

Exterkate, F.A. (1977) Pyrrolidone carboxylyl peptidase in *Streptococcus cremoris*: dependence on an interaction with membrane components. *J. Bacteriol.* **129**: 1281–1288.

Exterkate, F.A. (1984) Location of peptidases outside and inside the membrane of *Streptococcus cremoris*. *Appl. Env. Microbiol.* **47**: 177–183.

Exterkate, F.A. (1990) Differences in short peptide-substrate cleavage by two cell envelope-located serine proteinases of *Lactococcus lactis* subsp. *cremoris* are related to secondary binding specificity. *Appl. Microbiol. Biotechnol.* **33**: 401–406.

Exterkate, F.A. and de Veer, G.J.C.M. (1987) Purification and some properties of a membrane-bound aminopeptidase A from *Streptococcus cremoris*. *Appl. Environ. Microbiol.* **53**: 577–583.

Exterkate, F.A., Alting, A.C. and Slangen, C.J. (1991) Specificity of two genetically related cell-envelope proteinases of *Lactococcus lactis* subsp. *cremoris* towards α_{s1}-casein-(1–23)-fragment. *Biochem. J.* **273**: 135–139.

Exterkate, F.A., Alting, A.C. and Bruinenberg, P.G. (1993) Diversity of cell-envelope proteinase specificity among strains of *Lactococcus lactis* and the relation with charge characteristics of the substrate binding region. *Appl. Environ. Microbiol.*, in press.

Ezzat, N., Zevaco, C., El Soda, M. and Gripon, J.C. (1987) Partial purification and characterization of a cell wall-associated proteinase from *Lactobacillus bulgaricus. Milchwissenschaft* **42**: 95–97.

Fischetti, V.A., Pancholi, V. and Schneewind, O. (1990) Conservation of a hexapeptide sequence in the anchor region of surface proteins from Gram-positive bacteria. *Mol. Microbiol.* **4**: 1603–1605.

Gasson, M.J., Hill, S.A. and Anderson, P.H. (1987) Molecular genetics of metabolic traits in lactic streptococci. In: *Streptococcal Genetics* (Ferretti, J., and Curtiss, R., eds.), American Society for Microbiology, Washington DC, pp. 242–246.

Geis, A., Bockelmann, W. and Teuber, M. (1985) Simultaneous extraction and purification of a cell wall-associated peptidase and β-casein specific proteinase from *Streptococcus cremoris* AC1. *Appl. Microbiol. Biotechnol.* **23**: 79–84.

Gilson, E., Alloing, G. Schmidt, T., Claverys, J. P., Dudler, R. and Hofnung, M. (1988) Evidence for high affinity binding protein-dependent transport systems in Gram-positive bacteria and in Mycoplasma. *EMBO J.* **7**: 3971–3974.

Goodell, E.W. and Higgins, C.F. (1987) Uptake of cell wall peptides by *Salmonella typhimurium* and *Escherichia coli. J. Bacteriol.* **169**: 3861–3865.

Haandrikman, A.J., Kok, J., Laan, H., Soemitro, S., Ledeboer, A.M., Konings, W.N. and Venema, G. (1989) Identification of a gene required for the maturation of an extracellular serine proteinase. *J. Bacteriol.* **171**: 2789–2794.

Haandrikman, A.J., Meesters, R., Laan, H., Konings, W.N., Kok, J. and Venema, G. (1991a) Processing of the lactococcal extracellular serine proteinase. *Appl. Environ. Microbiol.* **57**: 1899–1904.

Haandrikman, A.J., Kok, J. and Venema, G. (1991b) Lactococcal proteinase maturation protein PrtM is a lipoprotein. *J. Bacteriol.* **173**: 4517–4525.

Hickey, M.W., Hillier, A.J. and Jago, G.R. (1983) Peptidase activities in lactobacilli. *Aust. J. Dairy Technol.* **38**: 118–123.

Higgins, C.F. (1990) The role of ATP in binding protein-dependent transport systems. *Res. Microbiol.* **141**: 353–360.

Higgins, C.F., Hyde, S.C., Mimmack, M.M., Gileadi, U., Gill, D.R. and Gallagher, M.P. (1990) Binding protein-dependent transport systems. *J. Bioenergetics and Biomembranes* **22**: 571–592.

Hiles, I.D., Gallagher, M.P., Jamieson, D.J. and Higgins, C.F. (1987) Molecular characterization of the oligopeptide permease of *Salmonella typhimurium. J. Mol. Biol.* **195**: 125–142.

Hugenholtz, J., Exterkate, F.A. and Konings, W.N. (1984) The proteolytic systems of *S. cremoris*. An immunological analysis. *Appl. Environ. Microbiol.* **48**: 1105–1110.

Hugenholtz, J., Dijkstra, M. and Veldkamp, H. (1987) Amino acid limited growth of starter cultures in milk. *FEMS Microbiol. Ecol.* **53**: 149–155.

Hwang, I.K., Kaminogawa, S. and Yamauchi, K. (1981) Purification and properties of a dipeptidase from *Streptococcus cremoris. Agric. Biol. Chem.* **45**: 159–165.

Hwang, I.K., Kaminogawa, S. and Yamauchi, K. (1982) Kinetic properties of a dipeptidase from *Streptococcus cremoris. Agric. Chem. Biol.* **46**: 3049–3053.

Hyde, S.C., Emsley, P., Hartshorn, M. Mimmack, M.M., Gileadi, U., Pearce, S. R., Gallagher, M.P., Gill, D.R., Hubbard, R. and Higgins, C. F. (1990) Structural model of ATP-binding proteins associated with cystic fibrosis, multidrug resistance and bacterial transport. *Nature* **346**: 362–365.

Kaminogawa, S., Azuma, N., Hwang, I.K., Susuki, Y. and Yamauchi, K. (1984a) Isolation and characterization of a prolidase from *Streptococcus cremoris* H61. *Agric. Biol. Chem.* **48**: 3035–3040.

Kaminogawa, S., Ninomiya, T. and Yamauchi, K. (1984b) Aminopeptidase profiles of lactic streptococci. *J. Dairy Sci.* **67**: 2483–2492.

Khalid, N.M. and Marth, E.M. (1990a) Proteolytic activity by strains of *Lactobacillus plantarum* and *Lactobacillus casei. J. Dairy Sci.* **73**: 3068–3076.

Khalid, N.M. and Marth, E.M. (1990b) Purification and partial characterization of a prolyl-dipeptidyl aminopeptidase from *Lactobacillus helveticus* CNRZ 32. *Appl. Environ. Microbiol.* **56**: 381–388.

Khalid, N.M. and Marth, E.H. (1990c) Purification and partial characterization of an aminopeptidase from *Lactobacillus helveticus* CNRZ 32. *Syst. Appl. Microbiol.* **13**: 311–319.

Kiefer-Partsch, B., Bockelmann, W., Geis, A. and Teuber, M. (1989) Purification of an X-prolyl-dipeptidyl aminopeptidase from the cell wall proteolytic system of *Lactococcus lactis* subsp. *cremoris. Appl. Microbiol. Biotechnol.* **31**: 75–78.

Kiwaki, M., Ikemura, H., Shimizu-Kadota, M. and Hirashima, A. (1989) Molecular characterization of a cell wall-associated proteinase gene from *Streptococcus lactis* NCDO 763. *Mol. Microbiol.* **3**: 359–369.

Kok, J. (1990) Genetics of the proteolytic system of lactic acid bacteria. *FEMS Microbiol. Rev.* **87**: 15–42.

Kok, J. (1991) Proteinase genes of cheese starter cultures. *Biochem. Soc. Trans.* **19**: 670–674.

Kok, J. and Venema, G. (1988) Genetics of proteinases in lactic acid bacteria. *Biochimie* **70**: 475–488.

Kok, J., van der Vossen, J.M.B.M. and Venema, G. (1984) Construction of plasmid cloning vectors for lactic streptococci which also replicate in *Bacillus subtilis* and *Escherichia coli. Appl. Environ. Microbiol.* **48**: 726–731.

Kok, J., J.M. van Dijl, J.M.B.M. van der Vossen and Venema, G. (1985) Cloning and expression of a *Streptococcus cremoris* proteinase gene in *Bacillus subtilis* and *Streptococcus lactis. Appl. Environ. Microbiol.* **50**: 94–101.

Kok, J., Leenhouts K.J., Haandrikman, A.J., Ledeboer, A.M. and Venema, G. (1988a) Nucleotide sequence of the cell wall proteinase gene of *Streptococcus cremoris* Wg2. *Appl. Environ. Microbiol.* **54**: 231–238.

Kok, J., Hill, D., Haandrikman, A., de Reuver, M.J.B., Laan, H. and Venema, G. (1988b) Deletion analysis of the proteinase gene of *Streptococcus cremoris* Wg2. *Appl. Environ. Microbiol.* **54**: 239–244.

Kolstad, J. and Law, B.A. (1985) Comparative specificity of cell wall, membrane and intracellular peptidases of group N streptococci. *J. Appl. Bacteriol.* **58**: 449–456.

Konings, W.N., Poolman, B. and Driessen, A.J.M. (1989) Bioenergetics and solute transport in lactococci. *CRC Critical Reviews in Microbiology* **16**: 419–476.

Kontinen, V.P. and Sarvas, M. (1988) Mutants of *Bacillus subtilis* defective in protein export. *J. Gen. Microbiol.* **134**: 2333–2344.

Kontinen, V.P., Saris, P. and Sarvas, M. (1991) A gene (*prsA*) of *Bacillus subtilis* involved in a novel, late stage of protein export. *Mol. Microbiol.* **5**: 1273–1283.

Kuipers, O.P., Rollema, H.S., de Vos, W.M. and Siezen, R.J. (1993) Characterization of the nisin gene cluster *nisABCTCIPR* of *Lactococcus lactis*: requirement of expression of the *nisA* and *nisI* genes for producer immunity. *Eur. J. Biochem.*, **216**, 281–291.

Kunji, E.R.S., and Smid, E.J., Plapp, R., Poolman, B. and Konings, W.N. (1993) Ditripeptides and oligopeptides are taken up via distinct transport mechanisms in *Lactococcus lactis. J. Bacteriol.* **175**: 2052–2059.

Laan, H., Bolhuis, H., Poolman, B., Abee, T. and Konings, W.N. (1993) Regulation of proteinase synthesis in *Lactococcus lactis. Acta Biotechnol.* **13**: 95–101.

Laan, H. and Konings, W.N. (1989) Mechanism of proteinase release from *Lactococcus lactis* subsp. *cremoris* Wg2. *Appl. Environ. Microbiol.* **55**: 3101–3106.

Laan, H. and Konings, W.N. (1990) Autoproteolysis of the extracellular serine proteinase of *Lactococcus lactis* subsp. *cremoris* Wg2. *Appl. Environ. Microbiol.* **57**: 2586–2590.

Laan, H., Smid, E.J., De Leij, L., Schwander, E. and Konigs, W.N. (1988) Monoclonal antibodies to cell-wall-associated proteinase of *Lactococcus lactis* subsp. *cremoris* Wg2. *Appl. Environ. Microbiol.* **54**: 2250–2256.

Laan, H., Smid, E.J., Tan, P.S.T. and Konings, W.N. (1989) Enzymes involved in the degradation and utilization of casein in *L. lactis. Neth. Milk Dairy J.* **43**: 327–345.

Laan, H., Kok, J., Haandrikman, A.J., Venema, G. and Konings, W.N. (1992) Localization and accessibility of antigenic sites of the extracellular proteinase of *Lactococcus lactis. Eur. J. Biochem.* **204**: 815–820.

Laloi, P., Atlan, D., Blanc, B., Gilbert, C. and Portalier, R. (1991) Cell-wall associated proteinase of *Lactobacillus delbrueckii* subsp. *bulgaricus* CNRZ 397: differential extraction, purification and properties of the enzyme. *Appl. Microbiol. Biotechnol.* **36**: 196–204.

Law, B.A. (1979) Extracellular peptidases in group N streptococci used as cheese starters. *J. Appl. Bacteriol.* **46**: 455–463.

Law, B.A. and Kolstad, J. (1983) Proteolytic systems in lactic acid bacteria. *Antonie van Leeuwenhoek* **49**: 225–245.

Law, J., Vos, P., Hayes, F., Daly, C., de Vos, W.M. and Fitzgerald, G.F. (1992) Cloning and partial sequencing of the proteinase gene complex from *Lactococcus lactis* subsp. *lactis* UC317. *J. Gen. Microbiol.* **138**: 1–10.

Leenhouts, J.K., Kok, J. and Venema, G. (1991) Chromosomal stabilization of the proteinase genes in *Lactococcus lactis*. *Appl. Environ. Microbiol.* **57**: 2568–2575.

Machuga, E.J. and Ives, D.H. (1984) Isolation and characterization of an aminopeptidase from *Lactobacillus acidophilus* R–26. *Biochem. Biophys. Acta* **789**: 26–36.

Martin-Hernandez, M.C., Alting, A.C. and Exterkate, F.A. (1993) Purification and characterization of the mature membrane-associated cell-envelope proteinase of *Lactobacillis helveticus* L89. *Appl. Microbiol. Biotechnol.*, in press.

Mayo, B., Kok, J., Venema, K., Bockelmann, W., Teuber, M., Reinke, H. and Venema, G. (1991) Molecular cloning and sequence analysis of the X-prolyl dipeptidyl aminopeptidase gene from *Lactococcus lactis* subsp. *cremoris. Appl. Environ. Microbiol.* **57**: 38–44.

Mayo, B., Kok, J., Bockelmann, W., Haandrikman, A., Leenhouts, K.J. and Venema, G. (1993) The effect of X-prolyl dipeptidyl aminopeptidase deficiency in *Lactococcus lactis. Appl. Environ. Microbiol.* **59**: 2049–2055.

McKay, L.L. (1983) Functional properties of plasmids in lactic streptococci. *Antonie van Leeuwenhoek* **49**: 259–274.

McKay, L.L. and Baldwin, K.A. (1978) Stabilization of lactose metabolism in *Streptococcus lactis* C2. *Appl. Environ. Microbiol.* **36**: 360–367.

Meyer, J. and Jordi, R. (1987) Purification and characterization of X-prolyl-dipeptidyl-aminopeptidase from *Lactobacillus lactis* and from *Streptococcus thermophilus. J. Dairy Sci.* **70**: 738–745.

Mierau, I., Tan, P.S.T., Haandrikman, A.J., Mayo, B., Kok, J., Konings, W.N. and Venema, G. (1993) Cloning and sequencing of the gene for a lactococcal endopeptidase, an enzyme with sequence similarity to mammalian enkephalinase *J. Bacteriol.* **175**: 2087–2096.

Mills, O.E. and Thomas, T.D. (1978) Release of cell-wall-associated proteinase(s) from lactic streptococci. *N.Z.J. Dairy Sci. Techol.* **13**: 209–215.

Moon, Y.-I., Kim, M.-B. and Kim, Y-K. (1988) A study on the proteolytic action of extracellular protease of *Lactobacillus bulgaricus* CH–18. *Korean J. Anim. Sci.* **30**: 498–504.

Monnet, V., Ley, J.P. and Gonzalez (1992) Substrate specificity of the cell envelope-located proteinase of *Lactococcus lactis* subsp. *lactis* NCDO 763. *Int. J. Biochem.* **24**: 707–718.

Mou, L., Sullivan, J.J. and Jago, G.R. (1975) Peptidase activities in group N streptococci. *J. Dairy Res.* **42**: 147–155.

Muset, G., Monnet, V. and Gripon, J.-C. (1989) Intracellular proteinase of *Lactococcus lactis* subsp. *lactis* NCDO 763. *J. Dairy Res.* **56**: 765–778.

Naes, H. and Nissen-Meyer, J. (1992) Purification and N-terminal amino acid sequence determination of the cell-wall-bound proteinase from *Lactobacillus paracasei* subsp. *paracasei. J. Gen. Microbiol.* **138**: 313–318.

Nardi, M., Chopin, M.-C., Chopin, A., Cals, M.-M. and Gripon, J.-C. (1991) Cloning and DNA sequence analysis of an X-prolyl dipeptidyl peptidase gene from *Lactococcus lactis* subsp. *lactis* NCDO 753. *Appl. Environ. Microbiol.* **57**: 45–50.

Neviani, E., Boquien, C.Y., Monnet, V., Phan Thanh, L. and Gripon, J.-C. (1989) Purification and characterization of an aminopeptidase from *Lactococcus lactis* ssp. *cremoris* AM2. *Appl. Environ. Microbiol.* **55**: 2308–2314.

Nissen-Meyer, J. and Sletten, K. (1991) Purification and characterization of the free form of the lactococcal proteinase and the autoproteolytic cleavage products. *J. Gen. Microbiol.* **137**: 1611–1618.

Niven, G.W. (1991) Purification and properties of aminopeptidase A from *Lactococcus lactis* subsp. *lactis* NCDO 712. *J. Gen. Microbiol.* **137**: 1207–1212.

Nowakowski, C.M., Bhowmik, T.K. and Steele, J.L. (1993) Cloning of peptidase genes from *Lactobacillus helveticus* CNRZ 32. *Appl. Microbiol. Biotechnol.* **39**: 204–210.

Otto, R., de Vos, W.M. and Gavrieli, J. (1988) Plasmid DNA in *Streptococcus cremoris*: influence of pH on selection in chemostats of a variant lacking a protease plasmid. *Appl. Environ. Microbiol.* **43**: 1272–1277.

Perego, M., Higgins, C.F., Pearce, S.R., Gallagher, M.P. and Hoch, J.A. (1991) The oligopeptide transport system of *Bacillus subtilis* plays a role in the initiation of sporulation. *Mol. Microbiol.* **5**: 173–186.

Pugsley, A.P. (1988) In: *Protein Transfer and Organelle Biosynthesis* (Das, R.A., and Robbins, P.W., eds), Academic Press, Orlando, F1, pp. 607–652.

Rabier, D. and Desmazeaud, M.J. (1973) Inventaire des différentes activites peptidasiques intracellulaires des *Streptococcus thermophilus*. Purification et proprietés d'une dipeptide-hydrolase et d'une aminopeptidase. *Biochimie* **55**: 389–404.

Rauch, P.J.G. and de Vos, W.M. (1992) Characterization of the novel nisin-sucrose conjugative transposon Tn*5276* and its insertion in *Lactococcus lactis*. *J. Bacteriol.* **174**: 1280–1287.

Reid, J.R., Moore, C.H., Midwinter, G.G. and Pritchard, G.G. (1991a) Action of a cell wall proteinase from *Lactococcus lactis* subsp. *cremoris* SK11 on bovine α_{s1} and β-casein. *Appl. Microbiol. Biotechnol.* **35**: 222–227.

Reid, J.R., Ng., K.H., Moore, C.H., Coolbear, T. and Pritchard, G.G. (1991b) Comparison of bovine α-casein hydrolysis by PI and PIII-type proteinases from *Lactococcus lactis* subsp. *cremoris*. *Appl. Microbiol. Biotechnol.* **36**: 344–351.

Siezen, R.J. (1993) Modelling and engineering of enzyme/substrate interactions in subtilisin-like enzymes of unknown 3-dimensional structures. In: *Proceedings International Symposium on Subtilisin Enzymes* (Betzel, Ch., and Bott, R., eds), Plenum Press, in press.

Siezen, R.J., de Vos, W.M., Leunissen, J.A.M. and Dijkstra, B.W. (1991) Homology modelling and protein engineering strategy of subtilases, the family of subtilisin-like proteinases. *Prot. Engin.* **4**: 719–737.

Siezen, R.J., Bruinenberg, P.G., Vos, P., van Alen- Boerrigter, I.J., Nijhuis, M., Alting, A.C., Exterkate, F.A. and de Vos, W.M. (1993) Engineering of the substrate binding region of the subtilisin-like, cell envelope proteinase of *Lactococcus lactis*. *Prot. Engin.*, in press.

Smid, E.J. (1991) Physiological implications of peptide transport in lactococci. Thesis, University of Groningen, Groningen, The Netherlands.

Smid, E.J., Driessen, A.J. and Konings, W.N. (1989a) Mechanism and energetics of dipeptide transport in membrane vesicles of *Lactococcus lactis*. *J. Bacteriol.* **171**: 292–298.

Smid, E.J., Plapp, R. and Konings, W.N. (1989b) Peptide uptake is essential for growth of *Lactococcus lactis* on the milk protein casein. *J. Bacteriol.* **171**: 6135–6140.

Sørhaug, T. and Solberg, P. (1973) Fractionation of dipeptidase activities of *Streptococcus lactis* and dipeptidase specificity of some lactic acid bacteria. *Appl. Microbiol.* **25**: 388–395.

Stadhouders, J., Toepoel, L. and Wouters, J.T.M. (1988) Cheesemaking with Prt- and Prt+ variants of N-streptococci and their mixtures. Phage sensitivity, proteolysis, and flavour development during ripening. *Neth. Milk Dairy J.* **42**: 182–193.

Strøman, P. (1992) Sequence of a gene (*lap* encoding a 95.3 kDa aminopeptidase from *Lactococcus lactis* ssp. *cremoris* Wg2. *Gene* **113**: 107–112.

Tan, P.S.T. (1992) The biochemical, genetic and physiological properties of aminopeptidase N from *Lactococcus lactis*. PhD Thesis, University of Groningen, Groningen, The Netherlands.

Tan, P.S.T. and Konings, W.N. (1990) Purification and characterization of an aminopeptidase from *Lactococcus lactis* subsp. *cremoris* Wg2. *Appl. Environ. Microbiol.* **56**: 526–532.

Tan, P.S.T., Pos, K.M. and Konings, W.N. (1991) Purification and characterization of an endopeptidase from *Lactococcus lactis* subsp. *cremoris* Wg2. *Appl. Environ. Microbiol.* **57**: 3539–3599.

Tan, P.S.T., Chapot-Chartier, M.-C., Pos, K.M., Rousseau, M., Boquien, C.-Y., Gripon, J.-C. and Konings, W.N. (1992a) Location of peptidases in lactococci. *Appl. Environ. Microbiol.* **58**: 285–290.

Tan, P.S.T., van Alen-Boerrigter, I.J., Poolman, B., Siezen, R.J., de Vos, W.M. and Konings, W.N. (1992b) Characterization of the *Lactococcus lactis pepN* gene encoding an aminopeptidase homologous to mammalian aminopeptidase N. *FEBS* **306**: 9–16.

Tan, P.S.T., van Kessel, T.A.J.M., van de Veerdonk, F.L.M., Zuurendonk, P.F., Bruins, A.P. and Konings, W.N. (1993) Degradation and debittering of a tryptic digest from β-casein by aminopeptidase N from *Lactococcus lactis* subsp. *cremoris* Wg2. *Appl. Environ. Microbiol.* **59**: 1430–1436.

Thomas, T.D. and Mills, O.E. (1981) Proteolytic enzymes of starter bacteria. *Neth. Milk Dairy J.* **35**: 255–273.

Thomas, T.D. and Pritchard, G.G. (1987) Proteolytic enzymes of dairy starter cultures. *FEMS Microbiol. Rev.* **46**: 245–268.

Tynkkynen, S. and von Wright, A. (1988) Characterization of a cloned chromosomal fragment affecting the proteinase activity of *Streptococcus lactis* ssp. *lactis*. *Biochimie* **70**: 531–534.

Tynkkynen, S., von Wright, A. and Syväoja, E-L. (1989) Peptide utilization encoded by *Lactococcus lactis* subsp. *lactis SSL135 chromosomal DNA. Appl. Environ. Microbiol.* **55**: 2690–2695.

Tynkkynen, S., Buist, G., Kunji, E., Kok, J., Poolman, B., Venema, G. and Haandrikman, A. (1993) Genetic and biochemical characterization of the oligopeptide transport system of *Lactococcus lactis. J. Bacteriol.*, in press.

Van Alen-Boerrigter, I.J., Baankreis, R. and de Vos, W.M. (1991) Characterization and overexpression of the *Lactococcus lactis pepN* gene and localization of its product, aminopeptidase N. *Appl. Environ. Microbiol.* **57**: 2555–2561.

Van Boven, A., Tan, P.S.T. and Konings, W.N. (1988) Purification and characterization of a dipeptidase from *Streptococcus cremoris* Wg2. *Appl. Environ. Microbiol.* **54**: 43–49.

Van der Guchte, M., Kodde, J., van der Vossen, J.M.B.M., Kok, J. and Venema, G. (1990) Heterologous gene expression in *Lactococcus lactis* subsp. *lactis*: Synthesis, secretion, and processing of the *Bacillus subtilis* neutral protease. *Appl. Environ. Microbiol.* **56**: 2606–2611.

Van der Meer, J.R., Polman, J., Beerthuyzen, M.M., Siezen, R.J., Kuipers, O.P. and de Vos, W.M. (1993) Characterization of the *Lactococcus lactis* Nisin A operon genes, *nisP*, encoding a subtilisin-like serine protease involved in the precursor processing, and *nisR*, encoding a regulatory protein involved in nisin biosynthesis. *J. Bacteriol.* **175**: 2578–2588.

Van der Vossen, J.M.B.M., Kodde, J., Haandrikman, A.J., Venema, G. and Kok, J. (1992) Characterization of transcription initiation and termination signals of the proteinase genes of *Lactococcus lactis* Wg2 and enhancement of proteolysis in *L. lactis. Appl. Environ. Microbiol.* **58**: 3142–3149.

Visser, S. (1993) Proteolytic enzymes and their relation to cheese ripening and flavor: An overview. *J. Dairy Sci.* **76**: 329–350.

Visser, S., Hup., G., Exterkate, F.A. and Stadhouders, J. (1983) Bitter flavour in cheese. Model studies on the formation and degradation of bitter peptides by proteolytic enzymes from calf rennet, starter cells and starter cell fractions. *Neth. Milk Dairy J.* **37**: 169–175.

Visser, S., Exterkate, F.A., Slangen, K.J. and de Veer, G.J.C.M. (1986) Comparative study of the action of cell wall proteinases from various strains of *Streptococcus cremoris* on bovine α_{s1}-β- and 2-casein. *Appl. Environ. Microbiol.* **52**: 1162.

Visser, S., Robben, A.J.P.M. and Slangen, C.J. (1991) Specificity of a cell envelope-located proteinase (PIII-type) from *Lactococcus lactis* subsp. *cremoris* AM1 and its action on bovine β-casein. *Appl. Microbiol. Biotechnol.* **35**: 477–483.

Von Wright, A., Tynkkynen, S. and Suominen, M. (1987) Cloning of a *Streptococcus lactis* subsp. *lactis* chromosomal fragment associated with the ability to grow in milk. *Appl. Environ. Microbiol.* **53**: 1584–1588.

Vos, P., Simons, G., Siezen, R.J. and de Vos, W.M. (1989a) Primary structure and organization of the gene for a procaryotic, cell envelope-located serine proteinase. *J. Biol. Chem.* **264**: 13579–13585.

Vos, P., van Asseldonk, M., van Jeveren, F., Siezen, R., Simons, G. and de Vos, W.M. (1989b) A maturation protein is essential for the production of active forms of *Lactococcus lactis* SK11 serine proteinase located in or secreted from the cell envelope. *J. Bacteriol.* **171**: 2795–2802.

Vos, P., Boerrigter, I.J., Buist, G., Haandrikman, A.J., Nijhuis, M., de Reuver, M.B., Siezen, R.J., Venema, G., de Vos, W.M. and Kok, J. (1991) Engineering of the *Lactococcus lactis* proteinase by construction of hybrid enzymes. *Protein Eng.* **4**: 479–484.

Xu, F., Pearce, L.E. and Yu, P.-L. (1990) Molecular cloning and expression of a proteinase gene from *Lactococcus lactis* subsp. *cremoris* H2 and construction of a new lactococcal vector pFX1. *Arch. Microbiol.* **154**: 99–104.

Yan, T-R., Azuma, N., Kaminogawa, S. and Yamauchi, K. (1987a) Purification and characterization of a substrate-size-recognizing metalloendopeptidase from *Streptococcus cremoris* H61. *Appl. Environ. Microbiol.* **53**: 2296–2302.

Yan, T-R., Azuma, N., Kaminogawa, S. and Yamauchi, K. (1987b) Purification and characterization of a novel metalloendopeptidase from *Streptococcus cremoris* H61. *Eur. J. Biochem.* **163**: 259–265.

Zevaco, C., Monnet, V. and Gripon, J.-C. (1990) Intracellular X-prolyl dipeptidyl peptidase from *Lactococcus lactis* ssp. *lactis*: purification and properties. *J. Appl. Bacteriol.* **68**: 357–366.

5 Bacteriocins of lactic acid bacteria

H.M. DODD and M.J. GASSON

5.1 Introduction

The production of bacteriocins by lactic acid bacteria has been known for many years. The potential value of this property in the preservation of fermented food products has been recognized and has prompted many research groups to direct their studies to an investigation of these antagonistic proteinaceous compounds. In the past, analysis of bacteriocins has in general concentrated on biochemical properties, i.e. inhibition spectra, crude purification and size determination. More recently advances in the techniques available for genetic study of lactic acid bacteria have allowed the characterization of these antimicrobial agents at the molecular level. The genetic determinants for the production of and immunity to a number of bacteriocins have been cloned and sequenced. Progress made in studying the molecular biology of these molecules has led to a greater understanding of the mechanisms by which they exert their antimicrobial effect. The aim of this chapter is to review information currently available on bacteriocins produced by lactic acid bacteria. The focus is on more recent developments in their genetic and biochemical characterization.

It is becoming apparent that, whilst there is a wide range of bacteriocins produced by many different lactic acid bacteria, a number of common features exist. In a recent review of bacteriocins produced by the genus *Lactobacillus*, Klaenhammer *et al.* (1992) have assigned antimicrobial proteins into one of three groups based on size, stability to heat and presence of modified amino acids. This classification system can be extended to include the many antagonistic proteins and peptides generated by other genera and on this basis the various bacteriocins discussed in this chapter are divided into three groups:

(i)   small heat-stable peptides;
(ii)   large heat-labile proteins;
(iii) modified peptides i.e. lantibiotics.

In some cases only rudimentary genetic or biochemical data is available making inclusion in a particular group provisional. However, it is anticipated that current interest in alternative bacteriocins and the rapid technical advances in the analysis of lactic acid bacteria will soon permit their definite assignment to one or other of the above groups.

A report of a novel bacteriocin which may represent a fourth group has recently been described. Lactococcin G from *Lactococcus lactis* subsp. *lactis* LMG 2081 involves two distinct peptides (Nissen-Meyer *et al.*, 1992). Both are small and hydrophobic and display properties that might suggest they are members of the group of small heat-stable proteins. However, unlike bacteriocins that have been described previously, lactococcin G depends on the complementary activity of the two antimicrobial peptides to exert its inhibitory effect (Nissen-Meyer *et al.*, 1992). This novel property suggests that lactococcin G should be classed in a separate group of bacteriocins that require the complementary action and cooperation of more than one peptide.

Those bacteriocins which have attracted interest and for which investigation has yielded some molecular data are listed in Table 5.1. Five different genera are represented from the lactic acid bacteria with most entries under *Lactobacillus* and *Lactococcus*. This is presumably a reflection of the long-term interest in those two industrially important organisms and the fact that their molecular analysis is more technically advanced than that of the other genera listed. *Pediococcus, Leuconostocs* and the newly reclassified *Carnobacterium* (Collins *et al.*, 1987) are currently attracting interest in the food industry and this has led to a more intensive investigation of the antagonistic properties of these organisms.

Until relatively recently the only bacteriocin from lactic acid bacteria to have undergone characterization in any depth was the lantibiotic nisin from *Lactococcus* (section 5.4). Over the last few years a number of bacteriocins representing the other two groups have been investigated at the molecular level. Significant advances have been made in the genetic and biochemical analysis of lactococcins A, B and M from *Lactococcus lactis* subsp. *cremoris* (van Belkum *et al.*, 1989; 1991a,b; Holo *et al.*, 1991), lactacin F from *Lactobacillus acidophilus* (Muriana and Klaenhammer, 1987; 1991a,b), leucocin A UAL-187 from *Leuconostoc gelidum* (Hastings and Stiles, 1991; Hastings *et al.*, 1991) and pediocin PA-1 from *Pediococcus acidilactici* (Marugg *et al.*, 1992; Henderson *et al.*, 1992). Unlike nisin the latter bacteriocins were all found to be plasmid encoded. Other similarities were evident, both biochemically and genetically, justifying their grouping together in the class of small heat-stable peptides which is discussed in detail in section 5.2.

A large number of other bacteriocins have been described in the literature which have undergone preliminary analysis and which are referred to in Table 5.1. In the majority of cases bacteriocin production appears to be associated with the presence of plasmids in the host strain. For example, the genetic determinants for both production of and immunity to a lactococcin from *L. lactis* subsp. *lactis* ADRIA 85L030 have been linked to a 10 kb region of plasmid, pOS5 (Dufour *et al.*, 1991). Other bacteriocins have been tentatively associated with plasmids on the basis of curing or conjugation experiments. These include diplococcin from *L. lactis* subsp. *cremoris* 346 (Davey and Richardson, 1981; Davey, 1984), and bacteriocin S50 from *L. lactis* subsp. *diactylactis* S50 (Kojic *et al.*, 1991). Bacteriocin producing strains of *Lactobacillus* have recently attracted interest and there is good evidence that the genetic determinants for lactocin S from *Lactobacillus sake* L45 are encoded by a 50 kb unstable plasmid (Mortvedt and Nes, 1990; Mortvedt *et al.*, 1991; section 5.4). Preliminary genetic analysis has also indicated that sakacin A from *Lactobacillus sake* LB706 (Schillinger and Lucke, 1989; Holck *et al.*, 1992) and sakacin P from *Lactobacillus sake* LTH673 (Tichaczek *et al.*, 1992) are plasmid-encoded.

Until relatively recently, the molecular characterization of bacteriocins from *Pediococcus* strains was limited to plasmid-encoded pediocin PA-1 produced by *Pediococcus acidilactici* PAC1.0 (section 5.2). However, several other strains of *Pediococci* have been found to produce bacteriocins as a consequence of reportedly plasmid-encoded genetic determinants (Daeschel and Klaenhammer, 1985; Graham and McKay, 1985; Ray *et al.*, 1989; Table 5.1). They display similar characteristics being small, heat stable proteins with a relatively broad spectrum of activity. A recent report on the purification and characterization of pediocin AcH, produced by *Pediococcus acidilactici* H, revealed that the amino acid sequence was identical to that of pediocin PA-1 (Motlagh *et al.*, 1992).

Within the genus of *Carnobacterium* a number of strains have been identified as producers of different bacteriocins (Shaw and Harding, 1985; Ahn and Stiles, 1990a, Stoffels *et al.*, 1992a,b). Genetic evidence has only been presented for plasmid-encoded bacteriocins produced by *Carnobacterium piscicola* LV17 (Ahn and Stiles, 1990a,b). Recently it has been reported that this strain contains two plasmids (40 and 49 MDa) each of which mediate the production of 3 independent bacteriocins (Worobo *et al.*, 1991).

Where convincing physical data is lacking the genetic determinants for the majority of the above mentioned bacteriocins can only provisionally be classed as plasmid-encoded. A degree of care is required when assigning bacteriocin determinants to a particular genomic location. This is well illustrated by the case of the genes for nisin biosynthesis in *L. lactis* subsp. *lactis* which, because of their transmissible and curable nature (Gasson, 1984; Tsai and Sandine, 1987), were though to be plasmid encoded. More recent molecular analysis has established that these genes are located on the chromosome (Dodd *et al.*, 1990) and are carried by a large conjugative transposon (Horn *et al.*, 1991; Rauch and De Vos, 1992; see chapter 1). Genetic analysis of nisin has progressed rapidly over recent years with a number of research groups investigating the biosynthesis of this post-translationally modified peptide. Because of its unique properties the genetic and structural analysis of nisin will be dealt with in a separate section together with other lantibiotics from lactic acid bacteria (section 5.4).

The only other identified bacteriocins thought to be chromosomally encoded are lactacin B, lactocin 27 and helveticin J, all produced by *Lactobacillus* strains (Table 5.1). With the former two bacteriocins these conclusions are based on an inability to physically isolate plasmid DNA from the producing strains (Barefoot and Klaenhammer, 1983; 1984; Upreti and Hinsdill, 1975). Apart from nisin determinants only genes for helveticin J biosynthesis and immunity have been

Table 5.1

Producer strain	Bacteriocin	Genetic determinants		Genes characterized			Sequence		Size kDa	Lanti-biotic	References
		Plasmid	Ch'some	Structural	Immunity	Other	nt	aa			
Lactobacillus											
L. helveticus LP27	lactocin 27	+							12.4		Upretti and Hindsdill (1975)
L. helveticus 487	helveticin J	+	+	*hlvJ*		ORF2	+		37		Joerger and Klaenhammer (1986; 1990), Klaenhammer et al. (1992)
L. acidophilus N2	lactacin B	+							6.3		Barefoot and Klaenhammer (1988; 1984)
L. acidophilus 11088	lactacin F	110 kb		*laf*		ORFs 23,4	+	+[a]	6.3		Muriana and Klaenhammer (1987; 1991a,b), Klaenhammer et al. (1992)
L. sake L45	lactocin S	50 kb		*lasA*			+	+[a]	3.8	+	Mortvedt and Nes (1990), Mortvedt et al. (1991), Nes et al. (1993)
L. sake LB706	sakacin A	28 kb		*sakA*			+	+	4.3		Schillinger and Lucke (1989), Holck et al. (1992)
L. sake LTH673	sakacin P							+[a]			Tichaczek et al. (1992)
L. curvatus LTH1174	curvacin A							+[a]			Tichaczek et al. (1992)
Lactococcus											
L. lactis subsp. *lactis* ATCC11454, 6F3, F15876, NIZO R5	nisin A	(+)[b]	+[c]	*nisA*	*nisI*	*nisB.T. C.P.R*	+	+	3.35	+	Buchman et al. (1988; Kaletta and Entian (1989); Dodd et al. (1990); Steen et al. (1991); Engelke et al. (1992); van der Meer et al. (1993); Kuipers et al. (1993)
L. lactis subsp. *lactis* NIZO 22186, N8, SIK83	nisin Z		+	*nisZ*			+	+		+	Graefe et al. (1991); Mulders et al. (1991)
L. lactis subsp. *lactis* ADRIA 85LO30	lactococcin	70 kb		*bac*	*imm*				2.4		Dufour et al. (1991)
L. lactis subsp. *lactis* CNRZ 481	lacticin 481	70 kb		*lct*			+	+[a]	2.9	+	Piard et al. (1992, 1993)
L. lactis subsp. *lactis* LMG 2081	lactococcin G							+	4.3, 4.1		Nissen-Meyer et al. (1992)
L. lactis subsp. *lactis* biovar diacetyllactis WM4	lactococcin A	131 kb		*lcnA*	*lciA*	*lcnC, lcnD*	+				Scherwitz-Harmon and McKay (1987), Stoddard et al. (1992)

	Bacteriocin	Plasmid							pI		Reference
L. lactis subsp. *lactis* biovar diacetyllactis S50	bacteriocin S50	+									Kojic *et al.* (1991)
L. lactis subsp. *lactis* 9B4	lactococcin A	60 kb	*lcnA*	*lciA*		+		+			van Belkum *et al.* (1991a,b)
L. lactis subsp. *cremoris* 9B4	lactococcin B	60 kb	*lcnB*	*lciB*		+		+			van Belkum *et al.* (1992); Venema *et al.* (1993)
L. lactis subsp. *lactis* 9B4	lactococcin M	60 kb	*lcnMa* *lcnMb*	*lciM*		+		+			van Belkum *et al.* (1991a)
L. lactis subsp. *lactis* LMG2130	lactococcin A	55 kb	*lcnA*			+		+	5.8		Holo *et al.* (1991)
L. lactis subsp. *lactis* 346	diplococcin	81 kb							5.3		Davey (1984); Davey and Richardson (1981)
Leuconostoc											
L. gelidum UAL187	leucocin UAL-1	11.6 kb	*lcnA*		ORF2	+			3.9		Hastings and Stiles (1991); Hastings *et al.* (1991)
L. mesenteroides Y105	mesentericin Y105										Hechard *et al.* (1992)
Pediococcus											
P. acidilactici PAC1.0	pediocin PA.1	9.3 kb	*pedA*		*ped B* *ped C* *ped D*	+		+	4.6		Gonzalez and Kunka (1987); Henderson *et al.* (1992); Marugg *et al.* (1992); Nieto Lozano *et al.* (1992)
P. acidilactici H	pediocin AcH	8.9 kb	*pap*			+		+	2.7		Ray *et al.* (1989, 1992); Motlagh *et al.* (1992)
P. acidilactici SJ-1	pediocin SJ-1	7.8 kb							4.0		Schved *et al.* (1993)
P. pentosaceus L7230, FBB63	pediocin A	20 kb									Daeschel and Klaenhammer (1985)
P. cerevisae FBB63	bacteriocin	15.8 kb									Graham and McKay (1985)
Carnobacterium											
C. piscicola LV17	bacteriocins A1,A2,A3 B1,BM1,BM2	61.2 kb 75 kb						+[a]	4.6	+	Ahn and Stiles (1990a,b); Worobo *et al.* (1991)
Carnobacterium ssp. U149	carnocin U149										Stoffels *et al.* (1992a,b)

[a] partial amino acid sequence has been determined (see Figure 5.2).

[b] Kaletta and Entian (1989) reported that the *nisA* gene from *L. lactis* subsp. *lactis* 6F3 was located on a large plasmid.

[c] Nisin determinants from *L. lactis* subsp. *lactis* strains F15876 and NIZO R5 are carried by the chromosomally located conjugative elements Tn*5301* (Horn *et al.*, 1991) and Tn*5276* (Rauch and De Vos, 1992) respectively.

isolated and shown unequivocally to be chromosomally encoded (Joerger and Klaenhammer, 1986; 1990; section 5.3.1).

5.2 Small heat-stable bacteriocins

In a number of cases the initial genetic studies implying association of bacteriocin determinants with plasmids have been backed up by physical data. This has included the cloning of genes involved in bacteriocin production and immunity, nucleotide sequence determination, and in certain cases expression of the isolated gene(s) has been demonstrated. Figure 5.1 shows six plasmid-encoded bacteriocins for which molecular data is available enabling the structural organization of genetic determinants to be elucidated. These are lactococcins A, B and M from *Lactococcus lactis* subsp. *cremoris*, lactacin F from *Lactobacillus acidophilus*, leucocin A UAL-187 from *Leuconostoc gelidum* and pediocin PA-1 from *Pediococcus acidilactici*. From these maps (Figure 5.1) it is clear that a pattern is emerging in which two or more genes are arranged in an operon-like structure with the first gene specifying the structural gene for the particular bacteriocin. The grouping together of these peptides into the class

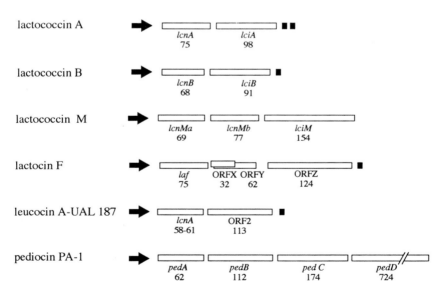

Figure 5.1 Genetic organization of small heat-stable bacteriocin determinants in operons. Open reading frames (ORFs) are represented by open boxes with the number of amino acids in the primary translation product given under the gene. Potential promoters (solid arrows) and terminator regions (black boxes) are indicated. The diagonal lines intersecting ORF3 of pediocin PA-1 indicates that only the ends of this large gene are represented.

of small heat-stable bacteriocins is based on both genetic and biochemical properties.

5.2.1 Genetic analysis

The most intensively studied small heat-stable bacteriocins from lactococcal strains are the lactococcins A, B and M all produced by the same strain, *L. lactis* subsp. *cremoris* 9B4. Extensive work carried out by van Belkum *et al.*, (1989; 1991a; 1992) has established that the genetic determinants for bacteriocin production (*lcn*) and immunity (*lci*) for all three bacteriocins are located on a 60 kb conjugative plasmid p9B4-6. In each case the genes are arranged in an operon structure with the immunity gene located downstream of the structural gene for lactococcin production (Figure 5.1). The lactococcin M operon contains a third gene, *lcnMb*, the function of which is unknown, but its expression is required for antimicrobial activity. Although the specificities of the three lactococcins are distinct, they share similar nucleotide sequences upstream of the structural genes for lactococcin production. Primer extension analysis has identified promoter active sequences in this region (van Belkum *et al.*, 1991a). The apparent similarity between sequences in this region of the three operons has led to the conclusion that observed differences in antagonistic activity are not due to differences in promoter strength (van Belkum *et al.*, 1991a). The expression of cloned genes from all three operons was achieved in *L. lactis* subsp. *lactis* IL1403 (van Belkum *et al.*, 1989; 1991a; 1992).

In a parallel study, carried out by Holo *et al.*, (1991), a bacteriocin apparently identical to lactococcin A has been characterized and the genes cloned and introduced into a number of *lactococcal* strains. This work demonstrated that although strain IL1403 could be successfully transformed to a *lcnA*+ phenotype, albeit with reduced levels of production, other strains became immune to the lactococcin A, but did not produce detectable amounts of the bacteriocin (Holo *et al.*, 1991; van Belkum *et al.*, 1991a).

More recently a third lactococcin A operon, this time originating from *L. lactis* subsp. *lactis* biovar diacetylactis WM4 (Scherwitz-Harmon and McKay, 1987), has been described in the literature (Stoddard *et al.*, 1992). The genes responsible for bacteriocin production and immunity are located on a 131 kb plasmid, pNP2 (Scherwitz-Harmon and McKay, 1987). Nucleotide sequence analysis has revealed that they are identical to the *lcnA* and *lciA* genes previously described (van Belkum *et al.*, 1991a; Holo *et al.*, 1991). However, in this case the operon included two other upstream genes, *lcnC* and *lcnD*, which played a role in lactococcin A expression and secretion (Stoddard *et al.*, 1992; see below). The gene cluster is preceded by a putative promoter which could initiate transcription of a polycistronic message encompassing all four genes, *lcnC, lcnD, lcnA* and *lciA*.

The most intensively analysed bacteriocin from *Lactobacillus* is lactacin F produced by *Lactobacillus acidophilus* 11088 (Muriana and Klaenhammer,

1987; 1991a,b). The genetic determinants necessary for its production are carried on a large conjugative plasmid (Muriana and Klaenhammer, 1987). A 2.2 kb restriction fragment was cloned into a shuttle vector and the recombinant plasmid introduced into non-producing *Lactobacillus acidophilus* strains. From these experiments it was established that all the sequences necessary for lactacin F production and immunity were contained within this region of the plasmid (Muriana and Klaenhammer, 1991a). Nucleotide sequence determination revealed the presence of an operon structure containing 4 open reading frames (Muriana and Klaenhammer, 1991a; Klaenhammer *et al.*, 1992), the first of which (*laf*) encodes the structural gene for lactacin F, with the next two potential genes overlapping (Figure 5.1). The specific roles of the three downstream OFRs have not been evaluated although their involvement in expression of the *laf* phenotype has been implicated (Muriana and Klaenhammer, 1991a).

Leuconostoc species displaying antimicrobial activity against other lactic acid bacteria have been described in the literature, but the only in depth molecular analysis to date has been carried out on leucocin A-UAL 187 produced by *L. gelidum* (Hastings and Stiles, 1991; Hastings *et al.*, 1991). The N-terminal amino acid sequence of the antimicrobial protein was determined and used to clone, by reverse genetics, a fragment from a 7.6 MDa plasmid. Nucleotide sequence determinations of this fragment revealed an operon containing two ORFs with the first gene identified as the leucocin structural gene (Figure 5.1). On the basis of similar operon-like structures for other bacteriocins, it has been suggested that the second putative gene is responsible for immunity. However, no phenotypic expression of leucocin was achieved when several lactic acid bacteria, and *E. coli*, were transformed with a recombinant plasmid containing these sequences (Hastings *et al.*, 1991).

Cloning and nucleotide sequence determination of the plasmid-encoded pediocin PA-1 determinants identified four genes arranged in an operon-like structure (Marugg *et al.*, 1992; Nieto Lozano *et al.*, 1992; Figure 5.1). Expression of genes from this region, when cloned in a recombinant plasmid and introduced into *E. coli*, gave rise to pediocin production and secretion. These experiments also indicated that whereas *ped B* was apparently not required for antimicrobial activity *ped D* does play a role in bacteriocin production. The latter ORF is much larger than any of the genes identified in the operons described above and nucleotide sequence analysis suggested that the gene product is involved in protein translocation (see later, Marugg *et al.*, 1992).

5.2.2 *Common structural features*

From the class of small heat-stable bacteriocins only lactococcin A, lactacin F, leucocin A, pediocin PA-1 and sakacin A have as yet undergone independent nucleotide and amino acid sequence determination. The primary translation products of these five are shown in Figure 5.2 together with the predicted sequences of those bacteriocins which have been partially characterized (having

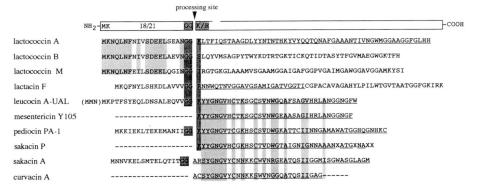

Figure 5.2 Primary amino sequences of small heat-stable bacteriocins. Regions of peptides that are underlined have been determined by amino acid sequence analysis. Other sequences have been predicted from nucleotide sequence data. Dashed lines indicate predicted N-terminal or C-terminal regions of the molecule for which the sequence has yet to be determined. Ambiguities in the amino acid sequence are shown as an X. A second possible initiation codon for leucocin A-UAL would result in an additional three residues at the start of the N-terminal region, shown in brackets (Hastings *et al.*, 1991). Common features of these peptides are summarized in the map above the sequences. Cleavage of the N-terminal leader occurs at the processing site shown as a gap between the two regions of the molecules. Conserved residues flanking the cleavage site of the peptides are shown by shading superimposed on the sequences. Homologies between sequences in the N-terminal leader regions of lactococcins A,B and M are indicated by lighter shading as are similarities between the C-terminal regions of the other peptide bacteriocins.

undergone either nucleotide or amino acid sequence analysis). From this data it is clear that members of this class of bacteriocin are transcribed as a precursor molecule with N-terminal extensions. Examination of homologies between predicted amino acid sequences with those of established N-terminal extensions enables the processing sites of lactococcins B and M to be positioned as shown in Figure 5.2. A comparison between the sequence of the different bacteriocins reveals a number of common features.

(i)   The short N-terminal sequences are either 18 or 21 amino acids in length.
(ii) The prepeptide starts with a methionine and is usually followed by a lysine residue.
(iii) Two adjacent glycine residues are found in the N-terminal leader at positions −1 and −2.
(iv) A lysine or arginine residue is usually located at the N-terminus (+1) of the mature bacteriocin.

The three lactococcins display particularly striking similarities having nearly identical N-terminal leaders (Figure 5.2). the nucleotide sequence homology extends into the 5′ region upstream of the respective coding regions such that the three operons are likely to be under the control of analogous promoters (van Belkum *et al.*, 1992). Sequence homology diverges quite markedly after the Gly–Gly dipeptide that marks the processing site (Figure 5.2) with neither the structural region of the mature lactococcins, nor the downstream immunity functions displaying any significant similarities (van Belkum *et al.*, 1991a; 1992). All three operons are encoded by the same large conjugative plasmid, p9B4, which thus contains regions of sequence duplication. It has been suggested that the individual bacteriocin determinants may have arisen from duplication of a common gene which subsequently underwent further recombinational rearrangements (van Belkum *et al.*, 1992). A conclusion drawn from the above observations is that as the promoter and N-terminal extensions of the three lactococcins are more or less equivalent the observed functional differences between these bacteriocins is not due to variation at the level of transcription (van Belkum *et al.*, 1991a; 1992).

The other well characterized N-terminal extensions display less homology. However, it has been noted that the same or similar residues are often found at identical positions within the sequence (Klaenhammer *et al.*, 1992). For example, in 5 out of 6 of the prepeptides the -11 and -12 positions are respectively Ser/Thr and Leu/Val residues (Figure 5.2).

A number of general rules have been found to apply to typical leader sequences (von Heijne, 1986). These include a positively charged N-terminus followed by a hydrophobic inner core region with a processing site obeying the -3,-1 law of von Heijne (1983). Despite being rather short the sequences within this region of the lactacin F precursor are consistent with those of a typical leader. Furthermore, secondary structure predictions for this molecule have identified an alpha-helical region at the N-terminus and a beta turn at the processing site which may correctly position the processing site for cleavage by a peptidase (Klaenhammer *et al.*, 1992). However, other bacteriocins of this class do not exhibit the same degree of hydropathicity in their equivalent N-terminal core region. Furthermore, the lactococcins A, B and M precursors all contain an asparagine in the -3 position (Figure 5.2). Such a large polar residue is not found in sites typically processed by leader peptidases (von Heijne, 1983). Holo *et al.* (1991) have suggested a stepwise processing of the lactococcin A precursor in which cleavage between the two glycine residues (positions -1 and -2) removing the N-terminal 20 residues is followed by the removal of the single N-terminal Gly (-1) to yield the mature 54 residue bacteriocin.

In those bacteriocins where the amino acid sequence of the processed peptide has been determined the N-terminal position was found to be occupied by a lysine residue or, in the case of lactacin F, an arginine residue (Figure 5.2). This preference for a positively charged amino acid at position +1 was also found in those bacteriocins for which only the amino acid sequence of the mature peptide was available (i.e. mesentericin Y105 and sakacin P, Figure 5.2). Indeed sequence homology between these latter peptides with pediocin PA-1, and to a lesser extent with leucocin A, extends throughout the molecules suggesting that these peptides represent a sub-group of small heat-stable bacteriocins which may have evolved from a common ancestor (Nieto Lozano *et al.*, 1992). The complete amino acid and nucleotide sequences data for mesentericin Y105, sakacin P and curvacin A awaits further molecular investigation, however, one might predict that the precursor genes encode N-terminal extensions sharing similar properties to those described above (Figure 5.2).

The presence of a charged residue at position +1 does not appear to be an absolute requirement for processing at the Gly–Gly site. Although the necessary amino acid sequence data is not available as yet the predicted N-terminal residues of lactococcins M and B are isoleucine and serine respectively (Figure 5.2). Curvacin A is also atypical in that the N-terminal of the processed peptide is an alanine residue (Figure 5.2).

From the growing volume of molecular data it can be concluded that, as with other secreted proteins, members of the class of small heat-stable bacteriocins are expressed as prepeptides which carry short N-terminal extensions. The first two residues of the precursor molecule are usually a methionine followed by a positively charged lysine residue. The internal processing site at which cleavage occurs is characterized by a Gly–Gly dipeptide at position -1 and -2 and often a positively charged residue is found at position +1, the N-terminus of the mature bacteriocin (Figure 5.2). These conserved regions suggest that processing of these small hydrophobic bacteriocins may occur via a common mechanism (Klaenhammer *et al.*, 1992; van Belkum *et al.*, 1992).

5.2.3 Bacteriocin translocation

It has been observed that the N-terminal leader sequences of small heat-stable bacteriocins do not conform to typical signal sequences involved in protein secretion. With the possible exception of lactacin F, the hydrophobic core common to secretory leaders, is lacking in the N-terminal regions shown in Figure 5.2. All these regions are short and contain a number of negatively charged residues. A novel secretory mechanism may exist for this class of bacteriocin which takes into account these differences in the primary sequence for this region of the precursor molecules.

The recent genetic investigation of lactococcin A production in *L. lactis* subsp. *lactis* biovar diacetylactis WM4 (Scherwitz-Harmon and McKay, 1987) identified two additional genes, *lcnC* and *lcnD*, upstream of the structural gene for lactococcin A (Stoddard *et al.*, 1992). These genes are thought to be involved in expression and secretion of lactococcin A (Stoddard *et al.*, 1992). This proposal provides a possible explanation for the observation of variable lactococcin A production in different *L. lactis* strains carrying the cloned *lacA* and *lciA* genes reported by van Belkum *et al.*, (1991a) and Holo *et al.*, (1991). The DNA sequences containing the latter lactococcin A operons did not include genes equivalent to *lcnC* and *lcnD* (Figure 5.1). However, it has been reported that the strain *L. lactis* subsp. *lactis* IL1403 displays homology with the lactococcin A gene cluster from pNP2 (Stoddard *et al.*, 1992). It is inferred from this observation that this particular strain encodes genes related to *lcnC* and *lcnD*. Hence when IL1403 is used as the production strain for clones carrying only the lactococcin A structural and immunity genes partial complementation of the *lcnC* and *lcnD* genes by related chromosomally encoded genes resulted in lactococcin A production. This interpretation explains why other host strains lacking these chromosomal determinants are incapable of producing active lactococcin A.

It is interesting to note that both *lcnC* and *lcnD* share significant homology with other proteins involved in secretory mechanisms of Gram-negative bacteria. Sequences within *lcnC* suggest that this gene, encoding a 716 amino acid protein LcnC, is a member of a family of ATP-binding proteins including several bacterial ATP-dependent transport proteins. The prototype of this family is HlyB, an *E. coli* membrane protein required for the signal peptide-independent export of haemolysin A (Randall *et al.*, 1987). As with other members of this group LcnC contains an ATP-binding domain in a conserved C-terminal stretch of approximately 200 amino acids and an N-terminal region containing three hydrophobic segments. The translation product of the *lcnD* gene is a 474 amino acid protein that shows structural similarities to other inner membrane proteins involved in transport systems, e.g. HlyD and related proteins (Stoddard *et al.*, 1992).

The large coding region identified in the pediocin PA-1 operon (Figure 5.1) and essential for production of this bacteriocin (Marugg *et al.*, 1992) specifies a protein of 724 amino acids. A search of protein sequence data bases revealed that this protein also belongs to the large HlyB family of ATP-dependent membrane translocators. The protein is of similar size to that encoded by *lcnC* in the lactococcin A operon and not surprisingly there is significant homology between these two amino acid sequences with particular sequence conservation found in the regions involved in ATP-binding and membrane spanning segments (Stoddard *et al.*, 1992; Marugg *et al.*, 1992).

The *L. lactis* subsp. *lactis* IL1403 chromosome-encoded secretion mechanism (Stoddard *et al.*, 1992) is capable of mediating the transport of lacto-coccins B and M as well as lactococcin A in the absence of *lcnC* and *lcnD* (van Belkum *et al.*, 1989; 1992). This suggests that the lactococcin A secretion apparatus can bring about the translocation of heterologous proteins. The fact that these three bacteriocins have almost identical N-terminal leaders (van Belkum *et al.*, 1992; Figure 5.2) may indicate that it is this region of the precursor molecule that is involved in recognition of the secretory apparatus. It may be predicted that the secretion and hence full activity of other bacteriocins of this class also requires the expression of translocation proteins. It will be interesting to see whether the functions of these analogous proteins are inter-changeable in bringing about the transport of small heat-stable bacteriocins in different lactic acid bacteria.

The involvement of similar secretory mechanisms has also been suggested for the group of lantibiotics. The *nisT* and *spaT* genes that are involved in the biosynthesis of nisin (Engelke *et al.*, 1992) and subtilin (Klein *et al.*, 1992; Chung *et al.*, 1992) respectively (section 5.4.4) are also members of the large family of proteins involved in membrane translocation typified by HlyB. A form of signal-sequence independent secretion thus appears to be a common mechanism employed by Gram-positive bacteria in the translocation of small heat-stable bacteriocins.

5.2.4 Mode of action

Detailed reports on how small heat-stable bacteriocins exert their antimicrobial activity have only been described for the well characterized bacteriocins lactococcin A and B. The mode of action of lactococcin A on whole cells and membrane vesicles of sensitive and immune lactococcal strains was studied by van Belkum et al., (1991b). They found that purified lactococcin A increased the permeability of the cytoplasmic membrane of sensitive L. lactis cells resulting in dissipation of the membrane potential. The bacteriocin induced the efflux of preaccumulated amino acids in whole cells and membrane vesicles, but had no effect on liposomes.

Lactococcin B appears to act in a similar manner to lactococcin A, causing dissipation of the membrane potential and pH gradient resulting in leakage of ions (Venema et al., 1993). Both lactococcin A and B are thought to assume a multipeptide complex in the membrane when involved in pore formation. Lactococcin B, however, is needed at high concentrations to achieve efflux of glutamate suggesting that at low levels of the bacteriocin the internal channel will allow the efflux of protons and other small molecules, but is too narrow for the passage of amino acids (Venema et al., 1993). The conclusion drawn from these studies was that both lactococcins cause pores to form in the membrane of susceptible cells with efflux of essential compounds resulting in inhibition of growth and ultimate death. In contrast to the mode of action of lantibiotics (section 5.4.7) lactococcin-induced pore-formation was found to be a voltage-independent process (van Belkum et al., 1991b; Venema et al., 1993).

Liposomes were found to be insensitive to both lactococcin A and B suggesting that the antimicrobial action of these bacteriocins requires a membrane-associated receptor protein (van Belkum et al., 1991b; Venema et al., 1993). Cells carrying the cloned immunity determinants were resistant to lactococcin B and lactococcin A which indicated that the location of the immunity proteins is probably the cytoplasmic membrane (van Belkum et al., 1991b; Venema et al., 1993). It has been proposed that the immunity protein may shield the receptor protein so that the target is no longer recognised in these cells. Alternatively pore-formation may be inhibited in the presence of the immunity protein (Venema et al., 1993).

Protein motifs often associated with membrane-related interactions include amphiphilic β-sheets, and α-helices. Sequences capable of forming the latter structure are present in the C-terminal region of lactococcin A (van Belkum et al., 1991; Holo et al., 1991) and have also been identified in the equivalent region of sakacin A (Holck et al., 1992). The lactacin F peptide contains two potential hydrophobic β-sheets that are predicted to form a transmembrane helix (Muriana and Klaenhammer, 1991a). Secondary structure predictions of this type have implicated the cell membrane as the primary target for others in this class of antimicrobial peptide.

5.3 Large heat-labile bacteriocins

Few bacteriocins of this class have been described in any detail in the literature. These proteins, which tend to be large (> 30 kDa), are usually inactivated within 30 minutes by temperatures of 100°C or less (see Klaenhammer, 1988; Klaenhammer *et al.*, 1992). The only member of this group to be characterized at the molecular level is helvetican J (see below). However, with the current interest in identifying new and novel bacteriocins it is likely that this group will increase in size. The recently described bacteriocins lacticin A and B, acidophilucin A and caseicin 80 (Rammelsberg and Radler, 1990; Toba *et al.*, 1991a,), all originating from *Lactobacillus* strains, can tentatively be assigned to this group on the basis of heat inactivation. How these proteins compare with other well characterized bacteriocins must await their further genetic and biochemical analysis.

5.3.1 Helvetican J

Helvetican J, produced by *Lactobacillus helveticus* 481, was identified by Joerger and Klaenhammer (1986) as a large (37 kDa) protein that exhibits a bactericidal activity against a narrow range of closely related *Lactobacillus* species. These initial experiments established that the genomic location of the biosynthetic determinants for helvetican J was the *Lactobacillus helveticus* 481 chromosome. An extension of this work resulted in the identification and cloning of these chromosomal sequences from a lambda library. Expression of the cloned bacteriocin genes in a heterologous host was also demonstrated (Joerger and Klaenhammer, 1990).

Nucleotide sequence analysis revealed the presence of two complete ORFs (ORF2 and ORF3) flanked by partially characterized upstream and downstream potential coding regions (Joerger and Klaenhammer, 1990; Klaenhammer *et al.*, 1992). The 999bp ORF3 has been implicated as the structural gene for helvetican J and accordingly has been designated *hlyJ*. OFR2 (315bp) is very close to the *hlyJ* gene with an intercistronic gap of only 30 bp. This upstream gene has the capacity to code for a protein of 11.8 kDa, but the function of this protein is not known at present. Expression signals have been located within the sequences. Upstream of OFR2 is a promoter-like region and both this gene and *hlyJ* are preceded by appropriately spaced ribosome binding sites. A rho-independent terminator is also found 37 bp after the end of the *hlyJ* gene. Two small inverted repeats are located between the proposed promoter and ribosome binding site of ORF2 and the suggestion has been made that they may play a regulatory role in helvetican J expression (Klaenhammer *et al.*, 1992). The organization of these genes together with the putative expression signals suggests that they are expressed as an operon.

The role of the additional ORFs that flank the proposed helvetican J operon are not known at present. As with other bacteriocins it might be expected that determinants specifying processing, export and immunity to helvetican J would have close genetic linkage to the structural gene for this bacteriocin. Helvetican J is hydrophilic in nature and the suggestion has been made that ORF2 may encode an immunity protein that binds to the bacteriocin molecule facilitating its secretion through the cell membrane (Joerger and Klaenhammer, 1990). It is interesting that expression of cloned helvetican J is host dependent. Failure to produce the active bacteriocin by certain hosts suggested that all the genetic information essential for processing and immunity was not present on the cloned chromosomal fragment. However, expression was achieved in an alternative host which indicated that the missing function could be provided in trans (Joerger and Klaenhammer, 1990). It is interesting that, although Laf⁻, the host strain which allowed production of helvetican J encodes certain genetic determinants involved in biosynthesis of lactacin F (section 5.2.1). It remains to be seen whether there are common steps in the processing or immunity mechanisms for these two very different types of bacteriocin.

5.4 Lantibiotics in lactic acid bacteria

Lantibiotics are bacteriocins that are synthesised ribosomally, but their distinguishing feature is the presence of a number of unusual amino acids residues, including lanthionine, β-methyllanthionine, dehydroalanine (Dha) and dehydrobutyrine (Dhb) in the active form of the molecule (reviewed recently by Jung, 1991). The presence of these non-coded amino acids in this class of bacteriocin indicates that enzymic modification of a primary translation product (prelantibiotic) must occur. Chemical reactions required to process a lantibiotic into the active molecule were first postulated by Ingram (1970). These involve the dehydration of serine and threonine residues to Dha and Dhb respectively (Figure 5.3). The dehydro-amino acids then undergo stereospecific nucleophilic addition by the sulfydryl group of a nearby cysteine residue to generate the sulfide bridge of lanthionine or β-methyllanthionine respectively (Figure 5.3). The complex biosynthesis of these lanthionine-containing antibiotics thus involves a post-translational modification step probably catalysed by the activity of at least two maturation enzymes.

An increasing number of lantibiotics have come to light over recent years and from molecular characterization it is evident that they can be divided into two subspecies, either linear or globularly shaped peptides (Jung, 1991). The most prominent member of this class of antimicrobial peptide is nisin, produced by

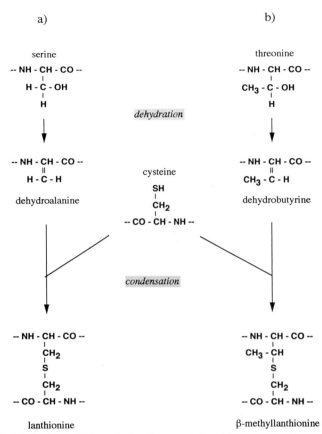

Figure 5.3 Mechanism of biosynthesis of non-coded residues present in mature nisin. (a) dehydroalanine and lanthionine and (b) dehydrobutyrine and β-methyllanthionine synthesis.

L. lactis subsp. *lactis* (Hurst, 1981; Fowler and Gasson, 1991). This linear molecule falls into the first group, together with the closely related lantibiotics subtilin (Gross and Kiltz, 1973), epidermin (Allgaier *et al.*, 1986), gallidermin (Kellner *et al.*, 1988) and Pep5 (Sahl and Brandis, 1981). Although these lantibiotics are produced by a diverse range of Gram-positive organisms, a comparison of their amino acid sequences indicate that they may well be related to a common ancestral peptide antibiotic.

For many years nisin was the only representative of this class of antibiotic produced by lactic acid bacteria (Hurst, 1981). More recently, a number of bacteriocins have been isolated from food grade organisms which, from preliminary analyses, can be classified as lantibiotics. Lactocin S from *Lactobacillus sake* (Mortvedt and Nes, 1990; Mortvedt *et al.*, 1991) and Lacticin 481 from *L. lactis* subsp. *lactis* (Piard *et al.*, 1992, 1993) have been purified to homogeneity and recent characterization at the molecular level indicated that both are highly likely to be lantibiotics (see below). The existence of a further lanthionine containing bacteriocin, Carnocin U149, from a *Carnobacterium piscicola* strain has recently been reported (Stoffels *et al.*, 1992 a,b). With the growing interest in food grade bacteriocins which exhibit a broad inhibitory spectrum, it may be anticipated that many more of this class of antagonistic peptide will come to light in lactic acid bacteria.

5.4.1 Lactocin S

Lactocin S was the first bacteriocin originating from a *Lactobacillus sake* strain to have been purified and biochemically characterized (Mortvedt and Nes, 1990; Mortvedt *et al.*, 1991). Partial sequence determination of the purified peptide indicated that the molecule was blocked at the N-terminal and also contained a number of unidentified amino acids. From the amino acid analysis it was thought likely that these residues may be components of lanthionine and β-methyllanthionine or dehydrated forms of serine or threonine (Mortvedt *et al.*, 1991; see Figure 5.3). This was confirmed by the recent identification and analysis of a genetic determinant for lactocin S (Nes *et al.*, 1993) proving that this bacteriocin can be classed as a lantibiotic. The authors state that nucleotide sequence analysis has identified an ORF that has the capacity to encode a 68 amino acid lactocin S precursor molecule (Nes *et al.*, 1993). This putative prepro-lactocin would contain a 33 residue N-terminal leader sequence and a number of serine, threonine and cysteine residues that may be subject to post-translational modification. From these results, together with the amino acid sequence analysis and mass spectroscopy data, a structure for mature lactocin S has been proposed as shown in Figure 5.4 (Nes *et al.*, 1993). The molecule, which contains two lanthionine rings and an additional four dehydro-residues, bears no similarities to the structures of other linear lantibiotics.

There is evidence that the genetic determinants responsible for lactocin S production and immunity are associated with an unstable 50 kb plasmid resident

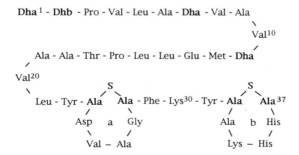

Figure 5.4 Proposed structure of lactocin S. Modified residues are dehydroalanine, Dha; dehydrobutyrine, Dhb and lanthionine, Ala-S-Ala.

in *Lactobacillus sake* L45 (Mortvedt and Nes, 1990). The genes involved in the biosynthesis of other lantibiotics, e.g. nisin and subtilin, are chromosomally encoded (see below) and plasmid-encoded bacteriocin genes tend to be more commonly associated with the family of small heat-stable bacteriocins (section 5.2). However, the involvement of plasmids is not unprecedented for lantibiotics, for example the genes required for epidermin biosynthesis are carried by a plasmid in *Staphylococcus aureas* (Schnell *et al.*, 1988). As with other lantibiotics a number of other genes would be expected to be involved in the production and secretion of active lactocin S. It is thought that genes encoding such functions as modification of precursor lactocin S and translocation of the active molecule through the cell membrane are located on the same plasmid as the structural gene (Nes *et al.*, 1993).

The carriage of genetic determinants on transmissible plasmids is known to facilitate their rapid spread throughout bacterial populations. It is interesting to note therefore, that the strain *Lactobacillus sake* 148 although independently isolated from a remote location (Sobrino *et al.*, 1991; 1992) to that of *Lactobacillus sake* L45 has recently been shown to produce a bacteriocin that is apparently identical to lactocin S (Nes, 1993).

5.4.2 Lacticin 481

Lacticin 481, produced by *L. lactis* subsp. *lactis* CNRZ 481, has been purified to homogeneity and molecular analysis indicated that its amino acid content included possible lanthionine residues (Piard *et al.*, 1992). The only previously described lantibiotic produced by *L. lactis* was nisin. However, partial amino acid sequence data indicated that lacticin 481 showed no similarity to nisin (Piard *et al.*, 1992) and thus, represented a new member of the group of lantibiotics in lactic acid bacteria. These preliminary studies have recently been backed up with genetic analysis in which the structural gene for lacticin 481 (*lct*) has been identified and cloned (Piard *et al.*, 1993). The nucleotide sequence revealed a ORF with the capacity to encode a peptide of 51 amino acids. Comparison of the deduced amino acid sequence with the N-terminal sequence indicated that the primary translation product represents a precursor molecule comprising a 24 residue N-terminal leader and a 27 residue C-terminal structural region. Furthermore, the predicted prelacticin sequence contained cysteine, serine and threonine residues (precursors of lanthionine and β-methyllanthionine, Figure 5.3) which were not detected in amino acid analysis of the mature molecule. Because of the arrangement of these potentially modified residues in the primary sequence there are two possible structures for lacticin 481 as shown in Figure 5.5 (Piard *et al.*, 1993).

The proposed molecular structures of both lacticin 481 (Figure 5.5) and lactocin S (Figure 5.4) radically differ from that of previously described lantibiotics, typified by nisin (Figure 5.6). The lanthionine rings, particularly the N-terminal rings a and b, are highly conserved in the linear lantibiotics nisin,

a)

b)

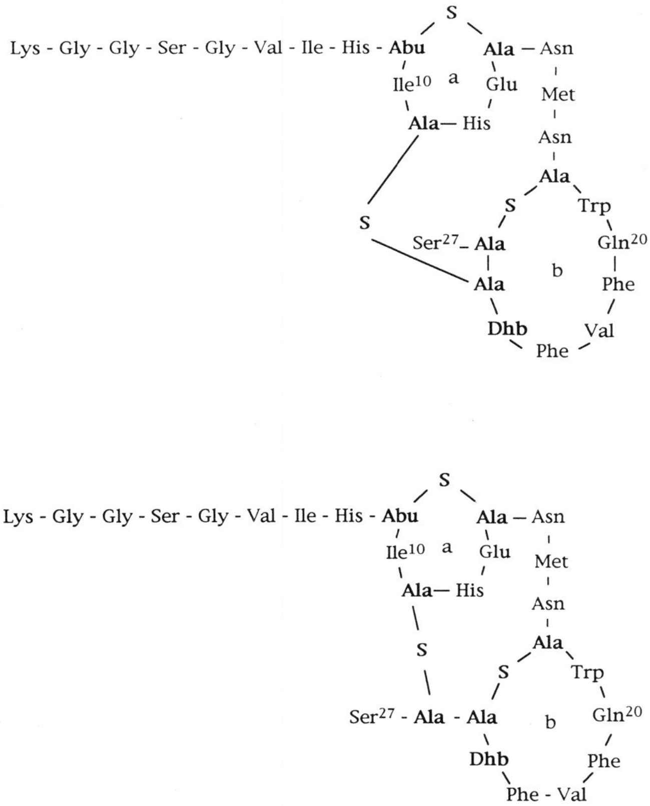

Figure 5.5 Potential structures of lacticin 481. Modified residues are dehydroalanine, Dha; lanthionine, Ala-S-Ala and β-methyllanthionine, Abu-S-Ala. Rings a and b are linked together by a further lanthionine residue. The alternative structures (a) and (b) depend on whether Ser₁₁, in ring a, reacts with either Cys₂₅ or Cys₂₆, in ring b, to generate this mono-sulfide bridge.

subtilin and epidermin (Jung, 1991) suggesting a common evolutionary ancestor. It will be interesting to see if the enzymes that introduce lanthionine rings in the newly described lantibiotics are interchangeable between the various lantibiotic-producing lactic acid bacteria.

5.4.3 Nisin structure and biosynthesis

Chemical analysis of nisin, carried out over two decades ago by Gross and Morrell (1967), established the complex chemical structure of the mature molecule as shown in Figure 5.6. Nisin is a 34 amino acid peptide that contains a high proportion of modified residues, including five lanthionine or β-methyllanthionine rings, two Dha residues and a Dhb residue. confirmation of the chemical structure of nisin in aqueous solution has been provided by

```
                    Dha                              Ala — Leu
                      \                               /      \
          Ile    Leu       S                     Gly       Met
           |   a  |       / \                      |    c   |
  Ile¹ — Dhb — Ala    Ala—Abu  b  Ala—Lys —Abu        Gly
           \   /        \   /                  \        /
            S          Pro —Gly¹⁰               S —Ala
                                                    |
                                                  Asn²⁰
                                                    |
                                             S      Met
                                           /   \     |
                                  His—Ala  d  Abu—Lys
                                  /      \    /
  Lys³⁴ — Dha — Val — His — Ile³⁰ — Ser—Ala   e  Abu—Ala
                                        \      /
                                          S
```

Figure 5.6 Structure of nisin A. Modified residues are Dha, Dhb, lanthionine and β-methyllanthionine. The α-carbon atoms of lanthionine and β-methyllanthionine are all in the D-configuration (from Gross and Morell, 1971).

sequence-specific assignment of its ^{1}H NMR spectrum (Chan et al., 1989b; Slijper et al., 1989). This work revealed that despite the rigidity imposed on the molecule by the presence of five lanthionine rings there also exists a degree of flexibility in the outermost C- and N-terminal regions of the peptide. From predictions of the three-dimensional structure of nisin, based on NMR data (Slijper et al., 1989; Van de Ven et al., 1992) it has been concluded that the internal region of the molecule contains a 'hinge' region with rings a, b and c on one side and rings d and e on the other side of a moveable joint (Figure 5.6). Lian et al., (1991) determined the structures for several parts of the nisin molecule in aqueous solution. In this study it was concluded that the molecule was very flexible with the only defined conformational features being those imposed by the presence of the lanthionine residues.

It has long been known that the antimicrobial activity of commercial samples of nisin deteriorate on storage and that a number of chemical components are found within such samples (Berridge et al., 1952). Chan et al., (1989a) have demonstrated that specific cleavage occurs as a result of acid treatment or long term storage. The chemical structures of the two major breakdown products were determined by ^{1}H NMR analysis and shown to arise by cleavage at the dehydroalanine residues in the mature molecule (Chan et al., 1989a). One degradation product, nisin$_{1-32}$, which had lost the C-terminal dipeptide, retained essentially wild-type activity. More severe acid treatment yields (desΔAla$_5$)nisin$_{1-32}$ in which cleavage at Dha5 resulted in loss of ring A (figure 5.6). The loss of biological activity by this molecule suggested that either this first lanthionine ring, or Dha$_5$ plays a key functional role. Protein engineering studies currently being carried out by a number of groups may help to elucidate the structural requirements of the molecule for full biological activity (see later, section 5.4.8).

The 34 residue nisin molecule originally characterized by Gross and Morell (1967) and more recently studied at the molecular level has been termed nisin A

(Figure 5.6). This distinguishes it from nisin variants that are derived either chemically, by protein engineering (see later) or through naturally occurring changes to the DNA coding sequence. An example of the last type of nisin variant was provided by the recent discovery of nisin Z (Graeffe *et al.*, 1991; Mulders *et al.*, 1991; De Vos *et al.*, 1992). This nisin molecule, produced by several strains of *L. lactis* has an asparagine residue in place of histidine at residue 27. DNA sequence analysis of the respective prenisin genes revealed the presence of a single base substitution, in the codon specifying residue 27 (Mulders *et al.*, 1991). The effect of this mutation on the properties of the peptide were found to be minimal with Nisin Z appearing to diffuse slightly more readily in agar than Nisin A (De Vos *et al.*, 1992). The staphylococcal lantibiotics epidermin and gallidermin are another example of peptides differing by one amino acid residue as a result of a single base pair substitution.

The presence of non-coded amino acids in the mature nisin molecule indicates that enzymic modification of a primary translation product (prenisin) must occur. The chemical reactions required to convert prenisin into an active form are outlined above (Figure 5.3). Based on these reactions it was possible to predict the sequence of the primary translation product for nisin. From this proposed prenisin sequence a DNA probe could then be designed that would encode the equivalent amino acid sequence. This strategy of 'reverse genetics' has been widely employed in the identification of a number of bacteriocin genes for which the antimicrobial protein has been purified (see above). A similar approach, first used successfully in the genetic analysis of the lantibiotic epidermin (Schnell *et al.*, 1988), confirmed the proposed mode of biosynthesis of lantibiotics.

Analysis of genes involved in nisin biosynthesis has been carried out by a number of different groups with Buchman *et al.*, (1988) first establishing the existence of a gene for prenisin. From the results of independent studies by Kaletta and Entian (1989) and Dodd *et al.*, (1990) similar conclusions were drawn which supported the above mechanism for nisin biosynthesis. Nisin is synthesized as a prepropeptide of 57 amino acids, the first 27 residues of which make up an N-terminal leader peptide. The 34 residue C-terminal region contains serine, threonine and cysteine residues in the positions predicted to generate the modified amino acids present in mature nisin. (Figure 5.6). The three independently isolated *nisA* genes (Buchman *et al.*, 1988; Kaletta and Entian, 1989; Dodd *et al.*, 1990), from distinct naturally occurring *L. lactis* strains, showed very little sequence divergence. The actual coding regions for the prenisin genes were identical although differences have been identified in flanking sequences (see below). The prenisin genes isolated by the latter two groups were both termed *nisA* in accordance with the genetic nomenclature suggested by De Vos *et al.*, (1991). This is based on the name of the antibiotic and is considered to be a more descriptive term than that of *spaN* (Buchman *et al.*, 1988) which refers to *s*mall *p*rotein *a*ntibiotic *n*isin. The term *nis* will be used throughout this review when discussing the *nisA* gene and other determinants involved in nisin biosynthesis.

5.4.4 Genetic analysis of the nisin biosynthetic genes

In addition to the structural gene for prenisin, *nisA*, a number of other genes are required for the biosynthesis of this complex lantibiotic. These include genes encoding enzyme(s) involved in dehydration and lanthionine formation in prenisin, proteolytic cleavage of the N-terminal leader, translocation/secretion of the mature peptide, nisin immunity determinants and possible regulatory functions. Analysis of the related peptides epidermin (Schnell *et al.*, 1992; Augustin *et al.*, 1992) and subtilin (Klein *et al.*, 1992; 1993; Chung *et al.*, 1992; Chung and Hansen, 1992) has revealed the presence of a number of genes, closely associated to the respective structural genes, which are essential for biosynthesis.

Approximately 800 bp upstream of the *nisA* gene a copy of the insertion sequence IS*904* was identified (Figure 5.7; Dodd *et al.*, 1990; Rauch *et al.*, 1990). A detailed analysis of the genomic relationship between this copy of IS*904* and the *nisA* gene in different nisin-producing transconjugants gave the first formal proof of the physical location of nisin determinants in the chromosome of *L. lactis* strains (Dodd *et al.*, 1990). Similar conclusions have been drawn from the analysis of nisin determinants encoded by independent nisin producing strains (Steen *et al.*, 1991; Rauch and De Vos, 1992; Gireesh *et al.*, 1992; de Vuyst and Vandamme, 1992). This work prompted a closer examination of the DNA sequences acquired as a result of conjugation which revealed

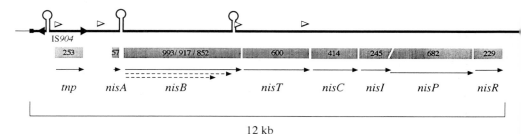

12 kb

Figure 5.7 Organization of nisin biosynthetic genes. The bold line represents the left end of the nisin transposon, Tn*5301*/Tn*5276*, integrated in the *L. lactis* subsp. *lactis* chromosome, shown by a thin line. A copy of IS*904*, close to one end of the transposon, is indicated by inverted arrows. Shaded boxes under the line represent genes with the number of amino acids in the translation products given and the direction of transcription shown by arrows. The *nisB* gene has been reported as three different sizes as indicated by the dashed arrows. The start of *nisP* overlaps with the end of the *nisI* open reading frame as shown by the staggered arrows. Putative expression signals identified from the sequence data are shown above the line. Promoters, open triangles and terminator sites, stem-loops. The map was compiled using data from Buchman *et al.* (1988), Dodd *et al.* (1990), Horn *et al.* (1991), Rauch and de Vos (1992), Steen *et al.* (1991), Engelke *et al.* (1992), van der Meer *et al.* (1993), Kuipers *et al.* (1993).

that nisin determinants were encoded by a novel chromosomal conjugative transposon, Tn5301 (Horn et al., 1991). In addition, the copy of IS904 upstream of nisA was located close to, but not precisely at, the terminus of Tn5301 and was not involved in the transposition of the much larger conjugative element (Horn et al., 1991). A more detailed description of the properties of these elements is given in chapter 1.

The discovery that the nisA gene was contained within a large 70 kb conjugative element was of some relevance to a subsequent search for other genetic determinants required for nisin biosynthesis. The points of fusion of the nisin transposon sequences with those of the chromosomal target site precisely defined the left and right termini of the element (Horn et al., 1991; Rauch and De Vos, 1992). IS904 lies upstream of nisA thus, an initial investigation of sequences required for nisin biosynthesis could be narrowed down to the region of the transposon lying to the right of nisA. This assumes that other nisin determinants are transferred along with the nisA gene rather than being resident in the genome of non-nisin producing strains and serving some other function which is also employed by the cell in the biosynthesis of mature nisin. The recent discovery of several genes involved in the production of the related lantibiotics subtilin (Klein et al., 1992; Chung et al., 1992; Chung and Hansen, 1992) and epidermin (Schnell et al., 1992) have shown them to be arranged in operon-like structures. This supports the proposal that other nisin determinants are likely to be encoded from sequences physically close to the nisA gene.

The start of a second ORF was discovered a short distance (108 bp) downstream from the end of the nisA gene (Buchman et al., 1988; Kaletta and Entian, 1989; Dodd et al., 1990). Extension of this sequence data by the same groups has, in each case, revealed the presence of a large coding region (Steen et al., 1991; Engelke et al., 1992; Dodd et al., unpublished data). While the sequences are largely identical throughout this region it is interesting to note that small differences at the 3' end of the independent ORFs leads to significant variations in the C-terminal ends of the respective gene products (Figure 5.7). The ORF described by Steen et al. (1991) is 2556 bp in length which would encode an 852 amino acid protein of 100.5 kDa. The equivalent ORF characterized by Engelke et al., (1992) has been termed nisB. A frameshift caused by an additional nucleotide at coordinate 2520 results in this gene having the capacity to encode a much larger 117 kDa protein of 993 amino acids. In the third nisB gene for which sequence data is available (Dodd et al., unpublished) a second frameshift mutation has introduced a stop codon after amino acid 917. No significant differences in nisin production by these three independent lactococcal isolates has been reported suggesting that the C-terminal end of the NisB protein does not have a vital functional role in the biosynthesis of this molecule.

The precise role of NisB in nisin biosynthesis has not yet been elucidated although it has been suggested that it may be involved in the enzymatic modification of prenisin (Steen et al., 1991; Engelke et al., 1992). It was found to share significant homology to the recently described proteins SpaB (Klein et al., 1992;

Chung and Hansen, 1992) and EpiB (Schnell *et al.*, 1992). The fact that these latter two proteins are absolutely required for the biosynthesis of the subtilin and epidermin respectively indicates that NisB plays an equivalent role in the production of active nisin. With the exception of these two lantibiotic genes no significant homology was found with other previously described proteins in various data bases. It has been suggested that the failure to find other proteins sharing similar sequences to those of NisB, SpaB and EpiB may indicate that these proteins have new biocatalytic functions (Engelke *et al.*, 1992), such as those reactions involved in the post-translational modification of prenisin.

Steen *et al.* (1991) detected homology between the *nisB* gene and ORFs encoding membrane-associated proteins from chloroplasts. a subsequent search for secondary structure motifs revealed a 14-residue trans-membrane helical segment near the C-terminal end of the molecule and a number of potential α-helices displaying amphipathic properties. From these observations it was inferred that *nisB* is a membrane associated protein with the lack of an N-terminal leader sequence suggesting that anchorage is on the cytoplasmic side of the membrane (Steen *et al.*, 1991). Engelke *et al.*, (1992) raised antibodies to a *nisB/trpE* fusion protein and used an immunoassay to detect *nisB* protein. The same group prepared *L. lactis* vesicles and, using the antibodies in Western blot analyses, showed that the *nisB* protein was tightly associated with the vesicle fraction. This result was consistent with a membrane location for the *nisB* protein and supported the proposal that nisin maturation, as with other lantibiotics, occurs while the protein is associated with the membrane (Schnell *et al.*, 1988).

The start of a third ORF is located immediately downstream of the *nisB* stop codon (Engelke 1992; Steen *et al.*, 1992; Dodd *et al.*, unpublished) the precise intercistronic distance ranging from 14 bp to 311 bp due to the variation in the 3' end-point of the *nisB* gene in the different strains analysed (Figure 5.7). The nucleotide sequence of this whole region has been presented (Engelke *et al.*, 1992) which indicates that this ORF has the capacity to code for a 600 amino acid protein with an expected molecular mass of 69 kDa. Database searches have revealed that the sequence is very similar (43.8% homologous) to that of the recently described *spaT* gene involved in subtilin biosynthesis (Klein *et al.*, 1992; Chung *et al.*, 1992). As the proteins encoded by these genes are likely to play similar roles in the production of the respective lantibiotics the ORF following *nisB* has been called *nisT* (see Engelke *et al.*, 1992).

Evidence for the involvement of the SpaT protein in translocation of the subtilin molecule comes from gene disruption studies (Klein *et al.*, 1992). This work demonstrated that loss of SpaT function resulted in a morphological change causing the cells to clump on solid media. This agglutination was attributed to the continued production of subtilin which accumulated intracellularly in cells deficient in a subtilin transport mechanism (Klein *et al.*, 1992). Similar experiments have not been reported for the equivalent nisin determinant, but it would be predicted that disruption to the *nisT* gene would adversely affect nisin secretion. It has been reported that expression of the *nisT* gene in *E. coli* is toxic and the suggestion made that this may be due to NisT interfering with the cellular metabolism of this particular host (Engelke *et al.*, 1992).

Significant homology has been found between the SpaT protein and several transport proteins including the protein involved in translocation of haemolysin B in *E. coli*, the ComA protein of *Streptococcus pneumoniae*, the mouse multidrug resistance protein and the protein responsible for human cystic fibrosis disease (see Klein *et al.*, 1992), all sharing approximately 20% homology over the entire protein. The NisT protein, as with all these secretory proteins, contains potential membrane-spanning helices within regions of hydrophobicity suggesting a membrane location. Both NisT and SpaT proteins can thus be considered to be members of the large family of proteins involved in membrane translocation (Blight and Holland, 1990).

The sequence analysis presented by Engelke *et al.* (1992) extends downstream of the *nisT* gene and reveals a further closely associated ORF. A potential coding region overlaps with the 3' end of the *nisT* gene (Figure 5.7) and has the capacity to encode a protein of 414 amino acids with a predicted molecular mass of 47.3 kDa. This ORF shared significant sequences homology to the *spaC* (Klein *et al.*, 1992) and the *epiC* (Schnell *et al.*, 1992; Augustin *et al.*, 1992) genes recently shown to be essential for biosynthesis of subtilin and epidermin respectively. Furthermore, hydrophobicity plots of the proteins encoded by these genes also revealed strong similarities throughout the molecules (Engelke *et al.*, 1992). These observations suggest that the three proteins carry out equivalent functions in the biosyntheses of their respective lantibiotics. Accordingly the gene was designated *nisC* (Engelke *et al.*, 1992).

With the exception of the SpaC and EpiC proteins no other previously described proteins were found in databases that showed any similarity to *nisC*. The class of proteins represented by NisC and also NisB (see above) thus appear to be unique to lantibiotic producing organisms. As such these proteins are obvious candidates for the role of lantibiotic maturation, however, as yet there is no formal proof of any involvement in the chemical reactions required for lanthionine production by these proteins.

Very recent reports by van der Meer *et al.*, (1993) and Kuipers *et al.*, (1993) have described the location of an additional three genes in the nisin gene cluster, termed *nisI, nisP* and *nisR* (Figure 5.7). Downstream of *nisC* is located *nisI* the product of which has features characteristic of a lipoprotein and appears to be involved in the expression of nisin immunity (Kuipers *et al.*, 1993). The proposed start of the next ORF in the gene cluster, designated *nisP*, overlapped with the end of *nisI* (Figure 5.7). This gene has the capacity to encode a protein of 682 residues. The deduced amino acid sequence of NisP contained an N-terminal signal sequence, a catalytic domain, which displayed homology to subtilisin-like serine proteases, and a C-terminal membrane anchor. Furthermore significant sequence homology was found between NisP and the putative epidermin leader peptidase EpiP from *Staphylococcus epidermis* (Schnell *et al.*, 1992). Sequence analysis indicated that the *nisP* gene encoded a serine protease. This proposal was supported by the finding that, *in vitro*, the NisP protein cleaved off the leader peptide sequence of purified precursor nisin to produce the active molecule (van der Meer *et al.*, 1993).

The last ORF in the nisin gene cluster is *nisR* (Figure 5.7). The putative nisR gene product (689 residues) shows similarities to the regulatory proteins of two component regulatory systems (van der Meer *et al.*, 1993). The coding region shares homology with the *spaR* gene which is involved in regulation of subtilin biosynthesis (Klein *et al.*, 1993). The finding that expression of *nisR* is essential for the production of precursor nisin indicated that this gene plays a similar regulatory role in nisin biosynthesis (van der Meer *et al.*, 1993). The conclusion drawn from these recent studies is that the organization of genes *nisABTCIPR*, located on a 10 kb region within the nisin transposon, represent the full complement of genes required for nisin biosynthesis and immunity (van der Meer *et al.*, 1993; Kuipers *et al.*, 1993).

5.4.5 *Gene expression of nisin determinants*

A search for expression signals in the nucleotide sequence has identified potentially promoter active sequences preceding *nisA, nisT* and *nisC* (Engelke *et al.*, 1992; Steen *et al.*, 1991). A number of inverted repeat sequences have also been located, often in intergenic regions, which are capable of forming stem-loop structures. These potential secondary structures may possibly serve a regulatory functions (Figure 5.7).

Northern hybridizations, using *nisA* specific probes, identified a small transcript of approximately 300 bp (Buchman *et al.*, 1988; Engelke *et al.*, 1992). From primer extension analyses the 5' end of this message was mapped 42 bases upstream of the start of the *nisA* gene (Engelke *et al.*, 1992). S1 mapping experiments had previously identified a similarly located, although 'ragged', end to the 5' end of the *nisA* transcript (Buchman *et al.*, 1988). The 3' end was located in the region of a large inverted repeat which occupies most of the intergenic gap between the *nisA* and *nisB* genes (Figure 5.7). This potential stem-loop possibly serves as a *nisA* terminator, however, the stretch of Ts which are characteristically found downstream of such a structure is absent from this sequence. Examination of sequences upstream of the 5' terminus did not reveal an obvious promoter and the suggestion has been made that this transcript may have resulted from processing a larger primary mRNA species (Buchman *et al.*, 1988). In support of this proposal Steen *et al.* (1991) have carried out primer extension analyses, using a *nisA* specific primer, which identified a number of larger mRNA species. Their data suggest that the *nisA* and *nisB* genes are transcribed as part of a polycistronic operon of greater than 8.5 kb and that the promoter for this proposed transcript lies approximately 4 kb upstream of the *nisA* gene.

Bearing in mind that the left termini of the nisin transposon, Tn*5301* has been mapped at 2.3 kb upstream up *nisA* (Horn *et al.*, 1991) transcription of a large polycistronic message would therefore be initiated from sequences outside the transposable element. The expression of nisin determinants would thus rely on the chromosomal target site for the nisin transposon containing a suitable promoter. Horn *et al.*, (1991) reported that insertion of Tn*5301* occurred at two preferred sites in the chromosome. It is possible that this site-specificity is a reflection of the requirement for external promoters for expression of nisin determinants. An alternative possibility is that the independently isolated nisin producing strains under investigation in the different groups are radically different in their genetic organisation in this region of the chromosome. In the similar primer extension experiments of Engelke *et al.* (1992), no such large transcripts were observed. Further analysis of this region of the chromosome is required, including nucleotide sequence determination, to define the precise location of the nisin promoter.

The absence of a characteristic promoter preceding the *nisB* gene argues in favour of transcriptional read through from a promoter upstream of the *nisA* gene. However, mRNA studies carried out by Engelke *et al.* (1992), using a *nisB* specific probe, failed to identify a large polycistronic transcript. This group only observed monocistronic transcripts for both *nisA* and *nisB* genes. It is however possible that a large unstable mRNA species, as proposed by Steen *et al.* (1991), may not be detected due to rapid processing at the inverted repeat structure identified in the *nisA*-*nisB* intercistronic gap.

At the end of the *nisB* gene a large stem-loop structure is located followed by a stretch of Ts. This potential rho-independent terminator is 180 nucleotides downstream of the end of the *nisB* gene described by Steen *et al.* (1991). This group make the proposal that this point marks the end of an operon encoding *nisA* and *nisB* in addition to the IS*904* transposase (Dodd *et al.*, 1990). The longer *nisB* genes characterized by Engelke *et al.* (1992) and Dodd *et al.*, (unpublished data) continue through the potential stem-loop structure. These results argue that the *nisB* transcript does not terminate at this point.

No transcriptional analysis has been reported for genes downstream of *nisB* however the region of putative secondary structure contains possible promoter active sequences and a ribosome binding site appropriately spaced for expression of the *nisT* gene (Steen *et al.*, 1991; Engelke *et al.*, 1992). the *nisC* gene immediately downstream of *nisT*, is also preceded by a potential promoter and ribosome binding site (Engelke *et al.*, 1992). Although read through from an upstream promoter may by under the control of regulatory factors, it appears that *nisT* and *nisC* transcription may be independent of upstream nisin determinants.

Indirect evidence for an operon-like arrangement of nisin genes was provided by the gene replacement studies reported by Dodd *et al.*, (1992). In this work the *nisA* gene was insertionally inactivated and attempts to complement the chromosomal mutation by the provision of plasmid encoded *nisA* initially proved unsuccessful. This was interpreted as the insertion in *nisA* exerting a polar effect on downstream genes, including those required for processing of prenisin.

5.4.6 Nisin immunity/resistance

In their recent characterization of nisin biosynthetic genes Kuipers *et al.* (1993) described the next ORF in the gene cluster, downstream of *nisC* (Figure 5.7). The gene product has been shown to provide a significant level of protection to *L. lactis* cells against exogenously added nisin. This indicated that the protein was involved in the cells' immunity mechanism and accordingly was designated *nisI*. The deduced amino acid sequence of the NisI protein contained a consensus liproprotein signal sequence suggesting that it was a lipid-modified extracellular membrane-anchored protein (Kuipers *et al.*, 1993).

Gene replacement studies, in which the *nisA* gene was insertionally inactivated revealed that the resulting Nis- strain exhibited lower levels of immunity to nisin than the parent Nis+ strain (Dodd *et al.*, 1992). This observation initially suggested that the nisin immunity determinants are co-transcribed with the *nisA* gene as part of a the same operon. An alternative explanation for the apparent polar affect of the *nisA* mutation on the *nisI* gene may be that the two genes are transcribed independently, but that expression of *nisA* is required for full expression of the nisin immunity determinants. Kuipers *et al.* (1993) observe a similar phenomenon using a Nis⁻ strain generated by substituting the *nisA* gene with a truncated gene. This strain was found to be ten-fold less resistant to nisin than the wild-type strain. Complementation of the nisin deficiency by providing a plasmid-encoded *nisA* gene restored this host's immunity to wild-type levels.

An understanding of the relationships between all the genes involved in nisin biosynthesis will no doubt come to light as the genes themselves are identified and characterized. However, early observations suggest that for optimal nisin production and host cell stability a fine balance is required between the levels of expression of these genes, including the structural gene, processing genes, secretory apparatus and immunity functions.

The incidence of nisin resistance, either naturally occurring or acquired, has been described for a range of organisms. The process by which these strains confer resistance to nisin is not known at present, but involves a mechanism distinct from that of nisin self-immunity exhibited by nisin producing strains. There have been a number of reports in which nisin resistance in lactococcal strains have been associated with plasmid DNA (McKay and Baldwin, 1984; Klaenhammer and Sanozky, 1985; Simon and Chopin, 1988; Froseth *et al.*, 1988; von Wright *et al.*, 1990). The most advanced genetic analysis of this form of resistance has been provided by Froseth and McKay (1991) who have identified and sequenced a gene (*nsr*) encoding a 319 amino acid protein. It has been predicted that this protein, which contains a hydrophobic region at the N-terminus, is located in the membrane (Froseth and McKay, 1991), although how it blocks nisin function has yet to be elucidated. The *nsr* gene, isolated from

L. lactis subsp. *lactis* biovar diactylactis, confers nisin resistance at a level one tenth of that of nisin immunity exhibited by nisin producing strains indicating that the two mechanisms are dissimilar. Furthermore, in hybridisations *nsr* sequences displayed no significant homology to genomic digests of a nisin producing strain (Froseth and McKay, 1991).

5.4.7 Mode of action

At present the mechanism by which nisin exerts its inhibitory effect on Gram-positive bacteria is not fully understood. It is thought that the primary cause of death is disruption of the cell membrane due to the detergent-like properties of nisin (Ramseier, 1960). More recently this disruption has been shown to result from channel formation due to the incorporation of nisin within the membrane (Henning *et al.*, 1986; Sahl *et al.*, 1987; Kordel *et al.*, 1989; Gao *et al.*, 1991). Efflux of small molecules and ions through these channels has the effect of dissipating the membrane potential (Ruhr and Sahl, 1985; Kordel and Sahl, 1986; Benz *et al.*, 1991; Gao 1991). Biophysical experiments involving artificial membranes have demonstrated that channel formation is a voltage dependent process (Benz *et al.*, 1991; Gao *et al.*, 1991). In these systems nisin dissipates both the membrane potential and pH gradient of liposomes (Gao *et al.*, 1991). This work also demonstrated that the phospholipid composition of the liposomal membrane affects the activity of nisin, an observation which may explain the variations in sensitivity between different bacteria. As a result of weakening the outer membrane of Gram-negative bacteria, by osmotic shock, treatment with chelating agents eg EDTA, or vesicle formation, access is gained to the cytoplasmic membrane. In such cells nisin has been shown to exert an inhibitory effect (Kordel and Sahl, 1986; Stevens *et al.*, 1991).

Nisin is proposed to generate channels in membranes as a consequence of its physical properties, i.e. a small hydrophobic, cationic peptide. However, it has been argued that this mechanism provides no obvious role for the dehydro-residues found in nisin and which are highly conserved features of all lantibiotics (Liu and Hansen, 1992). There is some evidence that nisin inhibition of outgrowing bacterial spores involves modification of membrane sulfydryl groups (Morris *et al.*, 1984). The unsaturated side-chains provided by the dehydro-residues may participate in these interactions however, no covalent linkages have been demonstrated. Structure/function studies with subtilin have shown that modification of the dehydroalanine residue at position 5, either chemically (Hansen *et al.*, 1991) or by protein engineering techniques (Liu and Hansen, 1992), was accompanied by the loss of its antimicrobial activity against outgrowing spores. These results indicate that the dehydroalanine-5 plays an important functional role in this type of inhibition. The suggestion has been made that inhibition of spore outgrowth involves the covalent attachment of the dehydroalanine residue to a sensitive cellular target. It is proposed that this inter-action is not required for the disruption of cell membranes as a result of channel formation and hence this latter mechanism does not require an equivalently sited dehydro-residue for activity. Recent progress in protein engineering the nisin molecule supports this proposal.

5.4.8 Protein engineered nisins

Until recently studies relating specific parts of the nisin molecule to a particular biological property were confined to the analysis of nisin variants that were generated by chemical means. With the increased knowledge of the genetics of lantibiotic biosynthesis it has become possible to develop protein engineering strategies allowing, in theory, unlimited modifications to the mature molecule. This approach will aid the understanding of structure/function relationships. Also the possibility of producing a nisin variant with advantageous properties such as enhanced stability, host range or greater biological activity may increase the application of this lantibiotic in the food industry.

Classical protein engineering involves small changes to a structural gene usually by site-directed mutagenesis so as to alter individual specific amino acids in an expressed protein. In the case of lantibiotic engineering the primary expressed protein is subject to post-translational modification before the active mature molecule is formed. Therefore, it is necessary to find a means of expressing a variant precursor gene in a lantibiotic-producing bacterial host such that the peptide is correctly modified and secreted from the cell. Two approaches to this problem have been successfully used to generate nisin variants. Kuipers et al. (1992) exploited the differences between nisin A and nisin Z to develop strains which expressed both molecules. The background strain was a plasmid-free L. lactis subsp. lactis harbouring an intact copy of the nisin transposon which encoded nisin A production. A plasmid encoding a copy of the nisin Z prenisin gene under control of the lactococcal lactose operon promoter was introduced into this strain leading to the production of both nisin A and nisin Z. It was possible to resolve the two molecule forms and thus to purify mature nisin Z using reverse phase HPLC (Kuipers et al., 1992). Changes introduced into the plasmid encoded nisZ gene could then be translated into variant mature nisin molecules and their properties could be investigated. This approach may be of value in the laboratory, however, it suffers from the need for chemical separation of the engineered variant nisins from normal nisin A. It would also be very limiting to use this strategy for the industrial scale production of a nisin variant should a commercially valuable molecule be designed.

An alternative protein engineering strategy requires a nisin producing strain in which the prenisin gene is inactivated, but which retains expression of genes for nisin maturation and immunity. Dodd et al. (1992) constructed such a strain by insertionally inactivating the nisA gene with an erythromycin resistance gene. The insertion which resulted in loss of the nisin phenotype was found to have also adversely affected immunity so that this Nis⁻ strain was now sensitive to reduced levels of nisin. Selection for nisin immunity at wild type levels generated a strain in which genes for immunity and nisin modification were sufficiently well expressed to convert plasmid encoded prenisin into the equivalent mature nisin (Dodd *et al.*, 1992). This strain was transformed with plasmid encoded variant prenisin genes expressed both under the control of the natural nisin promoter and the lactococcal lactose operon promoter. In this case only the variant nisin was produced facilitating immediate assessment of its properties and simplifying its purification and large scale production.

The expression of mutated *nisA* genes using the above strategies and the subsequent post-translational modification and production of specifically altered nisin variants opens the way to an investigation of the role of individual amino acids in the biology of the nisin molecule. In the naturally occurring nisin variant, nisin Z, the substitution of asparagine for histidine at position 27 introduces a more polar residue in ring D of the mature molecule (Figure 5.6). Nisin Z has normal biological activity but was found to produce larger inhibition zones than nisin A when used at higher concentrations. This effect has been explained by faster diffusion of the variant molecule (De Vos *et al.*, 1992). A site directed mutant of nisin A in which the same histidine residue in ring D was changed to glutamine was also found to exhibit normal biological activity (Dodd *et al.*, 1992). In the course of constructing this latter nisin variant, random mutations were co-introduced into the molecule resulting in two further chemically altered nisins. Their analyses indicated that substitution of a valine at position 32 with a similarly charged isoleucine residue did not affect biological activity. In contrast changing the threonine residue at position 23 of nisin A into a serine significantly reduced activity (Dodd *et al.*, 1992). Since this residue is involved in the closure of ring D correct maturation would generate a lanthionine bridge in place of the β-methyllanthionine of nisin A. From these experiments it was not possible to tell whether the reduced biological activity reflected this change to ring D or whether poor maturation of the variant prenisin was responsible.

Dehydro residues are highly conserved features of linear lantibiotics and their very presence throughout the mature molecules argues that these atypical residues play an important functional role. It has been suggested that they may contribute to the antimicrobial activity by their ability to react with free sulfydryl groups of bacterial cell-wall proteins (Morris *et al.*, 1984; Liu and Hansen, 1990). As such they are an obvious target for engineering the nisin molecule. The purification and analysis of spontaneous degradation products of nisin already indicated that the Dha at residue 5 was significant. A molecule in which this residue was lost together with Dha_{33} lacked biological activity (Chan *et al.*, 1989a). Kuipers *et al.* (1992) changed Ser_5 of prenisin Z to a threonine residue and expressed and purified the equivalent mature nisin. Analysis by two dimensional NMR revealed that the threonine residue was dehydrated to Dha and the molecule had 2–10 fold lower activity than nisin Z. It was concluded that this was due to the presence of the less reactive double bond of Dhb instead of Dha at position 5. This same group reported that replacing this Dha with an alanine in nisin Z did not yield an active molecule. In contrast to this observation a variant of nisin A has been produced in which the same substitution (i.e. Dha_5 to alanine) has been engineered (Dodd, Horn and Gasson unpublished data). In the expression system described by Dodd et al. (1992) this variant was found to have relatively normal antimicrobial activity. Similar techniques were used to substitute Dha_{33} for an alanine and this nisin A variant retains antimicrobial activity, albeit at slightly reduced levels. This result is in agreement with that of Chan et al. (1989a) who showed that the loss of the two C-terminal residues, including the dehydroalanine at position 33, by chemical degradation does not result in reduced activity. The substitution of both Dha_5 and Dha_{33} for alanine residues creating a double mutant, caused a significant loss in activity suggesting that the presence of at least one Dha residue is required for full antimicrobial activity (Dodd, Horn and Gasson unpublished data).

The closely related lantibiotic subtilin also contains a Dha residue at position 5 in ring A. However, loss of antimicrobial activity as a result of chemical modification of this residue (Hansen et al., 1991) occurs much more readily than was found for nisin (Chan et al., 1989a). Liu and Hansen (1992) demonstrated that the greater instability of subtilin was due to the molecular environment of Dha_5. It was postulated that this was primarily due to a chemical modification involving the adjacent glutamic acid residue at position 4 and by changing this amino acid to a less reactive isoleucine residue enhanced stability was achieved. These authors also changed Dha_5 of subtilin to alanine and found that this prevented the activity of subtilin against the outgrowth of Bacillus cereus spores. The suggestion has been made that these lantibiotics exhibit two distinct mechanisms of antimicrobial activity (Hansen et al., 1991), one directed against spore outgrowth and the other involving lysis of active growing cells. The observation that Dha_5-lacking mutants of subtilin are unable to inhibit spore outgrowth (Liu and Hansen, 1992), whereas the equivalent nisin variant retains its activity against growing cells (Dodd, Horn and Gasson, unpublished results) would strongly support this proposal.

In an attempt to change nisin Z to a molecule more closely resembling subtilin, Kuipers et al. (1992) changed residue 17 of nisin Z from methionine to glutamine and residue 18 from glycine to threonine. Expression of the variant prenisin resulted in a mixture of two mature nisins one with threonine at position 18 the other with the dehydrated form dehydrobutyrine at this position. Both molecules exhibited biological activity but failed to reproduce the antimicrobial activity of subtilin against nisin-producing strains of L. lactis suggesting that this property resided in other residues characteristic of subtilin.

In theory the number and type of alterations that can be made to nisin using the techniques of protein engineering are limitless. However, as post-translational modification is a fundamental part of lantibiotic formation the host cells maturation machinery must be able to specifically recognise the precursor molecule. Hence processing signals must be retained in the primary sequence if the precursor is to be modified to an active form. Those amino acids that are highly conserved in different lantibiotics are likely to be absolutely required for activity. As more lantibiotic variants are engineered and their biological properties assessed it may become apparent where molecular variation is most likely to yield improved nisins.

5.4.9 Applications

Nisin has a long history of use in food but following its original discovery in the 1920s it was first evaluated as a therapeutic agent. This potential application failed primarily because nisin was unstable at physiological pH and was sensitive to rapid enzymatic degradation. Also in therapeutic terms its relatively narrow antibacterial spectrum limited its value in human and vetinary medicine (Hurst, 1983). Following the end of the second world war nisin was developed as a food preservative, firstly in combatting the problem of clostridial spoilage in natural cheeses (Hirsch et al., 1951). This application was limited by the inhibition of vital cheese starter bacteria by nisin and whilst earlier attempts to circumvent this limitation by developing nisin resistant strains proved complex (Lipinska, 1977) the future understanding and exploitation of genes for nisin immunity may reawaken interest.

Processed cheese proved to be a more realistic target for the use of nisin and the prevention of cheese spoilage in such products was its first practical application (McClintock et al., 1952). It is of particular relevance that this type of spoilage often originates from contaminating spores of Clostridium and Bacillus and the capacity of nisin to inhibit spore out growth is thus an important property in its practical use in food. In some countries where milk shelf life is problematic nisin is used as an additive to prolong the maintenance of a usable product particularly where hot climates and lengthy transport are necessary (Delves-Broughton, 1990).

Another distinct use of nisin is in the production of canned vegetables. Here its ability to prevent thermophilic spoilage by heat resistant spores and to facilitate the safe use of lower heat treatment is of especial importance (Fowler and Gasson, 1991). More recently nisin has been used in Australia to prevent problems with Bacillus cereus spores in high moisture flour based products (Jensen, 1990). The use of nisin in meat based products has been less effective often requiring the addition of very high levels and its poor performance might be attributed to binding of a molecule to the food, uneven distribution, poor solubility or sensitivity to food enzymes (Fowler and Gasson, 1991). Nisin has been investigated for the protection of fish products from Clostridium botulinum spores particularly where vacuum or modified atmosphere packaging is used (Taylor et al., 1990).

The major spoilage problem in the brewing industry is associated with lactic acid bacteria and here nisin has been investigated as a novel protective strategy. The fact that the primary beer fermentation relies on yeasts which are quite insensitive to nisin is a particular advantage over the situation in dairying where both the starter organisms and the spoilage agents are nisin sensitive Gram-positive bacteria. Whilst nisin proved effective as an additive during the fermentation process the use of nisin during the washing of pitching yeast has proved particularly advantageous over acid treatments that are traditionally used to minimize the problem of bacterial contamination (Ogden, 1987). Nisin has also proved effective in bottle conditioned beers (Ogden *et al.*, 1988) and in pasteurized products it may allow a reduction in the intensity of heat treatment with analogous advantages to those in the canning industry. Additional applications have been investigated in both the wine-making (Radler, 1990a,b; Daeschel *et al.*, 1991) and the distilling industries (Henning *et al.*, 1986).

Nisin thus has a well established history of safe use in foods with a constantly growing range of future application opportunities. The advent of genetic control both over the *in vivo* transfer of the nisin transposon and the detailed manipulation of the nisin biosynthetic genes will open even more opportunities in the future. The capacity to protein engineer the nisin molecule may also provide variants that are more potent and effective in a wider rage of physical environments or against a wider ranger of spoilage and food-poisoning organisms.

5.5 Concluding remarks

Until relatively recently nisin was the only bacteriocin from lactic acid bacteria to have undergone characterisation in any depth. Now many more bacteriocins are being studied at the molecular level. Although these antimicrobial peptides have been isolated from a diverse range of organisms they can be categorized, on the basis of size, stability to heat and the presence of modified amino acids, into one of three main groups sharing common properties (section 5.1).

The molecular analysis of the lantibiotic nisin is still the subject of much intensive research as demonstrated by the recent work of Engelke *et al.* (1992), van der Meer *et al.* (1993) and Kuipers *et al.* (1993). Their studies have significantly advanced understanding of the biosynthesis of this unique bacteriocin. The discovery of two further lantibiotics in lactic acid bacteria (sections 5.4.1 and 5.4.2) should provide further insight into the mechanisms involved in processing, translocation and immunity to lanthionine containing molecules.

The molecular biology of a wide range of other bacteriocins are currently undergoing thorough genetic and biochemical investigation and advances made in recombinant DNA technology for lactic acid bacteria has seen a parallel increase in our knowledge of these antimicrobial proteins. A good example of this is the detailed genetic and biochemical analysis of the lactococcins by van Belkum *et al.*, 1989; 1991a,b; 1992 (section 5.2.2). These studies have led the way in investigating the mode of action of small heat-stable bacteriocins (section 5.3.4). More recently genetic analysis of lactococcin A has brought to light the existence of a system for translocation of these small heat-stable peptides (section 5.3.4). The proposed signal sequence-independent secretory pathway (Stoddard *et al.*, 1992) may well be a common feature of bacteriocins in Gram-positive organisms. Similarities between the lactococcal system and that of other bacteriocins, both non-modified peptides (e.g. pediocin PA-1, Marugg *et al.*, 1992) and lantibiotics (e.g. nisin, Engelke *et al.*, 1992) have been observed that suggest that very different antimicrobial peptides may utilize analogous secretory mechanisms.

Increased understanding of bacteriocins in lactic acid bacteria should lead to new biological preservatives for use in the food industry. The progress made in genetic and biochemical analyses of these molecules is likely to be exploited for yield improvement and the engineering of variant bacteriocins exhibiting altered and improved properties.

References

Ahn, C. and Stiles, M.E. (1990a) Antibacterial activity of lactic acid bacteria isolated from vacuum-packaged meats. *J. Appl. Bacteriol.* **69**, 302–310.

Ahn, C. and Stiles, M.E. (1990b) Plasmid-associated bacteriocin production by a strain of *Carnobacterium piscicola* from meat. *Appl. Envir. Microbiol.* **56**, 2503–2510.

Allgaier, H., Jung, G., Werner, R.G., Schneider, U. and Zahner, H. (1986) Epidermin: sequencing of a heterodet tetracyclic 21-peptide amide antibiotic. *Eur. J. Biochem.* **160**, 9–22.

Augustin, J., Rosenstein, R., Wieland, B., Schneider, U., Schnell, N., Engelke, G., Entian, K.-D. and Gotz, F. (1992) Genetic analysis of epidermin biosynthesis genes and epidermin-negative mutants of *Staphylococcus epidermis*. *Eur. J. Biochem.* **204**, 1149–1154.

Barefoot, S.F. and Klaenhammer, T.R. (1983) Detection and activity of lactacin B, a bacteriocin produced by *Lactobacillus acidophilus*. *Appl. Envir. Microbiol.***45**, 1808–1815.

Barefoot, S.F. and Klaenhammer, T.R. (1984) Purification and characterization of the *Lactobacillus acidophilus* bacteriocin lactacin B. *Antimicrob. Agents Chemother.* **26**, 328–334.

Benz, R., Jung, G. and Sahl, H.-G. (1991) Mechanism of channel-formation by lantibiotics in black lipid membranes. In *'Nisin and Novel Lantibiotics'* (eds. G. Jung and H-G. Sahl) ESCOM, Leiden, pp. 359–372.

Berridge, N.J., Newton, G.G.F. and Abraham, E.P. (1952) Purification and nature of the antibiotic nisin. *Biochem. J.* **52**, 529–535.

Blight, M.A. and Holland, I.B. (1990) Structure and function of haemolysin B, P-glycoprotein and other members of a novel family of membrane translocators. *Mol. Microbiol.* **4**, 873–880.

Buchman, W.B., Banerjee, S. and Hansen, J.N. (1988) Structure, expression, and evolution of a gene encoding the precursor of nisin, a small protein antibiotic. *J. of Biol. Chem.* **263**, 16260–16266.

Chan, W.C., Bycroft, B.W., Lian, L-Y. and Roberts, G. (1989a) Isolation and characterization of two degradation products derived from the peptide antibiotic nisin. *FEBS Lett.* **252**, 29–36.

Chan, W.C., Lian, L-Y., Bycroft, B.W. and Roberts, G.C.K. (1989b). Confirmation of the structure of nisin by complete ^{1}H N.m.r. resonance assignment in aqueous and dimethyl sulphoxide solution. *J. Chem. Soc. Perkin. Trans.* **1**, 2359–2367.

Chung, Y.J., Steen, M.T. and Hansen, J.N. (1992). The subtilin gene of *Bacillus subtilis* ATCC 6633 is encoded in an operon that contains a homolog of the hemolysin B transport protein. *J. Bacteriol.* **174**, 1417–1422.

Chung, Y.J. and Hansen, J.N. (1992). Determination of the sequence of *spaE* and identification of a promoter in the subtilin (*spa*) operon in *Bacillus subtilis*. *J. Bacteriol.* **174**, 6699–6702.

Collins, M.D., Ash, C., Farrow, A.E., Philips, B.A., Ferusus, S. and Jones, D. (1987). Classification of *Lactobacillus divergens*, *Lactobacillus piscicola* and some catalase-negative asporogenous, rod-shaped bacteria from poultry in a new genus *Carnobacterium*. *Int. J. Syst. Bacteriol.* **37**, 310–317.

Daeschel, M.A. and Klaenhammer, T.R. (1985). Association of a 13.6-megadalton plasmid in *Pediococcus pentosaceus* with bacteriocin activity. *Appl. Envir. Microbiol.* **50**, 1538–1541.

Daeschel, M.A., Jung, D.S. and Watson, B.T. (1991). Controlling wine malolactic fermentation with nisin and nisin-resistant strains of *Leuconostoc oenos*. *Appl. Envir. Microbiol.* **57**, 601–603.

Davey, G.P. (1984). Plasmid associated with diplococcin production in *Streptococcus cremoris*. *Appl. Envir. Microbiol.* **48**, 895–896.

Davey, G.P. and Richardson, B.C. (1981). Purification and some properties of diplococcin from *Streptococcus cremoris* 346. *Appl. Envir. Microbiol.* **41**, 84–89.

de Vuyst, L. and Vandamme, E.J. (1992). Localization and phenotypic expression of genes involved in the biosynthesis of the *lactococcus lactis* subsp. *lactis* lantibiotic nisin. In '*Bacteriocins, Microcins and Lantibiotics*.' (eds James, R., Ladzunski, C. and Pattus, F.), NATA ASI Series H, **65** Springer-Verlag, Berlin, Heidelberg, Germany, pp. 449–461.

De Vos, W.M., Jung, G. and H.-G. Sahl. (1991). Definitions and nomenclature of lantibiotics. In '*Nisin and Novel Lantibiotics*', (eds G. Jung and H.-G. Sahl), ESCOM, Leiden, pp. 457–463.

De Vos, W.M., Mulders, J.W.M., Siezen, R.J., Hugenholtz, J. and Kuipers, O.P. (1992). Properties of nisin Z and distribution of its gene, *nisZ* in *Lactococcus lactis*. *Appl. Envir. Microbiol.* **59**, 213–216.

Delves-Broughton, J. (1990). Nisin and its uses. *Food Technology* **44**, (11), 100–117.

Dodd, H.M., Horn, N. and Gasson, M.J. (1990) Analysis of the genetic determinant for production of the peptide antibiotic nisin. *J. Gen. Microbiol.* **136**, 555–566.

Dodd, H.M., Horn, N., Hao, Z. and Gasson, M.J. (1992) A lactococcal expression system for engineered nisins. *Appl. Envir. Microbiol.* **58**, 3683–3693.

Dufour, A.D., Thuault, D., Boulliou, A., Bourgeois, C.M. and Le Pennec, J-P. (1991) Plasmid-encoded determinants for bacteriocin production and immunity in a *Lactococcus lactis* strain and purification of the inhibitory peptide. *J. Gen. Microbiol.* **137**, 2423–2429.

Engelke, G., Gutochowski-Eckel, Z., Hammelman, M. and Entian, K.-D. (1992) Biosynthesis of the lantibiotic nisin: genomic organization and membrane localization of the NisB protein. *Appl. Envir. Microbiol.* **58**, 3730–3743.

Fowler, G.G. (1971) Nisin in combination with heat treatment of foods. Svenska Institutet for Konserveringsforskning. Report No. 292.

Fowler, G.G. and Gasson, M.J. (1991) Antibiotics – nisin. In '*Food Preservatives*,' (eds. N.J. Russel and G.W. Gould), Blackie, London, pp. 135–152.

Froseth, B.R. and McKay, L.L. (1991) Molecular characterization of the nisin resistance region of *Lactococcus lactis* subsp. *lactis* biovar *diacetylactis* DRC3. *Appl. Envir. Microbiol.* **57**, 804–811.

Froseth, B.R., Herman, R.E. and McKay, L.L. (1988) Cloning of nisin resistance determinant and replication origin on 7.6-kilobase *Eco*RI fragment of pNP40 from *Streptococcus lactis* subsp. diacetylactis DRC3. *Appl. Envir. Microbiol.* **54**, 2136–2139.

Gao, F.H., Abee, T. and Konings. W.N. (1991) Mechanism of action of the peptide antibiotic nisin in liposomes and cytochrome c oxidase-containing proteoliposomes. *Appl. Envir. Microbiol.* **57**, 2164–2170.

Gasson, M.J. (1984) Transfer of sucrose fermenting ability, nisin resistance and nisin production into *Streptococcus lactis* 712. *FEMS Microbiol Lett.* **21**, 7–10.

Gireesh, T., Davidson, B.E. and Hillier, A.J. (1992) Conjugal transfer in *Lactococcus lactis* of a 68-kilobase-pair chromosomal fragment containing the structural gene for the peptide bacteriocin nisin. *Appl. Envir. Microbiol.* **58**, 1670–1676.

Gonzalez, C.F. and Kunka, B.S. (1987) Plasmid associated bacteriocin production and sucrose fermentation in *Pediococcus acidilactici*. *Appl. Envir. Microbiol.* **53**, 2534–2538.

Graeffe, T., Rinalta, H., Paulin, L. and Saris, P. (1991) A natural nisin variant. In '*Nisin and Novel Lantibiotics*' (eds. G. Jung and H.-G. Sahl), ESCOM, Leiden, pp. 260–268.

Graham, D.C. and McKay, L.L. (1985) Plasmid DNA in strains of *Pediococcus cerevisiae* and *Pediococcus pentosaceus*. *Appl. Envir. Microbiol.* **50**, 532–534.

Gross, E., and Kiltz, H.H. (1973) The number and nature of α,β'-unsaturated amino acids in subtilin. *Biochem. Biophys. Res. Comm.* **50**, 559–565.

Gross, E. and Morell, J. (1967) The presence of dehydroalanine in the antibiotic nisin and its relationship to activity. *J. Amer. Chem. Soc.* **89**, 2791–2792.

Gross, E. and Morell, J. (1971) The structure of nisin. *J. Amer. Chem. Soc.* **93**, 4634–4635.

Hansen, J.N., Chung, Y.J., Liu, W. and Steen, M.T. (1991) Biosynthesis and mechanism of action of nisin and subtilin. In '*Nisin and Novel Lantibiotics*', (eds. G. Jung and H.-G. Sahl), ESCOM, Leiden, pp. 287–302.

Hastings, J.W., Sailer, M., Johnson, K., Roy, K.L., Vederas, J.C. and Stiles, M. (1991) Characterization of leucocin A-UAL 187 and cloning of the bacteriocin gene from *Leuconostoc gelidum. J. Bacteriol.* **173**, 7491–7500.

Hastings, J.W. and Stiles, M.E. (1991) Antibiosis of *Leuconostoc gelidum* isolated from meat. *J. Appl. Bacteriol.* **70**, 127–134.

Hechard, Y., Derijard, B., Letellier, F. and Cenatiempo, Y. (1992) Characterization and purification of mesentericin Y105, an anti-listeria bacteriocin from *Leuconostoc mesenteroides. J. Gen. Microbiol.* **138**, 2725–2731.

Henderson, J.T., Chopko, A.L. and Vanwassenaar, P.D. (1992) Purification and primary structure of pediocin PA–1 produced by *Pediococcus acidilactici* PAC1.0. *Arch Biochem Biophys* **295**, 5–12.

Henning, S., Metz, R. and Hammes, W.P. (1986) New aspects for the application of nisin to foods based on its mode of action. *Int. J. Food Microbiol.* **3**, 135–141.

Henning, S., Metz, R. and Hammes, W.P. (1986) Studies on the mode of action of nisin. *Int. J. Food Microbiol.* **3**, 121–134.

Hirsch, A., Grinsted, E., Chapman, H.R. and Mattick, A (1951) A note on the inhibition of an anaerobic sporeformer in Swiss-type cheese by a nisin-producing *Streptococcus. J. Dairy res.* **18**, 205–206.

Holck, A., Axelsson, L., Birkeland, S-E., Aukrust, T. and Blom, H. (1992) Purification and amino acid sequence of sakacin A, a bacteriocin from *Lactobacillus sake. J. Gen. Microbiol.* **138**, 2715–2720.

Holo, H., Nilssen, O. and Nes, I.F. (1991) Lactococcin A, a new bacteriocin from *Lactococcus lactis* subsp. *cremoris*: isolation and characterization of the protein and its gene. *J. Bacteriol.* **173**, 3879–3887.

Horn, N., Swindell, S., Dodd, H.M. and Gasson, M.J. (1991) Nisin biosynthesis genes are encoded by a novel conjugative transposon. *Mol. Gen. Genet.* **228**, 129–135.

Hurst, A. (1983) Nisin and other inhibitory substances from lactic acid bacteria. In *Antimicrobials in foods*, (eds. A.J. Branen and P.M. Davidson), Marcel Dekker Inc., New York, pp. 327–351.

Hurst, A. (1981) Nisin. *Adv. Appl. Microbiol.* **27**, 85–123.

Ingram, L. (1970) A ribosomal mechanism for synthesis of peptides related to nisin. *Biochem. Biophys. Acta.* **224**, 263–265.

Jensen, I. (1990) Nisin, *Bacillus cereus* and high moisture flour based products. Paper presented at Australian Society for Microbiology, Annual Scientific Meeting, Launceston, Tasmania. Paper No. S25.2.

Joerger, M.C. and Klaenhammer, T.R. (1986) Characterization and purification of helveticin J and evidence for a chromosomally determined bacteriocin produced by *Lactobacillus helveticus* 481. *J. Bacteriol.* **167**, 439–446.

Joerger, M.C. and Klaenhammer, T.R. (1990) Cloning, expression and nucleotide sequence of the *lactobacillus helveticus* 481 gene encoding the bacteriocin helveticin J. *J. Bacteriol.* **171**, 1597–1601.

Jung, G. (1991) Lantibiotics: a survey. In *Nisin and Novel Lantibiotics*, (eds G. Jung and H.-G. Sahl), ESCOM, Leiden, pp. 1–34.

Kaletta, C. and Entian, D.-D. (1989) Nisin, a peptide antibiotic: cloning and sequencing of the nisA gene and post-translational processing of its peptide product. *J. Bacteriol.* **171**, 1597–1601.

Kellner, R., Jung, G., Horner, T., Zahner, H., Schnell, N., Entian, K.-D. and Gotz, F. (1988) Gallidermin: a new lanthionine-containing polypeptide antibiotic. *Eur. J. Biochem.* **177**, 53–59.

Klaenhammer, T.R. (1988) Bacteriocins of lactic acid bacteria. *Biochimie.* **70**, 337–349.

Klaenhammer, T.R., Ahn, C., Fremaux, C. and Milton, K. (1992) Molecular properties of *lactobacillus* bacteriocins. In *Bacteriocins, Microcins and Lantibiotics* (eds James, R., Ladzunski, C. and Pattus, F) NATA ASI Series H, **65**, Springer-Verlag, Berlin, Heidleberg, Germany, pp. 37–58.

Klaenhammer, T.R. and Sanozky, R.B. (1985) Conjugal transfer from *Streptococcus lactis* ME2 of plasmids encoding phage resistance, nisin resistance and lactose-fermenting ability: Evidence for a high-frequency conjugative plasmid responsible for abortive infection of virulent bacteriophage. *J. Gen. Microbiol.* **131**, 1531–1541.

Klein, C., Kaletta, C. and Entian, K.-D. (1993) Biosynthesis of the lantibiotic subtilin is regulated by a histidine kinase/response regulator system. *Appl. Envir. Microbiol.* **59**, 296–303.

Klein, C., Kaletta, C., Schnell, N. and Entian, K-D. (1992) Analysis of genes involved in biosynthesis of the lantibiotic subtilin. *Appl. Envir. Microbiol.* **58**, 132–142.

Kojic, M., Svircevic, J., Banina, A. and Topisirovic, L. (1991). Bacteriocin-producing strain of *Lactococcus lactis* subsp. *diacitilactis* S50. *Appl. Envir. Microbiol.* **57**, 1835–1837.

Kordel, M., Schuller, F. and Sahl, H.-G. (1989) Interaction of the pore-forming peptide antibiotics Pep5, nisin and subtilin with non-energized liposomes. *FEBS Lett.* **244**, 99–102.

Kordel, M. and Sahl, H.-G. (1986) Susceptibility of bacterial eukaryotic and artificial membranes to the disruptive action of the cationic peptides Pep 5 and nisin. *FEMS Microbiol. Lett.* **34**, 139–144.

Kuipers, O.P., Rollema, H.S., Yap, W.M.G.J., Boot, H.J., Siezen, R.J. and De Vos, W.M. (1992) Engineering dehydrated amino acid residues in the antimicrobial peptide nisin. *J. Biol. Chem.* **267**, 2430–2434.

Kuipers, O.P., Yap, W.M.G.J., Rollema, H.S., Beerthuyzen, M.M., Siezen, R.J. and De Vos, W.M. (1991) Expression of wild-type and mutant nisin genes in *Lactococcus lactis*. In *Nisin and Novel Lantibiotics*, (eds. G. Jung and H.-G. Sahl), ESCOM, Leiden, pp. 250–259.

Kuipers, O.P., Beerthuyzen, M.M., Siezen, R.J. and De Vos, W.M. (1993) Characterization of the nisin gene cluster *nisABTCIPR* of *Lactococcus lactis*: requirement of expression of the *nisA* and *nisI* genes for producer immunity. *Eur. J. Biochem.* (in press).

Lian, L.Y., Chan, W.C., Morley, S.D., Roberts, G.C.K., Bycroft, B.W. and Jackson, D. (1991) Solution structures of nisin A and its two major degradation products determined by NMR. *Biochem. J.* **283**, 413–420.

Lipinska, E. (1977) Nisin and its applications. In *Antibiotics and Antibiosis in Agriculture*, (ed. M. Woodbine), Butterworth, London, pp. 103–130.

Liu, W. and Hansen, J.N. (1992) Enhancement of the chemical and antimicrobial properties of subtilin by site-directed mutagenesis. *J. Biol. chem.* **267**, 25078–25085.

Liu, W. and Hansen, J.N. (1990) Some chemical and physical properties of nisin, a small protein antibiotic produced by *Lactococcus lactis*. *Appl. Envir. Microbiol.* **56**, 2251–2258.

Marugg, J.D., Gonzalez, C.F., Kunka, B.S., Ledeboer, A.M., Pucci, M.J., Toonen, M.Y., Walker, S.A., Zoetmulder, L.C.M. and Vandenbergh, P.A. (1992) Cloning, expression and nucleotide-sequence of genes involved in production of pediocin PA–1, a bacteriocin from *Pediococcus acidilactici* PAC1.0. *Appl. Envir. Microbiol.* **58**, 2360–2367.

McClintock, M., Serres, L., Marzolf, J.J., Hirsch, A. and Mocquot, G. (1952) Action inhibitrice des streptrococques producteurs de nisine sur le developpement des sporules anaerobies dans le fromage de Gruyere fondu. *J. Dairy Res.* **19**, 187–193.

McKay, L.L. and Baldwin, K.A. (1984) Conjugative 50-megadalton plasmid in *Streptococcus lactis* subsp. *diacetylactis* DRC3 is associated with resistance to nisin and bacteriophage. *Appl. Envir. Microbiol.* **47**, 68–74.

Morris, S.L., Walsh, R.C. and Hansen, J.N. (1984) Identification and characterization of some bacterial membrane sulfhydryl groups which are targets of bacteriostatic and antibiotic action. *J. Biol. chem.* **259**, 13590–13594.

Mortvedt, C.I. and Nes, I.F. (1990) Plasmid associated bacteriocin production by a *Lactobacillus sake* strain. *J. Gen. Microbiol.* **136**, 1601–1607.

Mortvedt, C.I., Nissen-Meyer, J., Sletten, K. and Nes, I.F. (1991) Purification and amino acid sequence of Lactocin S, A bacteriocin produced by *Lactobacillus sake* L45. *Appl. Envir. Microbiol.* **57**, 1829–1834.

Motlagh, A.M., Bhunia, A.K., Szostek, F., Hansen, T.R., Johnson, M.C. and Ray, B. (1992) Nucleotide and amino acid sequence of *pap* gene (pediocin AcH production) in *Pediococcus acidilactici* H. *Letts. Appl. Microbiol.* **15**, 45–48.

Mulders, J.W.M., Boerrigter, I.J., Rollema, H.S., Siezen, R.J. and De Vos, W.M. (1991) Identification and characterization of the lantibiotic nisin Z, a natural nisin variant. *Eur. J. Biochem.* **201**, 581–584.

Muriana, P.M. and Klaenhammer, T.R. (1987) Conjugal transfer of plasmid encoded determinants for bacterion production and immunity in *Lactobacillus* acidophilus 88. *Appl. Envir. Microbiol.* **53**, 553–560.

Muriana, P.M. and Klaenhammer, T.R. (1991a) cloning, phenotypic expression and DNA sequence of the gene for lactacin F, an antimicrobial peptide produced by *Lactobacillus* spp. *J. Bacteriol* **173**, 1779–1788.

Muriana, P.M. and Klaenhammer, T.R. (1991b) Purification and partial characterization of lactacin F, a bacteriocin produced by *Lactobacillus acidophilus* 11088. *Appl. Envir. Microbiol.* **57**, 114–121.

Nes, I.F. (1993). Personal communication.

Nes, I.F., Mortvedt, C.I., Nissen-Meyer, J., Skaugen, M. (1993) Lactocin S, a lanthionine-containing bacteriocin isolated from *Lactobacillus sake* L45. In *Bacteriocins of Lactic Acid Bacteria* (eds de Vuyst, L. and Vandamme, E.J.) Elsevier, London, (in press)

Nieto Lozano, J.C., Nissen-Meyer, J., Sletten, K. and Nes, I.F. (1992) Purification and amino acid sequence of a bacteriocin produced by *Pediococcus acidilactici*. *J. Gen. Microbiol.* **138**, 1985–1990.

Nissen-Meyer, J., Holo, H., Havarstein, L.S., Sletten and K. Nes, I.F. (1992) A novel lactococcal bacteriocin whose activity depends on the complementary action of two peptides. *J. Bact.* **174**, 5686–5692.

Ogden, K. (1987) Cleansing contaminated pitching yeast with nisin. Inst. Brewing **91**, 302–307.

Ogden, K., Waites, M.J. and Hammond, J.R.M. (1988) Nisin and Brewing. J. Inst. Brewing **94**, 233–238.

Piard, J-C., Muriana, P.M., Desmazeaud, M.J. and Klaenhammer, T.R. (1992) Purification and partial characterization of lacticin 481, a lanthionine-containing bacteriocin produced by *lactococcus lactis* subsp. *lactis* CNRZ 481. *Appl. Envir. Microbiol.* **58**, 279–284.

Piard, J-C., Kuipers, O.P., Rollema, H.S., Desmazeaud, M.J. and De Vos, W.M. (1993) Structure, organization and expression of the *lct* gene for lacticin 481, a novel lantibiotic produced by *Lactococcus lactis*. *J. Biol. Chem.* (in press).

Radler, F. (1990a) Possible use of nisin in wine-making I. Action of nisin against lactic acid bacteria and wine yeasts in solid and liquid media. *Am. J. Enology and Viticulture* **41**, 1–6.

Radler, F. (1990b) Possible use of nisin in wine-making II. Experiments to control lactic acid bacteria in the production of wine. *Am. J. Enology and Viticulture* **41**, 7–11.

Rammelsberg, M. and Radler, F. (1990) Antibacterial polypeptides of *Lactobacillus* species. *J. Appl. Bacteriol.* **69**, 177–184.

Ramseier, H.R. (1960) Die Wirkung van Nisin auf *Clostridium Butyricum* prazm. *Archiv. fur Mikrobiologie* **37**, 57–94.

Randall, L.L., Hardy, S.J.S. and Thorn, J.R. (1987) Export of protein: a biochemical view. *Annu. Rev. Microbiol.* **41**, 507–541.

Rauch, P.J.G., Beerthuyzen, M.M. and De Vos, W.M. (1990) Nucleotide sequence of IS*904* from *Lactococcus lactis* subsp. *lactis* strain NIZO R5. *Nuc. Acids Res.* **18**, 4253.

Rauch, P.J.G. and De Vos, W.M. (1992) Characterization of the novel nisin-sucrose conjugative transposon Tn5276 and its insertion in *Lactococcus lactis*. *J. Bacteriol.* **174**, 1280–1287.

Rauch, P.J.G., Beerthuyzen, M.M. and de Vos, W.M. (1991) Molecular analysis and evolution of conjugative transposons encoding nisin production and sucrose fermentation in *Lactococcus lactis*. In *Nisin and Novel Lantibiotics*, (eds. G. Jung and H.-G. Sahl), ESCOM, Leiden, pp. 243–249.

Ray, B., Motlagh, A.M. Johnson, M.C. and Bozoglu, F. (1992) Mapping of pSMB74, a plasmid-encoding bacteriocin, pediocin AcH, production (Pap+) by *Pediococcus acidilactici* H. *Letts. Appl. Microbiol.* **15**, 35–37.

Ray, S.K., Kim, W.J., Johnson, M.C. and Ray, B. (1989) conjugal transfer of a plasmid encoding bacteriocin production and immunity in *Pediococcus acidilactici* H. *J. Appl. Bacteriol.* **66**, 393–399.

Ruhr, E. and Sahl, H.-G. (1985) Mode of action of the peptide antibiotic nisin and influence on the membrane potential of whole cells and on cytoplasmic and artificial membrane vesicles. *Antimicrob. Agents Chemother.* **27**, 841–845.

Sahl, H.-G., Kordel, M. and Benz, R. (1987) Voltage-dependant depolarization of bacterial membranes and artificial lipid bilayers by the peptide antibiotic nisin. *Arch. Microbiol.* **149**, 120–124.

Sahl, H.-G. and Brandis, H. (1981) Production, purification and chemical properties of an antistaphylococcal agent produced by *Staphylococcus epidermis*. *J. Gen. Microbiol.* **127**, 377–384.

Scherwitz-Harmon, K. and McKay, L.L. (1987) Restriction enzyme analysis of lactose and bacteriocin plasmids from *Streptococcus lactis* subsp. *diacetylactis* WM4 and cloning of *Bcl*I fragments coding for bacteriocin production. *Appl. Envir. Microbiol.* **53**, 1171–1174.

Schillinger, U. and Lucke, F.-K. (1989) Antibacterial activity of *Lactobacillus sake* isolated from meat. *Appl. Envir. Microbiol.* **55**, 1901–1906.

Schnell, N., Engelke, G., Augustin, J., Rosenstein, R., Ungermann, V., Gotz, F. and Jung, G. (1992)

Analysis of genes involved in the biosynthesis of lantibiotic epidermin. *Eur. J. Biochem.* **204**, 57–68.

Schnell, N., Entian, K.-D., Schneider, U., Gotz, F., Zahner, H., Kellner, R. and Jung, G. (1988) Prepeptide sequence of epidermin, a ribosomally synthesised antibiotic with four sulphide-rings. *Nature* (London). **333**, 276–278.

Schved, F., Lalazar, A., Henis, Y., and Juven, B.J. (1993) Purification, partial characterization and plasmid-linkage of pediocin SJ–1, a bacteriocin produced by *Pediococcus acidilactici. J. Appl. Bacteriol.* **74**, 67–77.

Shaw, B.G. and Harding, C.D. (1985) Atypical lactobacilli from vacuum-packaged meats: comparison by DNA hybridization cell composition and biochemical tests with a description of *Lactobacillus carnis* sp. nov. *Syst. Appl. Microbiol.* **6**, 291–297.

Simon, D. and Chopin, A (1988) Construction of a vector plasmid family and its use for molecular cloning in *Streptococcus lactis. Biochimie.* **70**, 559–566.

Slijper, M., Hilbers, C.W., Konings, R.N.H. and van den Ven. (1989) NMR studies of lantibiotics: assignment of the ^{1}H-NMR spectrum of nisin and identification of interresidual contacts. *FEBS Lett.* **252**, 22–28.

Sobrino, O.J., Rodriguez, J.M., Moreira, W.L., Fernandez, M.F., Sanz, B. and Hernandez, P.E. (1991) Antibacterial activity of *Lactobacillus sake* isolated from dry fermented sausages. *Int. J. Food Microbiol.* **13**, 1–10.

Sobrino, O.J., Rodriguez, J.M., Moreira, W.L., Clintas, L.M., Fernandez, M.F., Sanz, B. and Hernandez, P.E. (1992) Sakacin, M, a bacteriocin-like substance from *Lactobacillus sake* 148. *Int. J. Food Microbiol.* **16**, 215–225.

Steen, M.T., Chung, Y.J. and Hansen, J.N. (1991) Characterization of the nisin gene as part of a polycistronic operon in the chromosome of *Lactococcus lactis* ATCC 11454. *Appl. Envir. Microbiol.* **57**, 1181–1188.

Stevens, K.A., Sheldon, B.W., Klapes, N.A. and Klaenhammer, T.R. (1991) Nisin treatment for the inactivation of *Salmonella* species and other gram-negative bacteria. *Appl. Envir. Microbiol.* **57**, 3613–3615.

Stoddard, G.W., Petzel, J.P., van Belkum, M.J., Kok, J. and McKay, L.L. (1992) Molecular analysis of the lactococcin A gene cluster from *Lactococcus lactis* subsp. *lactis* biovar *diacetylactis* WM4. *Appl. Envir. Microbiol.* **58**, 1952–1961.

Stoffels, G., Nes, I.F. and Gudmundsdottir, A. (1992a) Isolation and properties of a bacteriocin producing *Carnobacterium piscicola* isolated from fish. *J. Appl. Bacteriol.* **73**, 309–316.

Stoffels, G., Nissen-Meyer, J., Gudmundsdottir, A., Sletten, K., Holo, H. and Nes, I.F. (1992b) Purification and characterization of a new bacteriocin isolated from a *Carnobacterium* sp. *Appl. Envir. Microbiol.* **58**, 1417–1422.

Tayler, L.Y., Cann, D.D. and Welch, B.J. (1990) Antibotulinal properties of nisin in fresh fish packaged in an atmosphere of carbon dioxide. *J. Food Protection*, **53**, 953–957.

Tichaczek, P.S., Nissen-Meyer, J., Nes, I.F., Vogel, R.F. and Hammes, W.P. (1992) Characterization of the bacteriocins curvacin A from *Lactobacillus curvatus* LTH1174 and sakacin P from *Lactobacillus sake* LTH673. *Systematic and Applied Microbiol.* **15**, 460–465.

Toba, T., Yoshioka, E. and Itoh, T. (1991a) Acidophilucin A, a new heat labile bacteriocin produced by *Lactobacillus acidophilus* LAPT 1060. *Letts. Appl. Microbiol.* **12**, 106–108.

Toba, T., Yoshioka, E. and Itoh, T. (1991b) Lacticin, a bacteriocin produced by *Lactobacillus delbrueckii* subsp. *lactis. Letts. Appl. Microbiol.* **12**, 43–45.

Tsai, H.-J. and Sandine, W.E. (1987) Conjugal transfer of nisin plasmid genes from *Streptococcus lactis* 7962 to *Leuconostoc dextranicum* 181. *Appl. Envir. Microbiol.* **53**, 352–357.

Upretti, G.C. and Hindsdill, R.G. (1975) production and mode of action of lactocin 27: bacteriocin from a homofermentative *Lactobacillus. Antimicrob. Agents Chemother.* **7**, 139–145.

van Belkum, M.J., Hayema, B.J., Geis, A., Kok, J. and Venema, G. (1989) Cloning of two bacteriocin genes from a lactococcal bacteriocin plasmid. *Appl. Envir. Microbiol.* **55**, 1187–1191.

van Belkum, M.J., Hayema, B.J., Jeeninga, R.E., Kok, J. and Venema, G. (1991a) Organization and nucleotide sequences of two lactococcal bacteriocin operons. *Appl. Envir. Microbiol.* **57**, 492–498.

van Belkum, M.J., Kok, J., Venema, G., Holo, H., Nes, I.F., Konings, W.N. and Abee, T. (1991b) The bacteriocin lactococcin A specifically increases permeability of lactococcal cytoplasmic membranes in a voltage-independent, protein-mediated manner. *J. Bacteriol.* **173**, 7934–7941.

van Belkum, M.J., Kok, J. and Venema, G. (1992) Cloning, sequencing and expression in

Escherichia coli of *lcnB*, a third bacteriocin determinant from the lactococcal bacteriocin plasmid p9B4–6. *Appl. Envir. Microbiol.* **58**, 572–577.

van de Ven, F.J.M., van den Hooven, H.W., Konings, R. and Hilbers C.W. (1991) NMR studies of lantibiotics. The structure of nisin in aqueous solution. *Eur. J. Biochem.* **202**, 1181–1188.

van der Meer, J.R., Polman, J., Beerthuyzen, M.M., Siezen, R.J., Kuipers, O.P. and de Vos, W.M. (1993) Characterization of the *Lactococcus lactis* Nisin A operon genes *nisP*, encoding a subtilisin-like serine protease involved in precursor processing, and *nisR*, encoding a regulatory protein involved in nisin biosynthesis. *J. Bacteriol.* **175**, 2578–2588.

Venema, K., Abee, T., Haandrikman, A.J., Leenhouts, K.J., Kok, J., Konings, W.N. and Venema, G. (1993) Mode of action of lactococcin B, a thiol-activated bacteriocin from *Lactococcus lactis*. *Appl. Envir. Microbiol.* **59**, 1041–1048.

von Heijne, G. (1986) A new method for predicting signal cleavage sites. *Nucleic Acids Res.* **14**, 4683–4690.

von Heijne, G. (1983) Pattern of amino acids near signal-sequence cleavage sites. *Eur. J. Biochem.* **133**, 17–21.

Von Wright, A., Wessels, S., Tynkkynen, S. and Saarela, M. (1990) Isolation of a replication region of a large lactococcal plasmid and use in cloning of a nisin resistance determinant. *Appl. Envir. Microbiol.* **56**, 2029–2035.

Worobo, R.W., Quadri, L., Stiles, M., Henkel, T., Sailer, M. and Vederas, J.C. (1991) Purification and characterization of bacteriocins produced by *Carnobacterium piscicola* LV17. Presented at the EMBO-FEMS-NATO Workshop on Bacteriocins, Microcins and Lantibiotics. Island of Bendor, France, 22–26, September 1991.

6 Genetic engineering of lactobacilli, leuconostocs and *Streptococcus thermophilus*

A. MERCENIER, P.H. POUWELS and B.M. CHASSY

6.1 Introduction

The lactic acid bacteria represent a large heterogeneous family of micro-organisms that share the property of converting fermentable carbohydrates primarily to lactic acid. The homofermentative lactic acid bacteria convert one mole of glucose to two moles of lactic acid, while the heterofermentative lactic acid bacteria convert one mole of glucose to one mole of lactic acid and a variety of other products such as acetic acid, ethanol and CO_2. The family includes very diverse species that naturally occupy quite different ecological niches. Lactic acid bacteria are found on plant surfaces, on external cavities of human and animal bodies, as commensal colonizers of the gastrointestinal systems of verte-brates, as well as in sewage and manure. They are used extensively for the manufacture of a variety of fermented food and feed products. Although they are best known for application in the preparation of fermented dairy products (i.e. cheeses, sour milks, yogurts), lactic acid bacteria are also used in the pickling of vegetables, baking, wine-making, curing of fish, meats and sausages, prepa-ration of silage, remediation of biowastes and the commercial manufacture of lactic acid (Kandler and Weiss, 1986). Due to the global economic importance of the agro-food sector, our knowledge of the lactic acid bacteria has benefited from an intensive research activity over the last decade. For economic, practical and historical reasons attention has been focused primarily on lactococci and the streptococci of medical importance. *Lactococcus lactis* subsp. *lactis, diacety-lactis* and *cremoris*, previously called Group N streptococci, were the first food starter cultures for which a genetic transfer system was devised (Gasson, 1990; McKay and Baldwin, 1990). This success opened the way for the application of recombinant DNA technology to these organisms.

Recently, other genera such as the lactobacilli, leuconostocs and other dairy streptococci have received increasing attention. The two last genera include a limited number of species of commercial interest (i.e. *Streptococcus thermo-philus* and *Leuconostoc mesenteroides* subsp. *mesenteroides*, *dextranicum* and *cremoris, Leuconostoc lactis, Leuconostoc paramesenteroides*, and *Leuconostoc oenos*. In contrast, the genus *Lactobacillus* contains about 50 species, some of which contain a number of subspecies (Kandler and Weiss, 1986). As a result of this diversity, the literature appears sparse and scattered for the lactobacilli, despite the fact that they have been the subject of active investigation during the last 15 years. The published data on lactobacilli can be roughly categorized into two research areas: one focuses on the rod-shaped micro organisms involved in industrial fermentation processes, and the other addresses the nutritional and human/animal health beneficial characteristics attributed to certain strains.

Paralleling the lactococci, genetic manipulation of *S. thermophilus*, *Lactobacillus* spp. and *Leuconostoc* spp. required the development of gene transfer methods and useful cloning vectors. Progress in this area was followed by a search for efficient gene transcription and translation signals that is still in progress. Recently acquired genetic knowledge was applied further to study important metabolic properties, to generate specific mutants, and to construct the first recombinant strains of potential utility. Research has focused on plasmid biology, naturally occurring phages and phage defense mechanisms and on biochemical studies of key enzymes. This chapter discusses advances in our understanding of the taxonomy, plasmid biology, cloning vectors, gene transfer methods, gene expression signals, phages, and significant metabolic traits associated with the lactobacilli, leuconostocs, and *S. thermophilus*.

6.2 Overview of taxonomy and health benefits

From a taxonomic perspective, *S. thermophilus*, the lactobacilli and the leuconostocs are grouped with the other lactic acid bacteria and *Bacillus* spp. within a supercluster of the clostridial sub-branch of the Gram-positive eubacteria, characterized by a low genomic G+C content (Stackebrandt *et al.*, 1983; Stackebrandt and Teuber, 1988). It should be noted, however, that the organisms discussed in this chapter diverged long ago from a common ancestor. The phylogenetic distances between the genera are great. Among the lactobacilli, ancient divergences have created a genus whose representative species are as distantly related as are the genera of Gram-negative bacteria (Stackebrandt *et al.*, 1983). The divergency among lactobacilli is reflected by the considerable differences found in G+C content (e.g. *Lactobacillus divergens* 33–35%; *Lactobacillus fermentum* 52%). As the techniques used in bacterial classification have evolved, the phylogenetic position of several species or strains has been subjected to frequent changes. The taxonomy of these bacteria has been the subject of recent specialized reviews (Collins *et al.*, 1991; Scheifler, 1987; Scheifler, *et al.*, 1992; Stackebrandt *et al.*, 1983; Stackenbrandt and Teuber, 1988). *S. thermophilus* offers a good example of constant evolution in taxonomic status. Since the first classification in 1937 by Sherman (1937), this micro organism has been moved several times within the genus *Streptococcus* (Hutkins and Morris, 1987). Although assigned in recent years as a subspecies of *S. salivarius* (Farrow and Collins, 1984), recent studies suggest its status should be elevated to that of a distinct streptococcal species (Bently *et al.*, 1991; Scheifler, *et al.*, 1991); therefore we have referred to *S. salivarius* subsp. *thermophilus* as *S. thermophilus* throughout this chapter. The organism appears related to *S. agalactiae*, *S. dysagalactiae*, and *S. acidominimus* (the viridans group of streptococci), but it occupies a different natural habitat and displays a higher optimal growth temperature.

Optimal growth temperatures for *S. thermophilus*, *Leuconostoc* spp. and *Lactobacillus* spp. vary over the range from 20–45°C. While *S. thermophilus* and the so-called 'thermophilic' lactobacilli (including *Lactobacillus helveticus* spp). grow well at 37–45°C, the majority of *Lactobacillus* spp. are of the mesophilic type displaying optimal development at 30–37°C. The optimal growth temperatures of the leuconostocs are even lower, ranging from 20–30°C. These three types of microorganisms also differ in shape. Cells of *S. thermophilus* are usually seen as large ovoid cocci, growing in variable chain lengths, depending on the strain and the growth conditions. The leuconostocs appear predominantly as spherical or lenticular cocci in pairs or chains. In contrast, the rod-shaped lactobacilli are more polymorphic, existing in straight, spiral or coccobacillary forms depending on physiological conditions and genetic shape determinants. The size, shape and degree of refringency of individual cells may differ widely among *Lactobacillus* species and even within subspecies (Kandler and Weiss, 1986).

Although the lactic acid bacteria, especially dairy starter strains, are generally recognized as nonpathogenic bacteria, exceptions have been noted in the literature (Friedland, *et al.*, 1990; Gilliland, 1990; Handwerger *et al.*, 1990). Unlike *Streptococcus* spp. (belonging to Lancefield's groups A, B and D), *Lactobacillus* spp. and *Leuconostoc* spp. have rarely been reported to act as opportunistic pathogens (Gilliland, 1990; Swenson, *et al.*, 1990). On the contrary, the health and nutritional benefits of lactic acid bacteria have been extensively examined during the last century (for reviews see Fernandes, *et al.*, 1987; Gilliland, 1990). These benefits include such non-controversial properties as the nutritional value of fermented dairy products for lactose-intolerant individuals and the production of growth-inhibitory metabolites (e.g. H_2O_2, antibiotics and bacteriocins). Together with an intense acidification of the growth medium, these properties lead to a prophylactic role for lactic acid bacteria against some Gram-positive or Gram-negative pathogens and other spoilage organisms. Other traits such as the antitumour activity or the serum cholesterol lowering effect of certain *S. thermophilus* or *Lactobacillus* spp. are less clearly established (Fernandes *et al.*, 1987); Gilliland, 1990; Kaklij *et al.*, 1991; Klebanoff and Coombs, 1991). An interesting property of lactobacilli is that a number of strains of this genus (e.g. *Lactobacillus acidophilus*, *Lactobacillus casei*, *Lactobacillus reuteri*, and *Lactobacillus plantarum* spp.) exhibit the capacity to efficiently colonize external cavities of the body, such as the mouth, the urogenital, and the intestinal tracts, where they appear to play a critical role in maintaining an equilibrated microflora. Strain and tissue-specific adherence to epithelial cells of the corresponding mucosal surfaces has been demonstrated both *in vitro* and *in vivo* (Redondo-Lopez, 1990; Reid *et al.*, 1990). The characteristics associated with potential health benefits to the host do not seem to apply to *Leuconostoc* spp. or *S. thermophilus*. Although the persistence of the two components of yogurt symbiosis, *S. thermophilus* and *Lactobacillus delbrueckii* spp. *bulgaricus* (subsequently referred to as *Lactobacillus bulgaricus*), has been long debated, current thinking maintains that these organisms are only a transiently retained in the gastrointestinal system after ingestion (Gilliland, 1990).

6.3 Biochemical traits

The lactic acid bacteria derive energy primarily through the anaerorobic glycolysis of carbohydrates. These pathways trap energy as ATP by non-oxidative substrate-level phosphorylation. The net energy yield of the homofermentative pathway is 2 molecules of ATP/glucose. Under strictly anaerobic conditions, the net energy yield of the heterofermentative pathway is 1 molecule of ATP/glucose; however, 2 molecules of ATP are generated in an aerobic or microaerophilic environment due to the presence of an alternative electron sink for cofactor regeneration. Some strains have also been shown to derive energy from a pH gradient established by carrier-mediated lactate efflux (Gaetje *et al.*, 1991).

6.3.1 The lac-PTS

Efficient metabolism of lactose is a common biochemical characteristic of strains used in dairy fermentations; *S. thermophilus*, leuconostocs and lactobacilli typically ferment lactose. Like *Lactococcus lactis* (McKay, 1982) and *Staphylococcus aureus* (Hengstenberg *et al.*, 1968; Morse *et al.*, 1968), *Lactobacillus casei* (Chassy and Alpert, 1989) transports and metabolizes lactose by a unique pathway that appears to be limited to certain (sub)species of streptococci, enterococci, lactococci, lactobacilli, and staphylococci. The lac-PTS operon encodes proteins that mediate the transport of lactose into the cell with concomitant phosphorylation (lactose:phosphoenolpyruvate-dependent phosphotransferase system; lac-PTS) and catalyse the hydrolysis of lactose 6-phosphate to glucose and galactose 6-phosphate. The lac-PTS is composed of EnzymeII[lac] (LacE), an integral 55 kDa membrane protein responsible for the vectorial phosphorylation and translocation of lactose into the cell as lactose 6-phosphate (Alpert and Chassy, 1990) and FactorIII[lac] (LacF), a 39 kDa trimeric peripheral membrane protein that phosphorylates EnzymeII[lac] (Alpert and Chassy, 1988). The cytoplasmic lactose 6-phosphate-hydrolyzing enzyme, β-D-phosphogalactoside galactohydrolase (P-β-gal), is designated LacG (Porter and Chassy, 1988). The genes encoding each of these proteins in staphylococci (Breidt and Stewart, 1986; Breidt and Stewart, 1987), streptococci (de Vos *et al.*, 1990; de Vos *et al.*, 1989), and lactobacilli have been isolated by molecular cloning and sequenced (Alpert and Chassy, 1988; Alpert and Chassy, 1990;

Porter and Chassy, 1988). The deduced sequences of the LacG, LacF, and LacE proteins *Lactobacillus casei* are more similar to those found in *Lactococcus lactis* and *S. aureus* than would be expected for such phylogenetically diverse organisms. It also appears that, based on protein sequence homology, LacG is encoded by a member of the same gene family as *bgl*B, that encodes the β-D-phosphoglucoside glucohydrolase of *E. coli* (de Vos *et al.*, 1989; Porter and Chassy, 1988). Based on a several limited regions of significant similarity, *Lac*E has been postulated to have an ancestral relationship with the cellobiose permease of *E. coli* (Reizer *et al.*, 1990). It is interesting to note that the lac-PTS gene order, *lac*EGF, is not the same as that found in lactococci and staphylococci (Alpert and Chassy, 1990; Breidt and Stewart, 1987; de Vos *et al.*, 1990).

The metabolism of galactose 6-phosphate requires the additional gene-products of the tagatose 6-phosphate pathway gene cluster (*lac*ABCD) encoding galactose 6-phosphate isomerase, tagatose 6-phosphate kinase, and tagatose 1,6-diphosphate aldolase. The genes of the tagatose-phosphate pathway of *Lactobacillus casei* have not been characterized to date.

6.3.2 The lac *operon*

Strains of *S. thermophilus* and the majority of *Lactobacillus* spp. contain a pathway for lactose metabolism that is functionally analogous to that encoded by the classical *E. coli* lac operon genes, *lac*Y and *lac*Z. Unlike LacZ of *E. coli* that is composed of a tetramer of four identical 116 kDa subunits, the β-galactosidase of *Lactobacillus bulgaricus* is composed of two identical subunits of 114 kDa (Schmidt *et al.*, 1989). The β-galactosidase of *S. thermophilus* most probably forms a tetramer with an apparent mass of 530 kDa (Chang and Mahoney, 1989). The monomeric molecular weight deduced from sequence data (117 kDa) agrees well with published biochemical data (Schroeder *et al.*, 1991). The β-galactosidase and lactose permease-encoding genes of *S. thermophilus* (Poolman *et al.*, 1989; Schroeder *et al.*, 1991) and *Lactobacillus bulgaricus* (Leong-Morgenthaler *et al.*, 1991; Schmidt *et al.*, 1989) have been isolated by molecular cloning in *E. coli* and the DNA sequence determined. The deduced LacZ-like gene products of *S. thermophilus* and *Lactobacillus bulgaricus*, share 48% sequence similarity. The 30–60% similarities found in comparison with the LacZ proteins of other microbes provide evidence of a distant relationship with the β-galactosidases of *Clostridium acetobutylicum* (Hancock *et al.*, 1991), *E. coli lac*Z (Kalnins *et al.*, 1983) and *ebg*A (Hall *et al.*, 1989), and *Klebsiella pneumoniae* (Buvinger and Riley, 1985) but not with a β-galactosidase found in *B. stearothermophilus* (*bga*B, Hirata *et al.*, 1986). Quite surprisingly for an organism used in milk fermentations, the *lac*Z gene of some strains of *Lactobacillus bulgaricus* has been observed to undergo spontaneous deletions that give rise to a Lac⁻ phenotype (Mollet and Delley, 1990). An unusual 72 bp direct duplication event has been observed to reverse the loss of β-galactosidase activity caused by a 30 bp deletion in *lac*Z (Mollet and Delley, 1991). The factors underlying this instability are as yet unclear.

*Lac*S, the gene encoding the lactose permease of *S. thermophilus* that is capable of encoding a 69.5 kDa protein, is found in an ORF upstream of the gene encoding β-galactosidase (Poolman *et al.*, 1989). The permease catalyses both lactose exchange and uptake reactions in intact cells and membrane vesicles. The translated protein sequence bears little resemblance to that of the *E. coli lac* permease, LacY; however, the gene encodes an N-terminal domain that appears distantly related to the melibiose carrier of *E. coli*, and the C-terminal domain has similarity to the EnzymeIII^{glc} of *E. coli* (Poolman *et al.*, 1989). It has been suggested that the N-terminal region functions as the transmembrane portion of the permease that facilitates lactose transport, and the C-terminal, FactorIII-like domain is a regulatory region. A gene analogous to *lac*S has been identified in *Lactobacillus bulgaricus* (Leong-Morgenthaler *et al.*, 1991). The lactose permeases of *Lactobacillus bulgaricus* and *S. thermophilus* exhibit about 80% DNA sequence conservation and 60% conservation in the deduced amino acid sequences (Leong-Morgenthaler *et al.*, 1991). For the *lac*S and *lac*Z genes of both organisms, the stop codon of *lac*S and the start codon of *lac*Z are separated by 3 bp. Northern blot analyses of *lac* mRNA levels have shown that *lac*S and *lac*Z constitute a single operon (Mercenier, 1990; Poolman *et al.*, 1990) and that *lac* gene expression is subject to glucose metabolite repression in *S. thermophilus* (Poolman *et al.*, 1990). It is interesting to note that the regulation of the *Lactobacillus bulgaricus lac* operon differs from that of *E. coli* and *S. thermophilus* since high levels of expression are seen during growth in media containing glucose (Schmidt *et al.*, 1989). During growth on lactose-containing media, many strains of *S. thermophilus* do not metabolize galactose but instead expel the galactose portion of lactose into the medium. Recently, it was demonstrated that *S. thermophilus* has a lactose-galactose antiporter activity that allows galactose efflux to drive lactose uptake (Hutkins and Ponne, 1991). Thus, it appears that these Gram-positive *lac* operons are only partly related to the one found in *E. coli*. In contrast to the *lac* operon of *E. coli*, the genes encoding two enzymes, mutarotase and UDP-glucose 4-epimerase, of the Leloir pathway for the metabolism of galactose 1-phosphate are located immediately before (5') the genes for *lac*S and *lac*Z in *S. thermophilus* (Poolman *et al.*, 1990).

Two unique *Lactobacillus casei* strains possess β-galactosidase activity in addition to the lac-PTS (Jimeno *et al.*, 1984). The plasmid-encoded β-galactosidase gene from the 28 kbp plasmid, pLZ15, resident in strain ATCC393 has been cloned into *E. coli* and sequenced (Chassy *et. al.* unpublished results). A plasmid-encoded (pNZ63) β-galactosidase gene has also been isolated from a strain of *Leuconostoc lactis*, and the nucleotide sequence determined (David *et al.*, 1992). Somewhat surprisingly, the β-galactosidase-encoding genes isolated from *Leuconostoc lactis* and *Lactobacillus casei* are nearly identical, displaying greater than 98% sequence identity (David *et al.*, 1992). It was also observed that the coding sequences each contained two genes which encode overlapping out of frame ORFs, *lac*L and *lac*M, whose expression appeared to be translationally coupled. Analysis of the *lac*L-derived polypeptide shows 30–50% similarity to the N-terminal two-thirds of *Lac*Z gene products isolated from Gram-positive and Gram-negative bacteria. *Lac*M encoded sequences correspond to the C-terminal regions of *lac*Z proteins. The two genes, *lac*L and *lac*M, are assumed to have arisen from a deletion event in a parental *lac*Z-like gene. In the process, about 8% of the sequence of *lac*Z was deleted. The region it encoded is apparently unnecessary for enzymatic activity. Active β-galactosidase isolated from *Lactobacillus casei* has a molecular weight around 220 000. The enzyme, an $\alpha_2\beta_2$ type, is comprised of two identical 38 kDa and two identical 70 kDa subunits (Chassy, unpublished observations). The plasmids pNZ63 and pLZ215 also encode non-contiguous *lac* permeases that remain uncharacterized.

6.3.3 Xylose metabolism

After isomerization and phosphorylation to yield xylulose-5-phosphate, a key intermediate compound in degradation of all pentoses and pentitols, D-xylose is catabolized in lactobacilli via the pentose-phosphate pathway, (Kandler, 1983). The *xyl* genes of *Lactobacillus pentosus* have recently been isolated by molecular cloning and the nucleotide sequences determined (Lokman *et al.*, 1991). The genes are organized in a regulon comprised of five or more contiguous genes arranged in three transcriptional units (Lokman *et al*, 1991; Lokman *et. al.*, manuscript in preparation). All of the *xyl* genes are transcribed in the same direction. The order of the genes in the regulon is *xyl*P/Q, *xyl*R, and *xyl*A/B. *Xyl*P/Q form an operon, *xyl*R is transcribed as a single gene, and *xyl*A/B constitute a second operon. Genes present downstream (3') of *xyl*B are not involved in xylose catabolism, but presumably function in ribose catabolism (Leer, unpublished results). *Xyl*P, probably encodes xylose permease; whereas, *xyl*R, *xyl*A and *xyl*B code for the xylose repressor, xylose isomerase and xylulose kinase, respectively. The function of *xyl*Q is, at present, unknown.

6.3.4 Protein metabolism

Although the specific amino acids required may be species and strain dependent, all lactic acid bacteria have an absolute growth requirement for amino acids. Milk contains a high concentration of amino acids bound in the form of protein; however, milk is not a rich source of free amino acids. The free amino acids present in milk are sufficient to support approximately 15–25% of the growth observed for *Lactococcus lactis* in milk (Mills and Thomas, 1981). The remaining required amino acids must be obtained through proteolysis of milk proteins. Strains of lactic acid bacteria that develop well in milk are usually proteolytic (for recent reviews see Kok, 1990; Kok and Venema, 1988; Thomas and Pritchard, 1987). However, *S. thermophilus* spp. are often found to be poorly proteolytic (Shankar and Davies, 1977). For this reason, they are often used in conjunction with a more proteolytic organism (e.g. *Lactobacillus bulgaricus* in yogurt fermentations). In addition to one or more protease activities, strains must possess a variety of peptidases and peptide transport proteins to obtain individual amino acids (Kok, 1990; Kok and Venema, 1988).

The extracellular proteases of the lactococci have been described at the molecular and sequence level (Kok, 1990; Kok and Venema, 1988). Although none has been studied in great detail, extracellular proteases have been reported in lactobacilli and *S. thermophilus*. In a recent survey of *S. thermophilus* spp., 3 of 19 strains studied were found to be as proteolytic as Prt$^+$ strains of *Lactococcus lactis* (Shahbal *et al.*, 1991). The cell envelope-associated proteases of these strains exhibited biochemical properties similar to those of the lactococcal proteases, but appeared genetically distinct (Shahbal *et al.*, 1993). A strain of *Lactobacillus casei* was found to possess a cell-bound proteinase activity with properties similar to the lactococcal serine proteinases; chromosomal DNA from this strain hybridized to fragments of the *prt*P and *prt*M genes of *Lactococcus lactis* (Kojic *et al.*, 1991). Preliminary mapping analysis points to some conservation of restriction endonuclease cutting sites (Kojic *et al.*, 1991). Recently, the genes encoding the cell-envelope-associated proteinase (*prt*P) and maturation protein (*prt*M) of *Lactobacillus paracasei* spp. *paracasei* were isolated, and the nucleotide sequence was determined (Holck and Naes, 1992; Naes and Nissen-Meyer, 1992). The polypeptides potentially encoded by the two genes showed extensive similarity to PrtP and PrtM of *Lactococcus lactis* spp. *cremoris* Wg2. The proteinase gene was expressed from its own regulatory and secretory signals in *Lactobacillus plantarum* but not in *Lactococcus lactis*.

Broad specificity aminopeptidases capable of hydrolysing di-, tri- and tetrapeptides have been reported to occur *Lactococcus lactis* spp. *cremoris* (Kaminogawa *et al.*, 1984), *Lactobacillus acidophilus* (Machuga and Ives, 1984) and *Lactobacillus delbrueckii* spp. *lactis* (Eggimann and Bachmann, 1980). These enzymes preferentially hydrolyse peptides with more than two amino acids. All of the enzymes characterized to date are metalloenzymes that lack carboxypeptidase or endopeptidase activity.

Proteolysis of casein produces proline-rich polypeptides that are not substrates for the aminopeptidases or the di- and tri-peptidases. Aminopeptidase P, proline iminopeptidase, prolinase, prolidase, and X-prolyl-dipeptidyl aminopeptidases serve the lactic acid bacteria for the degradation of proline-containing peptides. X-prolyl-dipeptidyl-aminopeptidase activity has been reported in *Lactobacillus delbrueckii* spp. *bulgaricus* (Atlan *et al.*, 1990) and spp. *lactis* (Meyer and Jordi, 1987), and *S. thermophilus* (Meyer and Jordi, 1987), and prolyl-dipeptidyl aminopeptidase has been observed in *Lactobacillus helveticus* (Khalid and Marth, 1990). The distinct physical and biochemical identity of each of these activities underscores complexity of proline-peptide metabolism in the lactic acid bacteria. This complexity is further demonstrated by the observation that mutants of *Lactobacillus bulgaricus* that lack X-prolyl-dipeptidyl-aminopeptidase grow at an only slightly reduced growth rate and produce a lower biomass (Atlan *et al.*, 1990).

6.4 Plasmid biology and cloning vectors

Many lactic acid bacteria have been reported to harbor endogenous plasmids varying in size from 2 to 100 kilobases; the presence of multiple plasmids is not uncommon. In the case of lactococci, plasmids have been associated with the important traits of proteinase production, lactose metabolism, citrate metabolism, bacteriocin production/resistance, and phage resistance (McKay, 1983; McKay and Baldwin, 1990).

6.4.1 Naturally occurring plasmids

Studies have described the analysis of the plasmid content of *Lactobacillus* spp. isolated from dairy products, meat, sour doughs, silage and gastrointestinal tracts (Table 6.1). Most of the plasmids reported have remained cryptic, despite much work in recent years to link these plasmids to phenotypic properties. A few endogenous plasmids have been linked to physiological traits such as carbohydrate metabolism (see Figure 6.1), proteolytic capacity, antibiotic resistance(s), phage defense mechanism(s), and bacteriocin production/ immunity. Slime production and adherence to mucosal surfaces are believed to be plasmid-encoded in some instances (see Table 6.1). Large linear plasmids

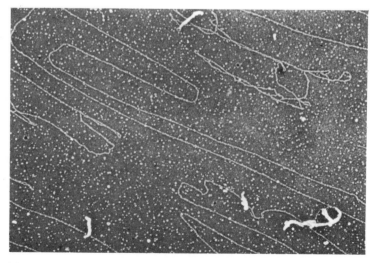

Figure 6.1   Electron micrograph of open circular form of the 35 kb lactose plasmid, pLZ64, isolated from *Lactobacillus casei* 64H.

Table 6.1 Plasmid-encoded traits in *Lactobacillus* spp.

Genus species	Metabolic function	Plasmid		Reference
		Size (kb)	Name	
Lb.casei	Lactose utilization	35	pLZ64	Chassy *et al.*, 1978
Lb. casei	Lactose	68.2	pLY101	Shimizu-Kadota, 1987
Lb. casei	Lactose/conjugal	52.5	pLZ18	Chassy and Rokaw, 1981
Lb. casei	β-galactosidase	28	pLZ15	Flickinger *et al.*, 1986
Lb. spp.	Maltose utilization	73.5	pLM291	Liu *et al.*, 1988
Lb. jugurti	N-acetyl-glucosamine utilization	13.2	–	Smiley and Fryder, 1978
Lb. helveticus	Proteolytic activity		pLHJ1	De Rossi *et al.*, 1989
Lb. sake	Cysteine assimilation	8.3,2.7	–	Shay *et al.*, 1988
Lb.ferementum	Em^R, Tc^R	15, 53	–	Ishiwa and Iwata, 1980
Lb. plantarum	Tc^R	8.5	pCAT	Ahn *et al.*, 1992
Lb. acidophilus	Antibiotic resistance	–	–	Vescovo *et al.*, 1982
Lb. reuteri	Antibiotic resistance	–	–	Vescovo *et al.*, 1982
Lb. reuteri	Em^R	10.5	pLUL631	Axelsson *et al.*, 1988
Lb. acidophilus	Em^R	–	–	Morelli *et al.*, 1983
Lb. acidophilus	Bacteriocin production/immunity	–	–	Muriana and Klaenhammer, 1987
Lb. sake	Antibacterial activity	–	–	Schillinger and Lücke, 1989
Lb. sake	Bacteriocin production	50	pCIMs	Mortvedt and Nes, 1990
Lb. spp.	Slime production	–	–	Ahm *et al.*, 1989
Lb. reuteri	Cell adhesion	–	–	Sarra *et al.*, 1989
Lb. helveticus	Restriction/modification	34	pLAR33	De Los Reyes-Gavilan *et al.*, 1990
Lb. spp. 100–33	MLS resistance	18	–	Rinckel and Savage, 1990.

(150 kb) have been detected by PFGE analysis in two strains of *Lactobacillus gasseri* (Roussel *et al.*, 1993).

The complete nucleotide sequences of eight small cryptic plasmids of *Lactobacillus* spp. have been determined: pLP-1 (Bouia *et al.*, 1989), pLB4 (Bates and Gilbert, 1989), p8014-2 (Leer *et al.*, 1992), pA1 (Vujcic and Topisirovic, 1993) and pC30i1 (Skaugen, 1989) of *Lactobacillus plantarum* strains; p353-2 of *Lactobacillus pentosus* (Leer *et al.*, 1992); pLAB1000 of *Lactobacillus hilgardii* (Josson *et al.*, 1990); and pLJ1 of *Lactobacillus helveticus* subsp. *jugurti* (Takiguchi *et al.*, 1989). The *Lactobacillus pentosus* plasmid p353-2 showed significant similarity to the *Lactobacillus plantarum* plasmids p8014-2, pC30i1, and pLP1, but little if any similarity to the other plasmids derived from lactobacilli. Most of the plasmids analysed contained elements (i.e. replication protein gene, plus and minus origins of replication) that are typical of plasmids that replicate through a single-stranded DNA rolling circle mechanism (Gruss and Ehrlich, 1989). Plasmids pLB4 and pLAB1000 also encode a site-specific recombinase and a potential recombinase target site (RS$_A$). Based on sequence comparisons, it appears that pLP1 and pAB1000 are related to the pC194 family of plasmids; while pLB4 and pA1 are related to members of the pE194 group. It has been demonstrated that the copy number of p353-2 is controlled by an ~70 bp RNA transcribed from the DNA strand opposite to that encodes RepA (van Luijk *et al.*, manuscript in preparation). A deletion that removed the promoter associated with the transcription of the 70 bp RNA increased plasmid copy number 5–10-fold.

Plasmids are apparently not common in the yogurt starter, *S. thermophilus*. About a fifth of the *S. thermophilus* strains examined thus far have been shown to contain a single small plasmid (2–25.5 kb). In a limited number of isolates, two or three replicons were found to coexist in the same cell. The function of the endogenous cryptic plasmids found in *S. thermophilus* remains uncertain (for a review, see Mercenier, 1990). DNA homology studies performed by cross-hybridization detected homology between certain single plasmids of individual strains. One plasmid (pER8; 2.2 kb) was also used to search for the presence of homologous DNA sequences in 12 *Lactococcus* and 4 *Lactobacillus* strains; no homologous sequences were detected in these Southern-blot hybridization experiments (Somkuti and Steinberg, 1991). A cryptic plasmid of *S. thermophilus* (pA33; 6.9 kb) has been sequenced (Mercenier, 1990). Analysis of its nucleotide sequence allowed the identification of 5 open reading frames and 22 bp directly repeated sequences. However, no other elements (i.e. *rep*A, plus and minus origins) often associated with small Gram-positive plasmids could be found. It is interesting to note that the presence of pA33 appears to affect a number of physiological traits of the host cell, such as chain length, antibiotic and phage resistance/sensitivity profile and milk clotting time (O'Regan *et al.*, manuscript in preparation). The nucleotide sequence of a second plasmid isolated from *S. thermophilus*, pST1, has recently been reported (Janzen *et al.*, 1992).

The absence of endogenous plasmids in the yogurt starter *Lactobacillus bulgaricus* spp. seems to be an almost general rule. This is in marked contrast to the other *Lactobacillus* spp., most of which contain plasmids. In this regard, it may be significant that *Lactobacillus bulgaricus* is one of a few (sub) species of lactobacilli that remain relatively refractory to transformation by autonomously replicating extrachromosomal elements (see below).

The plasmid content of leuconostocs has also been examined. Extrachromosomal elements have been observed in non-acidophilic *Leuconostoc* spp. (especially in dairy strains) and in a number of *Leuconostoc oenos* strains (O'Sullivan and Daly, 1982; Orberg and Sandine, 1984) even though plasmid-free isolates of this latter (sub)species are not uncommon. Although certain characteristics such as lactose and citrate fermentation are unstable in some *Leuconostoc* strains, only two plasmids have been linked to specific metabolic traits. The initial observation of a correlation between the presence of specific plasmids and the ability to use lactose and citrate (O'Sullivan and Daly, 1982) has been substantiated recently. The existence of a lactose plasmid, pNZ63 (23 kb) was demonstrated in *Leuconostoc lactis* NZ6009 (David, 1992; David *et al.*, 1989). This replicon carries at least two genes encoding components of the lactose system, the β-galactosidase (David *et al*, 1992) and the lactose permease (David, 1992). Interestingly, the two *lac* genes are not adjacent on pNZ63. In another strain, *Leuconostoc lactis* NZ6070, the citrate permease gene was shown to reside on a 23 kb plasmid, pCT71 (David and de Vos, personal communication). A 22 kb plasmid of *Leuconostoc mesenteroides* subsp. *mesenteroides* is implicated in citrate metabolism (Lin *et al.*, 1991). The plasmid linkages of both lactose and citrate fermentation were recently established in *Leuconostoc cremoris* and *Leuconostoc dextranicus* as well (Fantuzzi, 1991). The DNA sequence of a cryptic plasmid of *Leuconostoc oenos*, pL013 (3.9 kb), has been reported (Fremaux, 1990; see also GenBank: LEUMOBREP, Accession: M95954). Analysis of the nucleotide sequence revealed 4 open reading frames (ORFs). A putative protein encoded by one of the ORFs shows 20% homology with the Pre protein of pT181 of *S. aureus*, but no classical RS_A site or minus origin of replication could be identified (Fremaux, 1990).

6.4.2 *Insertion sequences*

The lactose plasmid pLZ15 of *Lactobacillus casei* carries multiple copies of a DNA sequence that shows homology (cross-hybridization reaction in Southern-blot experiments) to the *Lactobacillus casei* insertion element ISL*1* (Shimizu-Kadota *et al.*, 1988; Shimizu-Kadota *et al.*, 1985). pLZ15 also encodes a *lac*Z gene that is nearly identical in nucleotide sequence to the *lac*Z gene found on a plasmid in *Leuconostoc lactis*, pNZ63 (see section 6.3.2). DNA sequence homology between these two plasmids apparently exists only in the *lac*-coding regions, a finding that is suggestive of a recent transpositional event. However, it is not yet known if the *lac* region in either plasmid is bounded by IS-like elements such as ISL*1*. ISL*1* was the first transposable element to be discovered in lactic acid bacteria (Shimizu-Kadota *et al*., 1985). It was identified as being responsible for the conversion of a temperate phage of *Lactobacillus casei* (φFSW) to a lytic state (see section 3.9). Apparently, ISL*1* exhibits a relatively narrow host-range, having been found only in 3 out of 19 *Lactobacillus casei* strains studied (Shimizu-Kadota *et al*., 1988). The element contains no nucleotide sequences or gene products (based on translated ORFs) in common with known Gram-positive or Gram-negative IS sequences. Among the IS elements subsequently identified in lactococci, only one, IS*981*, shares homology with ISL*1* (for a review, see Gasson, 1990). In lactococci, such IS sequences may be found in the vicinity of genes coding for unstable or conjugally transferable phenotypic traits (lactose utilization, proteolytic capacity, nisin production/sucrose utilization) of plasmid or chromosomal origin. In several instances, it was elegantly demonstrated that they participate to DNA rearrangements involving plasmid–plasmid or plasmid–chromosome interactions (see Gasson, 1990).

An insertion sequence from *Leuconostoc mesenteroides* spp. *cremoris* strain DB*1165* has been characterized (Johansen and Kibenich, 1992). The 1553 bp element, designated IS*1165*, occurs with a copy number of 4 to 13 in *Leuconostoc mesenteroides* spp. *cremoris*. IS*1165*, contains an open reading frame of 1236 bp and is not related to any previously described insertion sequence. Sequences hybridizing to IS*1165* have been detected by hybridization in *Leuconostoc lactis*, *Leuconostoc oenos*, *Pediococcus* spp., *Lactobacillus helveticus*, and *Lactobacillus casei* but not in *Lactococcus* spp. As far as we are aware, insertion elements have not been described in *S. thermophilus*.

6.4.3 Cloning vectors

A number of potential cloning vectors have been constructed for use with lactic acid bacteria. Replicons functioning in *S. thermophilus* and lactobacilli have been reviewed (Chassy, 1987; Mercenier and Lemoine, 1989; Pouwels, *et al*., 1992); however, the list of useful vectors for these species and for leuconostocs (David *et al*., 1989) increases regularly. Many of the vectors are based upon replication functions and origins encoded on endogenous cryptic plasmids of lactococci, lactobacilli or streptococci; principally on those that have been sequenced (for a recent review see Kok, 1991 and also Bringel *et al*., 1989; Cocconcelli *et al*., 1991; De Rossi *et al*., 1991; de Vos; *et al*., 1987; Kok *et al*., 1984; Posno *et al*., 1991b; Pouwels *et al*., 1992). Other vectors are based on the origin of replication of broad host-range plasmids of Gram-positive bacteria, for example, pIL251 (Simon and Chopin, 1988) was derived from pAMβ1. Shuttle vectors able to replicate in both *E. coli* and lactic acid bacteria have also been constructed. Replication in *E. coli* is based either on a unique origin of replication originating from lactococcal cryptic plasmids and recognized by a large number of bacterial species; or on origins containing a DNA fragment that encodes an origin of replication derived from *E. coli* (i.e. pBR322 or pUC18). Typical examples of the broad host-range type are the 'prototype' vectors pGK12 (Kok *et al.*, 1984; Leenhouts *et al.*, 1991) and pNZ12 (de Vos, 1987), and the Gram-positive: *E. coli* bifunctional shuttles pLP825 (Posno *et al.*, 1991b) pSA3 (Dao and Ferretti, 1985) and pVA838 (Macrina *et al.*, 1982). The plasmids pNZ12 and pGK12 have been transformed successfully into lactococci (de Vos, 1987; Kok *et al.*, 1984), *S. thermophilus* (Mercenier and Lemoine, 1989), lactobacilli (Chassy and Flickinger, 1987; Luchansky *et al.*, 1988) and leuconostocs (David *et al.*, 1989). In addition, some of the vectors developed for *Bacillus* spp. have proven effective for lactobacilli (Chassy, 1987). Enhanced stability of such vectors was recently demonstrated in *Lactobacillus casei* when they contained a functionally active minus origin of replication like those found in plasmids that replicate through a single-stranded intermediate (Shimizu-Kadota *et al*, 1991). The presence of a functional minus origin in p353-2 has been shown to greatly enhance segregation stability in *Lactobacillus casei* and *Lactobacillus pentosus* (Leer *et al.*, 1992).

Most of the vectors currently used with lactic acid bacteria replicate via a single-stranded DNA intermediate rolling-circle mode, a feature that has been reported to be a potential source of segregational instability (Gruss and Erlich, 1989; Posno *et al.*, 1991b). Indeed, cloning of large DNA fragments in such replicons has been found often difficult, although it has been reported (Leer *et al.*, 1992; Posno *et al.*, 1991b). It has also been observed that spontaneous deletions can occur in recombinant plasmids introduced into *S. thermophilus* (Solaiman and Somkuti, 1992). From an analysis of deletion-prone regions, it was concluded that direct- and inverted-repeats contribute to the observed structural instability. Cloning vectors that do not replicate via a single-stranded rolling circle mechanism such as pLP825 (Posno *et al.*, 1991b) and pIL251 (Simon and Chopin, 1988) have been developed.

6.5 Genetic transfer methods

Bacterial genetics has traditionally relied on a variety of techniques for the introduction of exogeneous DNA into a target bacterium. Most of these methods are in the initial stages of development of their applicability to *S. thermophilus*, the lactobacilli and leuconostocs. Transformation by natural competence, protoplast formation and regeneration, and electroporation are the most common methods. Of these, electroporation has proven useful for many strains of lactic acid bacteria (Chassy *et al.*, 1988). Conjugal exchange of DNA has been reported; however, the method has not been used to practical advantage. Transduction has been demonstrated, but remains largely unexploited. This section describes transformation in detail and briefly points the reader to key references pertaining to alternative methods of genetic exchange (for reviews see Fitzgerald and Gasson, 1988; Gasson, 1990; Mercenier and Chassy, 1988).

6.5.1 Transformation by electroporation

The introduction of commercial high voltage pulse generators in the mid-1980s made possible the electrotransformation of the Gram-positive bacteria *Lactococcus lactis* (Harlander, 1987), *Lactobacillus casei* (Chassy and Flickinger, 1987), *Leuconostoc paramesenteroides* (David *et al.*, 1989) and *S. thermophilus* (Mercenier, 1990; Somkuti and Steinberg, 1988) by plasmid DNA (for reviews see Chassy *et al.*, 1988; Trevors *et al.*, 1992). The majority of species of Gram-negative bacteria as well as many strains of Gram-positive bacteria have been successfully transformed by electroporation. Electroporation replaces tedious, time-consuming and frequently unreliable protoplast transformation techniques that had taken years to develop for the lactic acid bacteria (Kondo and McKay, 1982; Mercenier and Chassy, 1988). The frequency of transformation observed with Gram-positive bacteria is generally lower than is observed with *E. coli*, but yields of 10^5 to 10^7 transformants/µg of DNA are achieved with some strains. Large differences in transformation frequency are often observed between two closely related strains. Other strains remain refractory to currently used techniques of electrotransformation.

Several factors may influence the outcome of electroporation attempts. The dense Gram-positive cell wall may present a barrier to entry of DNA in some stains. This possibility is suggested by the observation that the efficiency of transformation of *Listeria monocytogenes* can be raised from 300 to 8×10^5 transformants/µg DNA by incorporation of low concentrations of the cell wall synthesis inhibitor, penicillin G, in the growth medium (Park and Stewart, 1990). The incorporation of glycine in the growth medium, an agent which affects cell wall biosynthesis, can also lead to increases in electroporation frequency (Dunny *et al.*, 1991). In addition, it has been observed that treatment of cells with muralytic enzymes prior to the electrotransformation of lactococci can be used to raise the transformation efficiency (Powell *et al.*, 1988). It is also possible that incompatibility between transforming vectors and resident plasmids may lower frequencies of transformation. Incompatibility has been demonstrated in the attempted introduction of DNA by electroporation into *Lactococcus lactis* (Van der Lelie *et al.*, 1988) and lactobacilli (Posno *et al.*, 1991b). Other possible barriers include restriction systems, non-specific nucleases, non-expressed marker genes, non-expressed replication-essential functions, and interference with host-essential functions.

6.5.2 Liposome-mediated protoplast transfection, fusion and transformation

The value of liposome-mediated transfection of protoplasts with phage DNA is not at first apparent, since useful recombinants do not result. However, each plaque observed after transfection represents a single transfected protoplast. Transfection, therefore, is a rapid assay that measures the frequency of DNA entry into protoplasts. Transfection is not complicated by the frequently inefficient and slow regeneration of protoplasts. Protoplast transfection can be used to arrive at the optimal conditions for DNA uptake as an adjunct to development of a protopolast transformation system. The direct utility of phage transfection may increase when more is learned about the phages of these organisms. Transfection was first reported for *Lactobacillus casei* (Shimizu-Kadota and Kudo, 1984). The report also determined the need for liposomes, as no transfectants were observed when naked phage DNA was substituted for liposome-encapsulated DNA. Frequencies of 10^6–10^7 transfectants/µg phage DNA have been observed with *Lactobacillus casei* and *Lactobacillus bulgaricus* (Boizet *et al*, 1988).

Protoplasts of *Lactobacillus reuteri* (Morelli *et al.*, 1987) and *Lactobacillus acidophilus* (Lin and Savage, 1986) can be transformed without the use of liposomes. However, the transformation of *Lactobacillus casei* and *Lactobacillus plantarum* protoplasts required the incoming DNA to be trapped in liposomes (Chassy, 1987). Very fastidious strains of *Lactobacillus bulgaricus* can also be transformed by this method (Boizet and Chassy, unpublished obserations). Transformation of protoplasts of *S. thermophilus* has been reported (Mercenier, 1990; Mercenier *et al*, 1989). A critical prerequisite of this technique is the formulation of a suitable medium for the regneration of protoplasts. The protoplast transfection technique has also been applied to the development of regeneration media (Mercenier *et al*, 1989).

Protoplast fusion was employed as a gene transfer system for lactic streptococci prior to the introduction of a protoplast transformation system (Gasson, 1980). Intrageneric and intergeneric fusion of lactobacilli containing plasmids with selectable antibiotic resistance markers (Cocconcelli *et al.*, 1986; Iwata *et al.*, 1986) has been reported. Transfer of a chromosomal marker, trehalose fermentation, has been observed to occur during fusion of *Lactococcus lactis* and *Lactobacillus reuteri* (Cocconcelli *et al.*, 1986).

6.5.3 *Conjugation*

Conjugation of bacteria presents a natural means for gene exchange. A number of reports have demonstrated that broad host-range Gram-positive R-factors such as pAMβ1, can be transferred by conjugation into lactobacilli (Gibson *et al.*, 1979; Shrago *et al.*, 1986; Tannock, 1987; Vescovo *et al.*, 1983), leuconostocs (Pucci *et al.*, 1987) and *S. thermophilus* (Romero *et al.*, 1987). *Lactobacillus casei* ATCC4646 harbours a self-transmissible plasmid that confers on lactose-negative recipients the ability to metabolize lactose via the lac-PTS (Chassy and Rokaw, 1981). Bacteriocin production and resistance are transmissible plasmid-linked traits in certain strains of *Lactobacillus acidophilus* (Klaenhammer, 1988). Tn919, a conjugal transposon that determines tetracycline resistance, was shown to transfer from *S. sanguis* to *Lactobacillus plantarum* at very low frequency during matings on solid surfaces (for a review see Fitzgerald and Gasson, 1988). The *lac* genes and the nisin production/immunity determinants have been transferred from *Lactococcus lactis* to *Leuconostoc* spp. by conjugation (Tsai and Sandine, 1987; 1987b).

To date, conjugal plasmids have not been used directly as vectors for passenger DNA. However, during the conjugal transfer of plasmid pVA797 from *S. sanguis* to *Enteroccocus faecalis*, it was observed that a cointegrate formed with a second compatible resident plasmid that carried an insert of foreign DNA, pVA838 (Smith and Clewell, 1984). Subsequent to transfer, the cointegrate resolved precisely. After curing and segregation it was possible to select recombinant pVA838 containing cells that lacked pVA797. The technique was successfully used to introduce DNA into then untransformable strains of *S. thermophilus* (Romero *et al.*, 1987). Transfer of pIP501 from *S. sanguis* to *Lactobacillus plantarum* has been demonstrated to occur with mobilization of pSA3 or pVA838 (Shrago and Dobrogosz, 1988). Interestingly, the pVA797::pVA838 cointegrate resolved, but the pVA797::pSA3 cointegrate failed to resolve. The *lac*G gene of *Lactococcus lactis* has been transferred into *S. thermophilus* by the conjugal co-mobilization technique (Romero *et al.*, 1987), and the glucanase gene of *B. amyloliquefaciens* has been transferred into *Lactobacillus helveticus* by this technique (Thompson and Collins, 1989). The potential utility of conjugation warrants further research on the mechanism of conjugation.

6.5.4 Transduction

DNA transfer through transduction can be mediated by both temperate and lytic bacteriophage. Although transduction has proven to be a valuable genetic tool for some bacterial genera (e.g. *E. coli*), there are few examples of its application to the lactic acid bacteria (Gasson, 1990). Transduction of lactose and proteinase genes into *Lactococcus lactis* C2 resulted in a strain in which these plasmid-encoded traits were stabilized in the chromosome (McKay and Baldwin, 1978). This stabilized strain, produced by a transductional event, represents the first starter culture produced by genetic manipulation of lactococci. Transduction of plasmid and chromosomal genes into *S. thermophilus* mediated by lytic phages has been reported (Mercenier, 1990; Mercenier *et al.*, 1988). The lytic phage φadh found in *Lactobacillus acidophilus* ADH (now renamed *Lactobacillus gasseri* ADH; Raya and Klaenhammer, 1992) has been used to mediate transfer of plasmids (Luchansky *et al.*, 1989; Raya and Klaenhammer, 1992; Raya *et al.*, 1989). Interestingly, insertion of DNA fragments having homology to the host chromosome into plasmids used to transduce *Lactobacillus gasseri* ADH raised the transduction frequency 10^4–10^5-fold (Raya and Klaenhammer, 1992). A temperate phage of *Lactobacillus salivarius* was reported to be capable of generalized transduction of auxotrophic markers (Tohyama *et al.*, 1971). Transduction of lactobacilli by temperate phages usually occurs at very low frequencies (10^{-8}–10^{-10} transductants/PFU); however, transduction of plasmids effected with virulent phages of *S. thermophilus* has been reported to attain frequencies as high as 10^{-1}–10^{-2} transductants/PFU (Mercenier, 1990; Mercenier *et al.*, 1988). The recent characterization of the *att* loci of ϕadh may lead to intensified research on the application of transduction to the genetic manipulation of these organisms (Raya *et al.*, 1992, see also Ssection 6.9).

6.6 Gene expression in lactobacilli

6.6.1 Transcription

To date more than fifty *Lactobacillus* genes have been cloned and sequenced. All of the genes that were isolated prior to the development of genetic transformation systems for lactobacilli were cloned using *E. coli* as a host organism. Despite the fact that cloning of *Lactobacillus* DNA in *E. coli* occasionally can be frustrated by instability problems (Lokman *et al.*, 1991; Vanderslice *et al.*, 1986; Vidgrén *et al.*, 1992), nearly all cloning experiments carried out since transformation procedures for *Lactobacillus* became available, still involve the use of *E. coli* as a host organism. As a result of the improvement of transformation procedures (Trevors *et al.*, 1992), direct cloning of DNA fragments in specific strains of *Lactobacillus* has now become possible (Posno *et al.*, 1991b).

The value of a comparison of specific features of *Lactobacillus* genes such as promoters, ribosome binding sites and terminators is restricted by differences in the origin of these genes. The genes that have been sequenced are derived from nine or more different species, that are evolutionary related but taxonomically quite distinct. Initial evidence has been obtained that expression of *Lactobacillus* genes takes place when such genes are transferred to other *Lactobacillus* species or to *E. coli* (Lerch *et al.*, 1989; Natori *et al.*, 1990; Posno *et al.*, 1991; Toy and Bognar, 1990). These results have allowed a preliminary analysis of lactobacilli transcription and translation signals that are sufficiently similar to be recognized in other hosts, even though control of gene expression might be different.

Transcription start sites have been determined for eight *Lactobacillus* genes originating from five different species (Table 6.2). The length of the untranslated region for these transcripts is relatively small (< 100 nt), comparable to that found in *E. coli* and *Bacillus* spp. In the region of 5' to the transcription start site, −35 and −10 sequences have been identified that closely resemble the consensus sequences reported for the corresponding promoter regions in *E. coli* and *Bacillus* spp. The region 5' to the −35 sequence is rich in A residues, as has previously been observed for *Bacillus* spp. promoters that are recognized by σ-70 factors (Graves and Rabinowitz, 1986; Moran *et al.*, 1982). The distance between the −10 sequence and actual transcription start site varies considerably and no specifically required sequence element has been discerned.

With the exception of genes that are located in the promoter-distal region of an operon, −35 and −10-like sequences with a spacer of ~18 nucleotides have

Table 6.2 Promoters whose start sites have been determined

Gene	Species	-35/-10 sequence					Tss*	Reference
		-45	-35	-35/-10 sequence	-10			
xylR	Lb. pentosus	agtaagaat	TTGACA	ggtaattgagcggtcat	CACATT	cttgtgacgG	78	Lokman et al., unpub.
xylA	Lb. pentosus	agaaagcg	TTTACA	aaaataagccaatgccgc	TGTAAT	ctTaC	~45	Lokman et al., unpub.
L-HicDH	Lb. confusus	gttgttcag	TGGTGA	acaacgtctctttttg	TATAAT	gaacaT	26	Lerch et al., 1989a
D-HicDH	Lb. casei	aaagatgctt	TTCACA	gaggcatctgta	TACGGT	gagtgG	26	Lerch et al., 1989b
hdcA	Lactobacillus	aaaacgc	TTTACT	tttttattttgaagt	TATAAT	gaaaatcG	73	Vanderslice et al., 1986
fgs	Lb. casei	gctaag	CTCATA	agttagacccatttttctgg	TATGAT	gaagctA	59	Toy and Bognar, 1990
lacP	Lb. bulgaricus	aaaacatc	TTGGTT	tatttagtaaacaagtc	TATACT	gtaattaT	31	Leong-Morgenthaler et al., 1991
SP-P1	Lb. brevis	agcgtaatcc	TTGTAT	ctccttaaggaaatcgc	TATACT	tatcT	135	Vidgrén et al., 1992
SP-P2	Lb. brevis	aaataaaattc	TTAACA	aaagcgctaacttcggt	**TATACT**	attcT	48	Vidgrén et al., 1992
Consensus		aAAa . . c	**TT**gACA	tatTt	**TATAAT**	gG		
						GAaa		

Tss* means *transcription* start site. Consensus sequence is determined according to Graves and Rabinowitz (1986). The nucleotide given is the one occurring most frequently.
Bold capital :>75%; capital: 50%–75%; lower case: <50%.

been identified upstream of the coding region of all other *Lactobacillus* genes sequenced. A consensus sequence can be inferred for these genes that resembles that of promoters in *E. coli* and *Bacillus* spp. However, it should be noted that the transcription start sites for this group of *Lactobacillus* promoters has not been determined and that promoter sequences were postulated based on the assumption that they should conform to known promoter sequences. The presence of regions of high A/T content in the DNA of these organisms further complicates the identification of promoters through sequence analysis.

A functional analysis of putative promoter sequences has been carried for a small number of genes. Three-hundred base pair fragments of the *xyl*/R gene or *xyl*A/B operon were identified by primer extension analysis as promoter-containing fragments. The fragments were cloned in a promoter-probe vector with the chloramphenicol-resistance gene (*cml*) as marker and used to transform *E. coli*. Numerous chloramphenicol resistant colonies were found, indicating that the promoter sequences are functional in *E. coli* (Lokman *et al.*, manuscript in preparation). The promoter of *xyl*/R was also found to give rise to chloramphenicol resistance in *Lactobacillus casei*, indicating that expression from this promoter takes place in a heterologous, though related host organism. The promoter P*xyl*/R has also been used successfully to express foreign proteins in *Lactobacillus casei* (see Section 6.4).

DNA fragments containing promoter activity have also been isolated from *Lactobacillus curvatus* (Gaier, 1991) and *Lactobacillus acidophilus* (Posno, personal communication) using promoter-probe vectors. Nucleotide sequence analysis of such fragments has revealed the presence of promoter-like elements. The sequence of one fragment has been identified as a ribosomal RNA promoter (Posno, personal communication). Comparison of sequences of other portions of the cloned fragments with the nucleotide sequence databases has yielded no clue as to whether these sequences represent genuine promoters or merely sequences with promoter activity.

Rho-independent-like terminators, that are identified as stem-loop structures followed by a series of A or T residues (Platt, 1986), have been observed at the 3' end of many of the *Lactobacillus* genes thus far studied. The results of Northern blot analyses of several transcripts are consistent with the notion that these sites function as true terminators of transcription (Copeland *et al.*, 1989; Lokman *et al.*, manuscript in preparation). In addition, transcription analysis by Northern hybridization of hybrid transcriptional units that inserted the presumed terminator sequence of the *xyl*A/B operon of *Lactobacillus pentosus* downstream (3') of a marker gene, indicated that transcription is terminated at the site of the putative terminator.

Another DNA sequence resembling a rho-independent transcription terminator that might function as an attenuator controlling the level of RepA has been found in the untranslated region of the *rep*A gene of plasmid p353-2 (Leer *et al.*, 1992). This hypothesis is supported by Northern blot analysis showing that two transcripts of *rep*A can be formed. One transcript was found to have a length (~1100 nt) corresponding to transcription of the entire *rep*A gene. A second, more abundant message of 200 nt hybridizes to a probe derived from the untranslated region of the *rep*A gene. It may represent an RNA molecule that is the result of premature transcription termination. An RNA of opposite polarity, that is transcribed from a promoter located between the promoter and start codon of *rep*A, may facilitate premature termination at the attenuator site (Pouwels *et al.*, 1993).

In the *hdc*A/B (histidine decarboxylase) operon of *Lactobacillus* 30A (Vanderslice *et al.*, 1986 and the *xyl*A/B operon of *Lactobacillus pentosus* Lokman *et al.*, 1991), palindromic sequences have been found in intergenic regions that could function as transcription terminators and/or RNA processing sites. The function of such elements might be to reduce the efficiency with which the promoter-distal gene is transcribed relative to that of the promoter-proximal gene, or to allow for a difference in RNA turnover rates by dissection of the two transcripts (Higgins *et al.*, 1982).

6.6.2 Translation

Inspection of the nucleotide sequences around the translation-initiation site of all *Lactobacillus* genes sequenced, shows that ATG is the preferred start codon. Of more than fifty genes reported, four start with GTG, two with TTG and all of the others start with ATG. A highly conserved sequence (AGGAGG), with an internal triplet (GGA), is found in 90% of the sequences analysed at a distance 6–10 nucleotides 5' to the start triplet. The sequence most probably functions as Shine-Dalgarno sequence in binding to the 3' end of 16S RNA (Shine and Dalgarno, 1974). In agreement with this hypothesis is the finding that the sequence of the 3' terminal end of the 16S RNA of *Lactobacillus* is 3' CUUCUU 5' (Martinez-Murcia and Collins 1990, 1990b; Weisburg *et al.*, 1989).

Translational coupling of contiguous genes can increase the efficiency of translation of the downstream gene. Overlapping start/stop codons, or partially overlapping reading frames, were observed for the *Lactobacillus casei fgs* (folypoly-g-glutamate synthetase) gene (Toy and Bognar, 1990) and the *Lactobacillus casei trpF* (N-5'-phosphoribosyl-anthranilic acid isomerase) gene (Natori *et al.*, 1990). Interestingly, the *lac*L/*lac*M (Chassy *et al.*, unpublished observations) and *trp*B/A (Natori *et al.*, 1990) genes of *Lactobacillus casei* are partially overlapped, presumably ensuring the equimolar synthesis of the two subunits of the enzymes (Oppenheim and Yanofsky, 1980).

A computer-assisted statistical evaluation of codon usage of *Lactobacillus* genes shows a clear bias for specific codons. All possible codons are used by lactobacilli, but not by all species. The same holds true for stop codon usage. UAA is the most frequently used stop codon, and its usage has been found in the genes of all species thus far examined. The rare stop codon UAG is not present in 15 sequenced *Lactobacillus casei* genes, except for an unidentified open reading frame in the insertion element ISL*1* (Shimizu-Kadota *et al.*, 1985). The stop codon UGA is not used in any of the 13 sequenced genes from the two closely related species *Lactobacillus plantarum* and *Lactobacillus pentosus*. The arginine codon AGA and the asparagine codon GAC are rare or absent from some species, but are common in other species. The same applies to the aspartate codon GAC. Since the number of sequenced genes from *Lactobacillus* spp. is relatively small, it would be premature to conclude that specific tRNAs are absent from these species. In the two genes that code for surface proteins, translation termination appears to be effected by two consecutive amber codons (Vidgrén *et al.*, 1992; Boot *et al.*, 1993).

6.6.3 Regulation of transcription

Knowledge of the regulation of gene expression in lactobacilli is still in its infancy. Investigations of gene expression at the molecular level have only become possible since the introduction of techniques of genetic manipulation. Compared to the lactococci, there is a considerable gap in our knowledge of operon structure and the regulation of expression of *Lactobacillus* genes. Early biochemical and physiological data, indicated that the uptake and degradation of sugars are tightly controlled in these organisms (Thompson, 1987). Since a wealth of information concerning expression of *lac* genes from other organisms is available, and since lactose is one of the most common energy sources for a number of lactobacilli (Kandler, 1983), it is not surprising that studies on control of gene expression in lactobacilli started with an investigation of the genes involved in lactose catabolism.

The nucleotide sequence upstream (5') to the *lac*EGF genes of *Lactobacillus casei* has recently been determined (Alpert and Chassy, 1988; Alpert and Chassy, 1990; Porter and Chassy, 1988; Alpert and Siebers, personal communication). The sequence data reveal a potential promoter that conforms to the Gram-positive consensus sequence. The promoter-like element is followed at a distance of 50 nt by a sequence capable of forming a stem-loop structure in the mRNA that resembles a rho-independent transcription terminator. An 880 nt ORF is found 25 nt downstream of the terminator-like structure. The *lac*E gene is found 120 nt downstream of the ORF. Northern hybridization experiments using a *lac*E probe reveal mRNA species of 4.4 kb – sufficient to encode the ORF and the three known genes of the lac-PTS operon, *lac*EFG. These data suggest that the *tag* genes do not exist in a single transcriptional unit with the lac-PTS genes in *Lactobacillus casei*, as they do in *Lactococcus lactis* (Van Rooijen and de Vos, 1991; Van Rooijen *et al.*, 1991) and *S. aureus* (Oskouian and Stewart, 1990). The putative translation product of the ORF (34 kDa) shows significant similarity to the anti-terminator proteins, BglG, ArbG, SacT and SacY (Debarbouille *et al.*, 1991). Based on these observations, it may be that lac-PTS gene transcription in *Lactobacillus casei* is at least partially regulated by an anti-termination mechanism similar to that observed for the β-glucoside operon of *E. coli*.

The enzymes involved in catabolism of D-xylose in *Lactobacillus pentosus* are induced by xylose. Control of *xyl* gene expression takes place at the level of transcription. Expression is induced by xylose and is negatively controlled by a repressor protein. The N-terminal region of the repressor displays the helix-turn-helix motif characteristic of DNA-binding proteins (Lokman *et al.*, manuscript in preparation). In the intergenic region between *xyl*R and *xyl*A, there is an element with two-fold rotational symmetry that closely resembles the operator of the *xyl* operon of *B. subtilis* and *B. licheniformis* (Kreuzer *et al.*, 1989; Scheler *et al.*, 1991; Table 6.3). Cloning of this element into a multicopy vector and introduction of the resultant recombinant plasmid into *Lactobacillus pentosus* demonstrated that the element can function as an operator that titrates the repressor. In the presence of multiple copies of the operator, expression of the *xyl*A/B genes is constitutive (Lokman *et al.*, manuscript in preparation).

During growth on a medium containing xylose, high levels of *xyl*P/Q mRNA and *xyl*A/B mRNA are found; however, these levels are greatly reduced when glucose is used as an energy source, or when a mixture of the two sugars is present (Lokman *et al.*, manuscript in preparation). The intergenic region between *xyl*R and *xyl*A contains a second sequence element with a palindromic structure overlapping the −35 sequence that might be a target site for a protein involved in glucose repression of the *xyl* genes (Table 6.2). An element with similar structure has been found at the corresponding position in the untranslated region of catabolite-repressed genes of Gram-positive bacteria (Weicker and Chambliss, 1990). The hypothesis that this element is involved in glucose-mediated catabolite repression of *xyl*A/B is supported by the finding that introduction of a multi-copy vector harbouring the element into *Lactobacillus pentosus* alleviates glucose repression. This observation suggests that a protein interacting with the element is titrated (Lokman *et al.*, manuscript in preparation).

Table 6.3 Sequence of operator-like regions of *xyl* genes

Operator I				Reference
Lb. pentosus	TagGTTggt	tgc	cgaAACAAA	Lokman *et al.*, unpub.
B. subtilis	TtaGTTtgt	ttg	ggcAACAAA	Kreuzer *et al.*, 1989
B. licheniformis	GTtaGTTtaa	tgg	ttaAACAAAC	Scheler *et al.*, 1991
Consensus	T..GTT		AACAAA	
Operator II				
Lb. pentosus	ctaGA	aAg CG tTt	ACAaaat	Lokman *et al.*, unpub.
B. subtilis	ttTGg	aAg CG taa	aCAaaag	Jacob *et al.*, 1991
B. megaterium	ttTGa	aAg CG caa	ACAaa	Rygus and Hillen, 1992
Consensus	TGT/A	NAN CG NTN	T/ACA	Weickert and Chambliss, 1990

Operator I is the palindromic sequence that occurs between the −10 region and the start codon ATG. The consensus sequence is based on nucleotides present in all three elements. Operator II overlaps the −35 region in the promotor of *Lb. pentosus* but is present within the coding region *xyl*A of *B. subtilis* and *B. megaterium*. Underlined regions can form a potential stem structure.

6.6.4 *Heterologous gene expression*

A limited number of studies have appeared describing the expression of foreign genes in *Lactobacillus* spp. The first example of heterologous gene expression in lactobacilli was associated with the transfer of conjugal plasmids such as pAMβ1. Both replication-essential and antibiotic resistance-determining genes needed to be expressed in the heterologous host for conjugation to be observed. The markers that are most frequently used in the lactic acid bacteria, Cm^R (*cml*) from pC194 and Em^R (*erm*) from pE194 or pAMβ1, are expressed from their own regulatory signals in a wide variety of hosts. A number of enzymes originating from other bacterial sources also have been expressed in various lactobacilli. Examples of genes expressed from an endogenous promoter are given in Table 6.4. Both intracellular and extracellular enzymes were expressed in lactobacilli. Extracellular enzyme expression was observed in several instances, indicating that the secretion signals were also functional in heterologous hosts. In these studies, little information was presented regarding the efficiency or regulation of expression.

Table 6.4 Expression of heterologous proteins in *Lactobacillus*

Protein	Origin	Species	Location	Reference
Lipase	*S. hyicus*	*Lb. curvatus*	Plasmid	Vogel *et al.*, 1990
Cellulase	*C. thermocellum*	*Lb. plantarum*	Plasmid	Bates *et al.*, 1989
			Chromosome	Scheirlinck *et al.*, 1989
Xylanase	*C. thermocellum*	*Lb. plantarum*	Plasmid	Scheirlinck *et al.*, 1989
			Chromosome	
α-Amylase	*B. amyloliquefaciens*	*Lb. plantarum*	Plasmid	Leer, Unpub., [90]
α-Amylase	*B. amyloliquefaciens*	*Lb. casei*	Plasmid	Chassy, Unpub., Leer, Unpub.
α-Amylase	*B. stearothermophilus*	*Lb. plantarum*	Plasmid	Leer, Unpub.
α-Amylase	*C. thermosulforogenes*	*Lb. plantarum*	Plasmid	Leer, Unpub.
α-Amylase	*C. thermosulforogenes*	*Lb. casei*	Plasmid	Leer, Unpub.
Glucanase	*B. subtilis*	*Lb. acidophilus*	Plasmid	Baik and Pack, 1990
Glucanase	*B. amyloliquefaciens*	*Lb. helveticus*	Plasmid	Thompson and Collins, 1991
Cholesterol oxidase	*Streptomyces*	*Lb. casei*	Plasmid	Somkuti *et al.*, 1992
XylA/XylB	*Lb. pentosus*	*Lb. casei*	plasmid	Leer, Unpub.
lux	*Vibrio fischeri*	*Lb. casei*	plasmid	Ahmad and Stewart, 1988
lysostaphin	*S. staphylolyticus*	*Lb. casei*	plasmid	Gaier *et al.*, 1992

Expression of heterologous proteins under the transcriptional control of *Lactobacillus* promoters was recently demonstrated for a number of enzymes and fusion proteins (Table 6.5). Where the amount of protein synthesized was quantitated, relatively low levels of expression were detected. For example, expression of the *E. coli lacZ* gene under the control of the promoter of the xylose repressor gene, *xyl*R, amounted to approximately 1% of total protein. When an epitope of foot-and-mouth disease virus or an epitope of rotavirus capsid protein VP7 was fused to the β-galactosidase moiety, expression levels

Table 6.5 Expression of heterologous proteins in *Lactobacillus*

Protein	Origin	Species	Promoter	Signal sequence	Reference
β-galactosidase	*E. coli*	*Lb. casei*	*xyl*R		Kottenhagen, Unpub.
FMDV-β-gal	*E. coli*	*Lb. casei*	*xyl*R		Kottenhagen, Unpub.
Proteinase	*Lc. cremoris*	*Lb. casei*	Proteinase	*Lc. cremoris* Proteinase	Posno, Unpub.
α-amylase	*B. stearothermophilus*	*Lb. casei*	Proteinase	*Lc. cremoris* Proteinase	Leer, Unpub.
α-amylase	*Lb. amylovorus*	*Lb. casei*	α-amy	*Lb. amylovorus* α-amy	Jore, Unpub.

were even lower (Claassen *et al.*, 1993). No data is presently available on the protein levels that can ultimately be produced in lactobacilli. Although a more than ten-fold overexpression of a homologous protein in *Lactobacillus plantarum* has been reported, no data was reported as to how much protein was synthesized (Christiaens *et al.*, 1992).

The genetic modification of food, dairy or industrial bacterial starter cultures often will require an understanding of their protein secretory system. A major obstacle to further understanding in this area has been the paucity of well-described extra-cellular proteins in these organisms. Although they do not appear to be prodigious protein secretors, two species of lactobacilli, *Lactobacillus amylophilus* (Nakamura and Crowell, 1979) and *Lactobacillus amylovorus* (Nakamura, 1981), are known to secrete α-amylases and degrade insoluble starch (Imam *et al.*, 1991). *Lactobacillus paracasei* subsp. *paracasei* secretes a PrtP-like protease (Holck and Naes, 1992; Naes and Nissen-Meyer, 1992). It has recently been observed that certain lactobacilli efficiently secrete inulinases (Szilágyi *et al.*, personal communication). Four genes encoding proteins that are transported through the membrane and secreted into the culture fluid or remain cell wall-bound have been analysed at the molecular level. They contain signal sequences that are typical for secreted proteins. The α-amylase from *Lactobacillus amylovorus* has a 49 amino acid extension at its N-terminus, an extension longer than that of the α-amylase of *B. subtilis* (Jore *et al.*, manuscript in preparation). Two surface-layer proteins from *Lactobacillus brevis* (Vidgrén *et al.*, 1992) and *Lactobacillus acidophilus* (Boot *et al.*, 1993) contain signal sequences that are 30 and 24 amino acids long, respectively. It should be noted that the signal peptide sequences encoded by bacteriocin-encoding genes are sometimes atypical of those often associated with secreted proteins (Muriana and Klaenhammer, 1991).

Expression and subsequent secretion of a few extra-cellular proteins into the culture fluid has been reported to occur in lactobacilli (Table 6.5). The α-amylase from *Lactobacillus amylovorus* was efficiently expressed in *Lactobacillus casei* under the control of its own promoter (Jore *et al.*, manuscript in preparation). The regulation of expression was similar to that found in *Lactobacillus amylovorus*. Signal sequences from *Bacillus* spp. and *Lactococcus* spp. also seem to function in *Lactobacillus*, as expression and secretion was demonstrated for enzymes that were fused to these elements (Table 6.5). It is striking that expression and secretion signals isolated from *Lactococcus lactis* have been used to direct the expression and secretion of a TEM β-lactamase of Gram-negative origin in *Lactobacillus plantarum* (Sibakov *et al.*, 1991).

6.7 Gene expression in *S. thermophilus* and leuconostocs

Compared to the lactococci and lactobacilli, very little information is available about gene structure and gene expression in *S. thermophilus* and the leuconostocs. Genes related to lactose and galactose metabolism in *S. thermophilus* have been isolated and characterized (Poolman *et al.*, 1989; Poolman *et al.*, 1990; Schroeder *et al.*, 1991); and genes involved in lactose and citrate metabolism in leuconostocs have also been described (David, 1992; David *et al.*, 1992). The *gal*E gene product of *S. thermophilus* has significant sequence similarity to the UDP-glucose 4-epimerases of *E. coli* and *Streptomyces lividans*, while the *gal*M gene product of *S. thermophilus* exhibits similarity to the mutarotase of *Acinetobacter calcaoceticus* (Poolman *et al.*, 1989). Although separated by only 72 bp, the *gal*E and *gal*M genes appear to be transcribed from distinct promoters. A 10 to 20-fold repression in the expression of *gal*T, *gal*E and *gal*M is observed upon the addition of glucose to the growth media, indicative that transcription is regulated by glucose-mediated catabolite repression.

The plasmid-encoded citrate permease, *cit*P, of a *Leuconostoc lactis* strain has been cloned and the nucleotide sequence determined (David and de Vos, unpublished results). Somewhat surprising is the observation that the gene encodes a CitP protein that is virtually identical to that found in *Lactococcus lactis* (three amino acid substitutions, one amino acid deletion). This has led to the speculation that *cit*P has spread horizontally between the two lactic acid bacteria by IS-element-mediated transfer. A partial copy of an iso-ISS*1*-element was found on the lactococcal citrate plasmid pSL2 (Jans *et al.*, 1991).

The sucrose phosphorylase gene of *Leuconostoc mesenteroides* has been isolated and the nucleotide sequence reported (Kitao and Nakano, 1992). The protein potentially encoded by the gene has 68% similarity to the product of the *gtf*A gene isolated from *S. mutans*. The cloning of the gene and the amino acid sequence of the glucose 6-phosphate dehydrogenase from *Leuconostoc mesenteroides* has been reported (Lee *et al.*, 1991). The characterization of leucocin A-UAL 187 (Hastings *et al.*, 1991) and cloning of the bacteriocin genes from *Leuconostoc gelidum* has also been reported; DNA sequencing of the leucocin gene(s) revealed an operon consisting of two open reading frames flanked by a putative upstream promoter and a downstream terminator (Hastings, 1992).

Since few genes of *S. thermophilus* and leuconostoc origin have been characterized, it is difficult to draw conclusions about codon usage and expression signals in these bacteria. To date, only two reports have appeared that describe transcription initiation sites used by *S. thermophilus* that are based on mRNA mapping data (Poolman *et al.*, 1990; Slos *et al.*, 1991). Random promoters that were isolated from *S. thermophilus* have been used to express an α-amylase gene in *S. thermophilus, Lactococcus lactis, Enterococcus faecalis, B. subtilis,* and *Lactobacillus* spp. (Slos *et al.*, 1991). The *lac* promoter of *S. thermophilus* has been identified by transcription mapping (Poolman *et al.*, 1990). There are few reports available on the expression of heterologous genes by these organisms, although functional expression of foreign proteins in *S. thermophilus* has been detected (Mercenier, 1990; Slos *et al.*, 1991, Dutot *et al.*, unpublished results).

6.8 Chromosomal integration of genes

The construction of stable genetically modified strains that do not depend on the maintenance of plasmid-encoded traits requires a system for the insertion of mutant or heterologous genes into the host chromosome. Integration of genes has been described for a number of bacterial genera, including the lactic acid bacteria discussed here (see, for example Leer *et al.*, 1993; Mercenier, 1990; Scheirlinck *et al.*, 1989). The successful application of this technique will depend on the acquisition of more detailed knowledge of both the genome structure and the generalized and specific recombination system(s) present in the lactobacilli, leuconostocs and *S. thermophilus*. Initial mapping and characterization of the genomes of *Leuconostoc oenos* (1.8–2.1 mb), *Lactobacillus bulgaricus* (2.3 mb) and *S. thermophilus* (1.7–1.8 mb) have been described (Le Bourgeois *et al.*, 1993). Using degenerate primers for polymerase chain reaction amplification and subsequent cloning of the amplified fragments, DNA sequences that encode a fragment of a RecA-like protein have been isolated from *Lactobacillus bulgaricus, Lactobacillus helveticus, Leuconostoc mesenteroides,* and *S. thermophilus* (Duwat *et al.*, 1992). The integrative *att* sites specific for *Lactobacillus gasseri* and *Lactobacillus bulgaricus* bacteriophages have been identified (Lahbib-Mansais *et al.*, 1992; Raya *et al.*, 1992). Site-specific integration via high-frequency plasmid transduction has been reported in *Lactobacillus gasseri* (Raya and Klaenhammer, 1992). Transduction of DNA fragments into the chromosome of *Lactobacillus bulgaricus* has also been observed (Lahbib-Mansais *et al.*, 1992).

The integration into the chromosome and expression of α-amylase and endoglucanase genes in *Lactobacillus plantarum* were accomplished by recombination through a random chromosomal fragment (Scheirlinck *et al.*, 1989). Under selective pressure, unstable plasmids have been observed to integrate into the chromosome of *Lactobacillus plantarum* (Rixon *et al.*, 1990). Recently, it has been reported that a *cml* gene can be inserted into the chromosome of *Lactobacillus plantarum* through homologous recombination with DNA sequences that encode the conjugated bile salt hydrolase gene (*cbh*) (Leer *et al.*, 1993). The double cross-over event disrupted *cbh* expression and resulted in stable 'knock-out' mutants (Leer *et al.*, 1993).

Recombination through the lactose genes can also be employed for the insertion of heterologous DNA into the chromosome of *S. thermophilus* (Mercenier, 1990). Recombination appeared to occur when DNA was introduced via electroporation, but did not occur when the protoplast transformation technique was used, even though the efficiency of the latter process was high (*c.* 10^5 transformants/μg DNA). It was found that linear DNA functioned as well as, or better than, closed circular DNA. The high efficiency of the process (10^{-2} recombinations/transformant) theoretically would allow isolation of recombinants without the need for selection mediated by an antibiotic resistance marker. Recently, it has been observed that plasmids carrying DNA fragments isolated from *lac* operon of *S. thermophilus* combined with a selectable *erm* gene, integrate into the chromosome of *S. thermophilus* by homologous recombination (Mollet *et al.*, 1993). Growth in high concentrations of erythromycin caused amplification the inserted cartridge. In the absence of selective pressure, the cointegrate structure resolved. Lactose negative strains bearing a deletion in the *lacZ* gene were constructed. A promoterless *cml* gene was inserted between the *lacS* and *lacZ* genes, resulting in a gene fusion in which *cml* became a functional part of the *lac* operon. The gene was stably maintained in the absence of lactose, and was expressed coordinately with the *lac* genes upon growth in lactose-containing media. These findings represent the first report of the stable integration and expression of an heterologous gene in *S. thermophilus* using a site-specific procedure that might be useful for food, dairy and industrial fermentations.

6.9 Phage and phage resistance

Starter strain lysis and subsequent fermentation failure (slow or low-acid fermentation) due to phage attack represents a major economic loss in dairy manufacturing. This has led to extensive analysis of phages isolated from factories and of phage defense mechanisms naturally encountered in certain dairy starters. Since lactococci are the predominant organism used for cheese production, the current body of knowledge focuses on them (for reviews see, Klaenhammer, 1987; Sanders, 1988). However, phage characterization and classification studies have been performed on *S. thermophilus* and certain lactobacilli (see below). The information available for leuconostoc bacteriophages is very limited.

In contrast to the phages studied from other lactic acid bacteria, all the *S. thermophilus* phages analysed seem to fall into a single DNA homology group

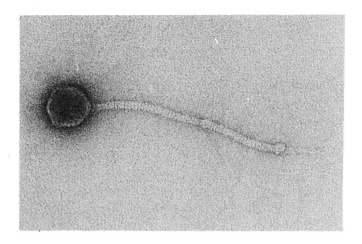

Figure 6.2   Lytic phage from *Streptococcus thermophilus*.

that can be subdivided into two or three subclasses. The vast majority of are lytic phages (see Figure 6.2), containing a double-stranded DNA genome of 35–45 kb, belonging to Bradley's morphology group type B1, and displaying a narrow host range (for reviews, see Jarvis, 1989; Mata and Ritzenthaler, 1988; Mercenier, 1990). Only recently (Mercenier, 1990; Neve *et al.*, 1990), has the lysogenic state been demonstrated in two *S. thermophilus* strains. Interestingly, these two lysogenic strains undergo autolysis at 45°C. Previous attempts to induce temperate phages from *S. thermophilus* had failed due to inadequate experimental conditions, lack of indicator strains and/or production of defective phage particles. It has since been observed that temperate phages could be induced from 12 of 120 strains of *S. thermophilus* analysed (Fayard, *et al.*, 1993). These phage isolates belonged to Bradley Group B1, possessed a DNA genome of 40–45 kb, and shared DNA homology with virulent phages of *S. thermophilus*. While phages attacking *Lactobacillus bulgaricus* have been isolated independently (see below), slow acid production during yogurt fermentation often appears to be linked to viral attack on the streptococcal component of the starter symbiosis. It should be noted that a majority of both *S. thermophilus* and *Lactobacillus bulgaricus* starter strains were resistant to all phages tested.

Phages of lactobacilli are quite diverse, as might be expected from the phylogenetic distance between their respective host species. As a consequence, the knowledge of *Lactobacillus* spp. phages is incomplete when compared to the phages of lactococci. A few phage isolates have been studied in detail, and the molecular mechanisms of their biosynthesis and the transition from a temperate state to a lytic state have been well documented. Due to the impact of industrial fermentation failures, most of the *Lactobacillus* phages studied thus far are isolates attacking dairy or meat starters, such as *Lactobacillus bulgaricus* and *Lactobacillus delbrueckii* subsp. *lactis, Lactobacillus helveticus, Lactobacillus casei* and *Lactobacillus acidophilus*. A few phages of *Lactobacillus plantarum* have also been described. Detailed descriptions of these bacteriophages are included in a number of recent reviews (Davidson *et al.*, 1990; Jarvis, 1989; Mata and Ritzenthaler, 1988; Séchaud, 1990; Séchaud *et al.*, 1988).

While most of the bacteriophages of *S. thermophilus* and *Lactobacillus* spp. have been isolated from defective dairy products, the few phages shown to infect *Leuconostoc* spp. were not all associated with major fermentation failures. However, phages attacking *Leuconostoc mesenteroides* or *Leuconostoc cremoris* were obtained from fermented milk products (see Jarvis 1989). They were only examined with respect to their morphological and growth characteristics. Although the occurrence of phages in wine does not seem very frequent, viruses specific for *Leuconostoc oenos* have been found when the initial pH of the beverage was higher than 3.2 or during the production of *Leuconostoc oenos* starter cultures (Arendt *et al.*, 1991; Davis *et al.*, 1985; Henick-Kling *et al.*, 1986; Nel *et al.*, 1987; Sozzi *et al.*, 1976). The malolactic fermentation of certain types of red wine was shown to be affected by phage attack in Australia (Davis *et al.*, 1985; Henick-Kling *et al.*, 1986). These Australian isolates had an hostrange limited to certain *Leuconostoc oenos* strains. These phages were unable to lyse *Lactobacillus* spp. or *Pediococci*. They all exhibited Bradley type B morphology. A virulent *Leuconostoc oenos* phage, φP581, was characterized in more detail (i.e. morphology, lytic spectrum, genomic DNA analysis, protein profile) by Arendt *et al.*, 1991b. Further analysis showed that this isolate was actually resident in the culture of the corresponding host strain (*Leuconostoc oenos* 58N) that exhibited all the characteristics of the phage-carrier state. The existence of lysogeny in *Leuconostoc oenos* has only been demonstrated recently (Arendt *et al.*, 1991). Mitomycin C treatment of 30 strains led to induction of temperate phages in 19 cases. Five of these phages were analysed more fully. The phages were found to be closely related and belonged to the same morphology group (isometric-headed). DNA–DNA hybridization detected phage DNA sequences in the genome of lysogenic strains. It is clear that many properties of the *S. thermophilus, Lactobacillus* spp. and *Leuconostoc* spp. phages remain to be investigated, though a few of the *Lactobacillus* phages have been extensively characterized.

Information on phage insensitivity or resistance mechanisms is more scarce than that available for lactococci. Very little molecular and biochemical data have been collected for *S. thermophilus*, lactobacilli and leuconostocs phages. For *S. thermophilus*, biological data indicative of the existence of endogenous restriction/modification system(s) in a number of strains (Benbadis *et al.*, 1990; Mercenier *et al.*, 1987) were substantiated recently by the isolation and characterization of Type II restriction endonucleases: *Sth*134I (C_mCGG; isoschizomer of *Hpa*II; (Solaiman and Somkuti, 1990), *Sth*117I (CC_mWGG; isoschizomer of *Bst*NI and *Eco*RII; Solaiman and Somkuti, 1991) and *Ssl*I (CC_mWGG; isoschizomer of *Eco*RII; Benbadis *et al.*, 1991). The optimal cleavage conditions for the two last enzymes were reported as different. In addition to R/M systems, at least two *S. thermophilus* strains possess an abortive infection-like defense mechanism (Larbi *et al.*, 1992). A 34 kb plasmid encoding for a R/M system has recently been identified in a *Lactobacillus helveticus* strain (De Los Reyes-Gavilan *et al.*, 1990). Fragments of this replicon, that does not seem to encode for lactose-fermenting ability and/or proteolytic capacity, allowed the detection of homologous plasmid-encoded R/M genes in two additional *Lactobacillus helveticus* strains. Although phage resistant strains of *Leuconostoc* spp. have been described, the mechanism(s) of resistance has not been reported.

6.10 Concluding remarks

The lactic acid bacteria discussed in this chapter are of major economic significance. They may be of value in maintaining and promoting the health of humans and animals. Improving existing strains with respect to traits such as susceptibility to phage attack, metabolic traits, and efficiency of proteolysis are all goals of research. The possibility also exists to use these bacteria for the production of value-added chemicals, for enhanced health-beneficial ways (i.e. as vectors for vaccination), and for biowaste remediation. All of these potential applications depend on efficient methods for genetically engineering strains.

The progress detailed in the previous sections demonstrates that we have learned much about these bacteria in the last 5–10 years. The majority of the literature citations in this chapter represent contributions since 1987. The advent of high-voltage electroporation greatly simplified the genetic manipulation of these organisms. Perhaps this has spurred the recent surge in our knowledge base of the molecular genetics of the lactic acid bacteria. One cannot help but note that the most recent issue of the NCBI (USA) Entrez Medline-Genbank CD-ROM (v.2.0, Dec. 15, 1992) included 81 references to sequences derived from lactobacilli, nine to sequences from leuconostocs, and seven to sequences from *S. thermophilus* in contrast to 62 sequences of lactococcal origin. This is a marked reversal of the earlier focus on the lactococci. However, the observation that the same database contains 2983 nucleotide sequences of *E. coli* origin and over 77 000 total nucleotide sequences should place in perspective the scope of our knowledge about the molecular biology of the lactic acid bacteria.

The tools for the genetic engineering of lactobacilli, leuconostocs and *S. thermophilus* are clearly available. Not only is transformation a reality for many strains, but useful second and third generation vectors have recently appeared. The catalogue of cloned genes and regulatory and expression signals available will allow thorough exploration of the expression and secretion signal sequence requirements and the potential of these bacteria. Additional research will be required to evaluate the possible use of these organisms for the efficient expression of heterologous genes. Recombinant DNA technology represents an invaluable tool for the generation of defined mutants with which to analyse the biochemistry of these organisms. Knowledge thus gained could be put to rational application in the design and modification of strains. We would predict that the next five years will see a considerable expansion of our knowledge of the molecular genetics of these organisms as well as a reasonable chance for the development of potentially useful strains.

Dedication and acknowledgements

This chapter is dedicated to the memory of Jean-Pierre Lecocq (1947-1992), Scientific Director, Transgene, SA, and dear friend whose keen mind, creative imagination, enthusiastic presence, and warmth of personality is sorely missed by all of us.

We would like to thank Ms Cynthia Murphy for thoughtful reading of the manuscript and constructive assistance in its preparation.

References

Ahmad, K.A. and Stewart, G.S.A.B. (1988) Cloning of the *lux* genes into *Lactobacillus casei* and *Streptococcus lactis*: phosphate-dependent light production. *Biochem. Soc. Trans.*, 1068.

Ahn, C., Collins-Thompson, C., Duncan, C. and Stiles, M.E. (1992) Mobilization and location of the genetic determinant of chloramphenicol resistance from *Lactobacillus plantarum* CAT2R. *Plasmid*, **27**, 169–176.

Ahrn, S., Molin, G. and Ståhl, S. (1989) Plasmids in *Lactobacillus* isolated from meat and meat products. *Syst. Appl. Microbiol.*, **11**, 320–325.

Alpert, C.A. and Chassy, B.M. (1988) Molecular cloning and nucleotide sequence of the factor III(lac) gene of *Lactobacillus casei*. *Gene*, **62**, 277–288.

Alpert, C.A. and Chassy, B.M. (1990) Molecular cloning and DNA sequence of *lacE*, the gene encoding the lactose-specific enzyme II of the phosphotransferase system of *Lactobacillus casei*: Evidence that a cysteine residue is essential for sugar phosphorylation. *J. Biol. Chem.*, **265**(36), 22560–22561.

Arendt, E.K., Lonvard, A. and Hammes, W.P. (1991) Lysogeny of *Leuconostoc oenos*. *J. Gen. Microbiol.*, **137**, 2135–2139.

Arendt, E.K., Neve, H. and Hammes, W.P. (1991b) Characterization of phage isolates from a phage-carrying culture of *Leuconostoc oenos* 58N. *Appl. Microbiol. and Biotech.*, **34**(2), 220–224.

Atlan, D., Laloi, P. and Portalier, R. (1990) X-Prolyl-dipeptidyl aminopeptidase of *Lactobacillus delbrueckii* spp. *bulgaricus*: characterization of the enzyme and isolation of deficient mutants. *Appl. Environ. Microbiol.*, **56**, 2174–2179.

Axelsson, L.T., Ahrné, S.E.I., Andersson, M.C. and Stahl, S.R. (1988) Identification and cloning of a plasmid-encoded erythromycin resistance determinant from *Lactobacillus reuteri*. *Plasmid*, **20**, 171–174.

Baik, B.H. and Pack, M.Y. (1990) Expression of a *Bacillus subtilis* endoglucanase gene in *Lactobacillus acidophilus*. *Biotechnol. Lett.*, **12**, 330–334.

Bates, E.E. and Gilbert, H.J. (1989) Characterization of a cryptic plasmid from *Lactobacillus plantarum*. *Gene*, **85**, 253–258.

Bates, E.E.M., Gilbert, H.J., Hazlewood, G.P., Huckle, J., Laurie, J.I. and Mann, S.P. (1989) Expression of a *Clostridium thermocellum* endoglucanase gene in *Lactobacillus plantarum*. *Appl. Environ. Microbiol.*, **55**., 2095–2097.

Benbadis, L., Faelen, M., Slos, P., Fazel, A. and Mercenier, A. (1990) Characterization and comparison of virulent bacteriophages of *Streptococcus thermophilus* isolated from yogurt. *Biochimie*, **72**(12), 855–862.

Benbadis, L., Garel, J.-R. and Hartley, D.L. (1991) Purification, properties and sequence specificity of *Ssl*I, a new type II restriction endonuclease from *Streptococcus salivarius* subsp. *thermophilus*. *Appl. Environ. Microbiol*, **57**, 3677–3678.

Bently, R.W., Leigh, J.A. and Collins, M.D. (1991) Intergeneric structure of *Streptococcus* based on comparative analysis of small- subunit rRNA. *Int. J. System. Bacteriol.*, **41**, 487–494.

Boizet, B., Flickinger, J.L. and Chassy, B.M. (1988) Transfection of *Lactobacillus bulgaricus* protoplasts by phage DNA. *Appl. Environ. Microbiol.*, **54**, 3014–3018.

Boot, H.J., Kolen, C.P.A.M., Van Noort, J.H. and Pouwels, P.H. (1993) 5–layer protein of *Lactobacillus acidophilus* ATCC 4356: Purification, expression in *Eschericha coli* and nucleotide sequence of the corresponding gene. *J. Bacteriol.*, in press.

Bouia, A., Bringel, F., Frey, L., Kammerer, B., Belarbi, A., Goyonvarch, A., and Hubert, J.C. (1989) Structural organization of pLP–1, a cryptic plasmid from *Lactobacillus plantarum* CCM1904. *Plasmid*, **22**(3), 185–192.

Breidt, F.J. and Stewart, G.C. (1986) Cloning and expression of the gene phospho-β-galactosidase of *Staphylococcus aureus* in *Escherichia coli*. *J. Bacteriol.*, **166**, 1061–1066.

Breidt, F.J. and Stewart, G.C. (1987) Nucleotide and deduced amino acid sequences of the *Staphylococcus aureus* phospho-β-galactosidase gene. *Appl. Environ. Microbiol.*, **53**, 969–973.

Bringel, F., Frey, L. and Hubert, J.C. (1989) Characterization, cloning, curing, and distribution in lactic acid bacteria of pLP1, a plasmid from *Lactobacillus plantarum* CCM 1904 and its use in shuttle vector construction. *Plasmid*, **22**(3), 193–202.

Buvinger, W.E. and Riley, M. (1985) Nucleotide sequence of *Klebsiella pneumoniae lac* genes. *J. Bacteriol.*, **163**, 850–857.

Chang, B.-S., and Mahoney, R.R. (1989) Purification and thermostability of β-galactosidase (lactase) from an autolytic strain of *Streptococcus salivarius* subsp. *thermophilus*. *J. Dairy Res.*, **56**, 117–127.

Chassy, B.M. (1987) Prospects for the genetic manipulation of lactobacilli. *FEMS Microbiol. Rev.*, **46**, 297–312.

Chassy, B.M. and Alpert, C.A. (1989) Molecular characterization of the plasmid-encoded lactose-PTS of *Lactobacillus casei*. *FEMS Microbiol. Rev.*, **63**(1–2), 157–166.

Chassy, B.M. and Flickinger, J.L. (1987) Transformation of *Lactobacillus casei* by electroporation. *FEMS Microbiol. Lett.*, **44**, 173–177.

Chassy, B.M., Gibson, E.V. and Giuffrida, A. (1978) Evidence for plasmid-associated lactose metabolism in *Lactobacillus casei* subsp. *casei*. *Current Microbiol.*, **1**, 141–144.

Chassy, B.M., Mercenier, A. and Flickinger, J. (1988) Transformation of bacteria by electroporation. *Trends Biotechnol.*, **6**, 303–309.

Chassy, B.M. and Rokaw, E. (1981) Conjugal transfer of plasmid-associated lactose metabolism in *Lactobacillus casei* subsp. *casei*. In *Molecular Biology, Pathogenesis and Ecology of Bacterial Plasmids*, New York: Plenum Press, p 590.

Christiaens, H., Leer, R.J., Pouwels, P.H. and Verstraete, W. (1992) Cloning and expression of a conjugated bile acid hydrolase gene from *Lactobacillus plantarum* using a direct plate assay. *Appl. Environ. Microbiol.*, **58**, 3792–3798.

Claassen, E., Pouwels, P.H., Posno, H. and Boersha, W. (1993) Development of safe oral vaccines based on Lactobacillus as a vector. In: *Recombinant Vaccines: New Vaccinology* (ed. E. Kurstak) Int. Comp. Virology Org., Montreal, in press.

Cocconcelli, P.S., Gasson, M.J., Morelli, L. and Bottazzi, V. (1991) Single-stranded DNA plasmid, vector construction and cloning of *Bacillus stearothermophilus* α-amylase in *Lactobacillus*. *Research in Microbiol.*, **142**(6), 643–652.

Cocconcelli, P.S. Morelli, L., Vescovo, M. and Bottazzi, V. (1986) Intergeneric protoplast fusion in lactic acid bacteria. *FEMS Microbiol. Lett.*, **35**, 211–214.

Collins, M.D., Rodrigues, U., Ash, C., Aguirre, M., Farrow, J.A.E., Martinez-Murcia, A., Phillips, B.A., Williams, A.M. and Wallbanks, S. (1991) Phylogenetic analysis of the genus *Lactobacillus* and related lactic acid bacteria as determined by reverse transcriptase sequencing of 16S rRNA. *FEMS Microbiol. Lett.*, **77**(1), 5–12.

Copeland, W.C., Domena, J.D. and Robertus, J.D. (1989) The molecular cloning, sequence and expression of the *hdc* B gene from *Lactobacillus* 30A. *Gene*, **85**(1), 259–266.

Dao, M.L. and Ferretti, J.J. (1985) *Streptococcus-Escherichia coli* shuttle vector pSA3 and its use in the cloning of streptococcal genes. *Appl. Environ. Microbiol.*, **49**, 115–119.

David, S. (1992) *Genetics of mesophilic citrate fermenting lactic acid bacteria*. Doctoral Dissertation, Rijksuniversiteit Wageningen.

David, S., Simons, G. and de Vos, W.M. (1989) Plasmid transformation by electroporation of *Leuconostoc paramesenteroides* and its use in molecular cloning. *Appl. Environ. Microbiol.*, **55**, 1483–1489.

David, S., Stevens, H., van Riel, M., Simons, G. and de Vos, W.M. (1992) *Leuconostoc lactis* β-galactosidase is encoded by two overlapping genes. *J. Bacteriol.*, **174**, 4475–4481.

Davidson, B.E., Powell, I.B. and Hillier, A.J. (1990) Temperate bacteriophages and lysogeny in lactic acid bacteria. *FEMS Microbiol. Rev.*, **87**, (1–2), 79–90.

Davis, C., Silveira, N.F.A. and Fleet, G.H. (1985) Occurrence and properties of bacteriophages of *Leuconostoc oenos* in Australian wines. *Appl. Environ. Microbiol.*, **50**, 872–876.

De Los Reyes-Gavilan, C.G., Limsowtin, G.K.Y., Sechaud, L. and Veaux, M. (1990) Evidence for a plasmid-linked restriction-modification system in *Lactobacillus helveticus*. *Appl. Environ. Microbiol.*, **56**(11), 3412–3419.

De Rossi, E., Brigidi, P., Rossi, M., Matteuzzi, D. and Riccardi, G. (1991) Characterization of Gram-positive broad host-range plasmids carrying a thermophilic replicon. *Research in Microbiol.*, **142**(4), 389–396.

De Rossi, E. *et al.* (1989) Preliminary studies on the correlation between the plasmid pIHJ1 and its proteolytic activity in *Lactobacillus helveticus* S36.2. Physical mapping and molecular cloning of the plasmid in *Escherichia coli*. *Microbiologica*, **12**, 273–276.

De Vos, W.M. (1987) Gene cloning and expression in lactic streptococci. *FEMS Microbiol. Rev.*, **46**, 281–295.

De Vos, W.M., Boerrigter, I., Van Rooyen, R.J., Reiche, B. and Hengstenberg, W. (1990) Characterization of the lactose-specific enzymes of the phosphototransferase system in *Lactococcus lactis*. *J. Biol. Chem.*, **265**(36), 22554–22560.

De Vos, W.M., Simons, G. and David, S. (1989) Gene organization and expression in the mesophilic lactic acid bacteria. *J. Dairy Sci.*, **72**, 3398–3405.

Debarbouillé, M., Martin-Verstraete, I., Arnaud, M., Klier, A. and Rapoort, G. (1991) Positive and negative regulation controlling expression of the *sac* genes in *Bacillus subtilis*. *Res. Microbiol*, **142**, 757–764.

Dunny, G.H., Lee, L.N. and LeBlanc, D.J. (1991) Improved electroporation and cloning vector system for Gram-positive bacteria. *Appl. Environ. Microbiol.*, **57**, 1194–1201.

Duwat, P., Ehrlich, D.S. and Gruss, A. (1992). Use of degenerate primers for polymerase chain reaction cloning and sequencing of the *Lactococcus lactis* spp. *lactis rec*A gene. *Appl. Environ. Microbiol.*, **58**, 2674–2678.

Eggimann, B. and Bachmann, H. (1980) Purification and partial characterization of an aminopeptidase from *Lactobacillus lactis*. *Appl. Environ. Microbiol.*, **40**, 876–882.

Fantuzzi, L. (1991) Instability of lactose and citrate metabolism of *Leuconostoc* strains. *Biotechnol. Lett.*, **13**, 433–436.

Farrow, J.A.E. and Collins, M.D. (1984) DNA base composition, DNA-DNA homology and long-chain fatty acid studies on *Streptococcus thermophilus* and *Streptococcus salivarius*. *J. Gen. Microbiol.*, **130**, 357–362.

Fayard, B., Haefliger, M. and Accolas, J.P. (1993) Interaction of temperate bacteriophages of *Streptococcus salivarius* subsp. *thermophilus* with lysogenic indicators affect phage DNA restriction patterns and host ranges. *J. Dairy Sci.*, in press.

Fernandes, C.F., Shahani, K.M. and Ames, M.A. (1987) Therapeutic role of dietary lactobacilli and lactobacillic fermented dairy products. *FEMS Microbiol. Rev.*, **46**, 343–356.

Fitzgerald, G.F. and Gasson, M.J. (1988) *In vivo* gene transfer systems and transposons. *Biochimie*, **70**(4), 489–502.

Flickinger, J.L., Porter, E.V. and Chassy, B.M. (1986) Molecular cloning of a plasmid-encoded β-galactosidase from *Lactobacillus casei*. In *86th Ann. Meeting Amer. Soc. Microbiol.*, ASM Publications.

Fremaux, C. (1990) *Application de la biologie moléculaire à la connaissance des bactéries lactiques du vin – identification par sondes nucléiques – étude de plasmides de Leuconostoc oenos.* Doctoral Dissertation, University of Bordeaux, France.

Friedland, I.R., Snipelisky, M. and Khoosal, M. (1990) Meningitis in a neonate caused by *Leuconostoc* sp. *J. Clin. Microbiol*, **28**(9), 2125–2126.

Gaetje, G., Mueller, V. and Gorrschalk, G. (1991). Lactic acid excretion via carrier-mediated facilitated diffusion in *Lactobacillus helveticus. Applied Microbiology and Biotechnology*, **34**(6), 778–782.

Gaier, W. (1991) *Untersuchungen zur transformation und gene expression bei Laktobazillen.* Doctoral Dissertation, University of Hohenheim, Germany.

Gaier, W., Vogel, R.F. and Hammes, W.P. (1992) Cloning and expression of the lysostaphin gene in *Bacillus subtilis* and *Lactobacillus casei. Lett. Appl. Microbiol.*, **14**, 72–76.

Gasson, M.J. (1980) Production, regeneration and fusion of protoplasts in lactic streptococci. *FEMS Microbiol. Lett.*, **9**, 99–102.

Gasson, M.J. (1990) *In vivo* genetic systems in lactic acid bacteria. *FEMS Microbiol. Rev.*, **87**(1–2), 43–60.

Gibson, E.M., Chace, N.M., London, S.B. and London, J. (1979) Transfer of plasmid-mediated antibiotic resistance from streptococci to lactobacilli. *J. Bacteriol.*, **137**, 614–619.

Gilliland, S.E. (1990) Health and nutritional benefits from lactic acid bacteria. *FEMS Microbiol. Rev.*, **87**(1–2), 175–188.

Graves, M.C. and Rabinowitz, J.C. (1986). *In vivo* and *in vitro* transcription of the *Clostridium pasteurianum* ferredoxin gene. *J. Biol. Chem.*, **261**, 11409–11415.

Gruss, A. and Erchlich, D. (1989) The family of highly interrelated single-stranded deoxyribonucleic acid plasmids. *Microbiol. Rev.*, **53**, 231–241.

Hall, B.G., Betts, P.W. and Wootton, J.C. (1989) DNA sequence analysis of artificially evolved *ebg* enzyme and *ebg* repressor genes. *Genetics*, **123**, 635–648.

Hancock, K.R., Rockman, E., Young, C.A., Pearce, L., Maddox, I.S. and Scott, D.B. (1991). Expression and nucleotide sequence of the *Clostridium acetobutylicum* β-galactosidase gene cloned in *Escherichia coli. J. Bacteriol.*, **173**, 3084–3095.

Handwerger, S., Horowitz, H., Coburn, K., Kolokathis, A. and Wormser, G.P. (1990) Infection due to *Leuconostoc* sp. – 6 cases and review. *Rev. of Infect. Diseases*, **12**(4), 602–610.

Harlander, S. (1987) Gene transfer systems in lactic streptococci. In J.J. Ferretti and R.C. Curtiss (eds.), *Streptococcal Genetics* Washington, D.C.: ASM Publications, pp. 229–233.

Hastings, J.W. (1992). Cloning and nucleotide sequence of a *Leuconostoc* bacteriocin operon. In *92nd General Meeting of the American Society for Microbiology*, **92**, New Orleans, Louisiana, USA, Abstr. Gen. Meet. Am. Soc. Microbiol, p. 333.

Hastings, J.W., Sailer, M., Johnson, K., Roy, K.L., Vederas, J.C. and Stiles, M.E. (1991) Characterization of Leucocin A-Val 187 and cloning of the bacteriocin gene from *Leuconostoc gelidum. J. Bacteriol.*, **173**, 7491–7500.

Hengstenberg, W., Penberthy, W.K., Hill, K.L. and Morse, M.L. (1968) Metabolism of lactose by *Staphylococcus aureus. J. Bacteriol.*, **96**, 2187–2188.

Henick-Kling, T., Lee, T.H. and Nicholas, D.J.D. (1986) Inhibition of bacterial growth and malolactic fermentation in wine by bacteriophage. *J. Appl. Bacteriol.*, **86**, 287–293.

Higgins, C.F., Ames, G.F.L., Barnes, W.M., Clement, J.M. and Hofnug, M. (1982). A novel intercistronic regulatory element of prokaryotic operons. *Nature*, **298**, 760–762.

Hirata, H., Fukazawa, T., Negoro, S. and Okada, H. (1986). Structure of a β-galactosidase gene of *Bacillus stearothermophilus. J. Bacteriol.*, **166**, 722–727.

Holck, A. and Naes, H. (1992) Cloning, sequencing, and expression of the gene encoding the cell-envelope-associated proteinase from *Lactobacillus paracasei* spp. *paracasei* NCDO 151. *J. Gen. Microbiol.*, **138**, 1353–1364.

Hutkins, R.W. and Morris, H.A. (1987) Carbohydrate metabolism by *Streptococcus thermophilus*: a review. *J. Food Protect.*, **50**, 876–894.

Hutkins, R.W. and Ponne, C. (1991) Lactose uptake driven by galactose efflux in *Streptococcus thermophilus* – Evidence for a galactose-lactose antiporter. *Appl. Environ. Microbiol.*, **57**(4), 941–944.

Imam, S.H., Burgess-Cassler, A., Cote, G.L., Gordon, S.H. and Baker, F.L. (1991) A study of cornstarch granule digestion by an unusually high molecular weight α-amylase secreted by *Lactobacillus amylovorus. Curr. Microbiol.*, **22**(6), 365–370.

Ishiwa, H. and Iwata, S. (1980) Drug resistance plasmids in *Lactobacillus fermentum. J. Gen Appl. Microbiol.*, **26**, 71–74.

Iwata, M., Mada, M. and Ishiwa, H. (1986) Protoplast fusion of *Lactobacillus fermentum. Appl. Environ. Microbiol.*, **52**, 392–393.

Jacob, S., Allmansberger, R., Gartner, D. and Hillen, W. (1991) Catabolite repression of the operon for xylose utilization from *Bacillus subtilis* W23 is mediated at the level of transcription and depends on a *cis* site in the *xyl*A reading frame. *Mol. Gen. Genet.*, **229**, 189–196.

Jahns, A., Schafer, A., Geis, A. and Teuber, M. (1991) Identification, cloning and sequencing of the replication region of *Lactococcus lactis* spp. *lactis* biovar. *diacetylactis* Bu2 citrate plasmid pSL2. *FEMS Microbiol. Lett.*, **80**, 253–258.

Janzen, T., Kleinschmidt, J., Neve, H. and Geis, A. (1992) Sequencing and characterization of pST1 a cryptic plasmid from *Streptococcus thermophilus. FEMS Microbiol. Lett.*, **95**, 175–180.

Jarvis, A.W. (1989) Bacteriophages of lactic acid bacteria. *J. Dairy Sci.*, **72**, 3406–3428.

Jimeno, J., Casey, M. and Hofer, F. (1984) The occurrence of β-galactosidase and P-β-galactosidase in *Lactobacillus casei* strains. *FEMS Microbiol. Lett.*, **25**, 275–278.

Johansen, E. and Kibenich, A. (1992) Isolation and characterization of IS*1165* an insertion sequence of *Leusonostoc mesenteroides* spp. *cremoris* and other lactic acid bacteria. *Plasmid*, **27**, 200–206.

Jones, S. and Warner, P.J. (1990) Cloning and expression of a-amylase from *Bacillus amyloliquefaciens* in a stable plasmid vector in *Lactobacillus plantarum. Lett. in Appl. Microbiol.*, **11**(4), 214–219.

Josson, K., Soetaert, P., Michiels, F., Joos, H. and Mahillon, J. (1990) *Lactobacillus hilgardii* plasmid pLAB1000 consists of two functional cassettes commonly found in other Gram-positive organisms. *J. of Bacteriol.*, **172**(6), 3089–3099.

Kaklij, G.S., Kelkar, S.M., Shenoy, M.A. and Sainis, K.B. (1991) Antitumor activity of *Streptococcus thermophilus* against fibrosarcoms: Role of T-cells. *Cancer Lett.*, **56**(1), 37–44.

Kalnins, A., Otto, K., Ruether, U. and Mueller-Hill, B. (1983) Sequence of the *lac*Z gene of *Escherichia coli. EMBO J*, **2**, 593–597.

Kaminogawa, S., Ninomiya, T. and Yamauchi (1984) Aminopeptidase profiles of lactic streptococci. *J. Dairy Sci.*, **67**, 2483–2492.

Kandler, O. (1983) Carbohydrate metabolism in lactic acid bacteria. *Antonie van Leeuwenhoek*, **49**, 209–224.

Kandler, O. and Weiss, N. (1986) Bergey's Manual of Systematic Bacteriology. In P.H.A. Sneath, N.S. Mair, M.E. Sharpe, and J.G. Holt (eds.), Baltimore: Williams & Wilkins, pp. 1208–34.

Khalid, N.M. and Marth, E.M. (1990) Purification and partial characterization of a prolyl-dipeptidyl aminopeptidase farm *Lactobacillus helveticus* CNRZ 32. *Appl. Environ. Microbiol.*, **56**, 381–388.

Kitao, S. and Nakano, E. (1992) Cloning of the sucrose phosphorylase gene from *Leuconostoc mesenteroides* and its overexpression using a 'sleeper' phage. *J. Ferment. and Bioeng.*, **73**, 179–184.

Klaenhammer, T.R. (1987) Plasmid-directed mechanisms for bacteriphage defense in lactic streptococci. *FEMS Micro. Rev.*, **46**, 313–325.

Klaenhammer, T.R. (1988) Bacteriocins of lactic acid bacteria. *Biochimie*, **70**, 303–316.

Klebanoff, S.J. and Coombs, R.W. (1991) Viricidal effect of *Lactobacillus acidophilus* on human immunodeficiency virus type 1: Possible role in heterosexual transmission. *J. Exp. Med.*, **174**(1), 289–292.

Kojic, M., Fira, D., Banina, A. and Topisirovic, L. (1991) Characterization of the cell wall-bound proteinase of *Lactobacillus casei* HN14. *Appl. and Environ. Microbiol.*, **57**(6), 1753–1757.

Kok, J. (1990) Genetics of the proteolytic system of the lactic acid bacteria. *FEMS Microbiol. Rev.*, **87**(1–2), 15–42.

Kok, J. (1991) Special-purpose cloning vectors for lactococci. In G.M. Dunny, P.P. Cleary, and L.L. McKay (eds.), *Genetics and Molecular Biology of Streptococci, Lactococci and Enterococci*, Washington, D.C.: ASM Publications, pp. 97–102.

Kok, J., van der Vossen, J.M.B.M. and Venema, G. (1984) Construction of plasmid cloning vectors for lactic streptococci which also replicate in *Bacillus subtilis* and *Escherichia coli. Appl Environ. Microbiol.*, **48**, 726–731.

Kok, J. and Venema, G. (1988) Genetics of proteinases of lactic acid bacteria. *Biochimie*, **70**, 475–488.

Kondo, J.K. and McKay, L.L. (1982) Transformation of *Streptococcus lactis* protoplasts by plasmid DNA. *Appl. Environ. Microbiol.*, **43**, 1213–1215.

Kreuzer, P., Gratner, D., Allmansberger, R. and Hillen, W. (1989) Identification and sequence analysis of the *Bacillus subtilis xyl*R gene and *xyl* operator. *J. Bacteriol.*, **171**, 3840–3845.

Lahbib-Mansais, Y., Boizet, B., Dupont, L., Mata, M. and Ritzenthaler, P. (1992) Characterization of a temperate bacteriophage of *Lactobacillus delbrueckii* spp. *bulgaricus* and its interactions with the host cell chromosome. *J. Gen. Microbiol.*, **138**, 1139–1146.

Larbi, D., Decaris, B. and Simonet, J.M. (1992) Different bacteriophage resistance mechanisms in *Streptococcus salivarius* subsp. *thermophilus*. *J. Dairy Res.*, **59**, 349–357.

Le Bourgeois, P., Lautier, M. and Ritzenthaler, P. (1993) Chromosome mapping in Lactic Acid Bacteria. *FEMS Microbiol. Rev.*, **12**, (1–3), 109–123.

Lee, W.T., Flynn, T.G., Lyons, C., and Levy, H.R. (1991) Cloning of the gene and amino acid sequence for glucose 6-phosphate dehydrogenase from *Leuconostoc mesenteroides*. *J. Biol. Chem.*, **266**, 13028–13034.

Leenhouts, K.J., Tolner, B., Bron, S., Kok, J., Venema, G. and Seegers, J. (1991) Nucleotide sequence and characterization of the broad-host-range lactococcal plasmid pWV01. *Plasmid*, **26**, 55–56.

Leer, R., Christiaens, H., Peters, L., Posno, M. and Pouwels, P. (1993) Gene-disruption in *Lactobacillus plantarum* strain 80 by site-specific recombination: isolation of a mutant strain deficient in conjugated bile salt hydrolase activity. *Mol. Gen. Genet.*, in press.

Leer, R.J., Van Luijk, N., Posno, M. and Pouwels, P.H. (1992) Structural and functional analysis of two cryptic plasmids from *Lactobacillus pentosus* MD353 and *Lactobacillus plantarun* ATCC 8014. *Mol. Gen. Genet.*, **234**, 265–274.

Leong-Morgenthaler, P., Zwahlen, M.C. and Hottinger, H. (1991) Lactose metabolism in *Lactobacillus bulgaricus*: Analysis of the primary structure and expression of the genes involved. *J. of Bacteriol.*, **173**(6), 1951–1957.

Lerch, H.-P., Frank, R. and Collins, J. (1989) Cloning, sequencing and expression of the L-2-hydroxyisocaproate dehydrogenase-encoding gene of *Lactobacillus confusus* in *Escherichia coli*. *Gene*, **83**, 263–270.

Lerch, H.P., Blocker, H., Kallwas, H., Hoppe, J., Tsai, H. and Collins, J. (1989b) Cloning, sequencing and expression in *Escherichia coli* of the D-2-hydroxyisocaproate dehydrogenase gene of *Lactobacillus casei*. *Gene*, **78**, 47–57.

Lin, J., Schmitt, P. and Divies, C. (1991) Characterization of a citrate-negative mutant of *Leuconostoc mesenteroides* spp. *mesenteroides*: Metabolic and plasmidic properties. *Appl. Microbiol. and Biotech.*, **34**(5), 628–631.

Lin, J.H.-C. and Savage, D. C. (1986) Genetic transformation of rifampicin resistance in *Lactobacillus acidophilus*. *J. Gen. Microbiol.*, **132**, 2107–2111.

Liu, M.L., Kondo, J.K., Barnes, M. B. and Bartholomeu, D.T. (1988) Plasmid-linked maltose utilization in *Lactobacillus* spp. *Biochimie*, **70**, 351–355.

Lokman, B.C., van Santen, P., Verdoes, J.C., Kruse, J., Leer, J.R., Posno, M. and Pouwels, P.H. (1991) Organization and characterization of three genes involved in D-xylose catabolism in *Lactobacillus pentosus*. *Mol. Gen. Genet.*, **230**, 161–169.

Luchansky, J.B., Kleeman, E.G., Raya, R.R. and Klaenhammer, T.R. (1989) Genetic transfer systems for delivery of plasmid DNA, deoxyribonucleic acid, to *Lactobacilus acidophilus* ADH: Conjugation, electroporation, and transduction. *J. of Dairy Sci.*, **72**(6), 1408–1417.

Luchansky, J.B., Muriana, P.M. and Klaenhammer, T.R. (1988) Application of electroporation for transfer of plasmid DNA to *Lactobacillus, Lactococcus, Leuconostoc, Listeria, Pediococcus, Bacillus, Staphylococcus, Enterococcus* and *Propionibacterium*. *Mol. Microbiol.*, **2**, 637–646.

Machuga, E.J. and Ives, D.H. (1984) Isolation and characterization of an aminopeptidase from *Lactobacillus acidophilus* R–26. *Biochem. Biophys. Acta*, **789**, 26–36.

Macrina, F.L., Tobian, J.A., Jones, K.R., Evans, R.P. and Clewell, D.B. (1982) A cloning vector able to replicate in *Escherichia coli* and *Streptococcus sanguis*. *Gene*, **19**, 345–353.

Martinez-Murcia, A.J. and Collins, M.D. (1990b) Nucleotide sequence of 16S ribosomal RNA from *Lactobacillus kandleri* and *Lactobacillus minor*. *Nucl. Acid Res.*, **18**, 3401.

Martinez-Murcia, A.J. and Collins, M.D. (1990) Nucleotide sequence of 16S ribosomal RNA from *Lactobacillus viridescens* and *Lactobacillus confusus*. *Nucl. Acid Res.*, **18**, 3402.

Mata, M. and Ritzenthaler, P. (1988) Present state of lactic acid bacteria phage taxonomy. *Biochimie*, **70**, 395–400.

McKay, L.L. (1982) Regulation of lactose metabolism in dairy streptococci. In R. Davies *Developments in Food Microbiology*, Essex, Applied Science Publishers Ltd., pp. 153–182.

McKay, L.L. (1983). Functional properties of plasmids in lactic streptococci. *Antonie van Leeuwenhoek J. Microbiol.*, **49**, 259–274.

McKay, L.L. and Baldwin, K. (1978) Stabilization of lactose metabolism in *Streptococcus lactis* C2. *Appl. Environ. Microbiol.*, **36**, 360–367.

McKay, L.L. and Baldwin, K.A. (1990) Applications for biotechnology: present and future improvements in lactic acid bacteria. *FEMS Microbiol. Rev.*, **87**(1–2), 3–14.

Mercenier, A. (1990) Molecular genetics of *Streptococcus thermophilus*. *FEMS Microbiol. Rev.*, **87**(1–2), 61–78.

Mercenier, A. and Chassy, B.M. (1988) Strategies for the development of bacterial transformation systems. *Biochimie*, **70**, 503–517.

Mercenier, A. and Lemoine, Y. (1989) Genetics of *Streptococcus thermophilus*: A review. *J. Dairy Sci.*, **72**, 3444–3454.

Mercenier, A. Robert, C., Romero, D.A., Castellino, I., Slos, P. and Lemoine, Y. (1989) Development of an efficient spheroplast transformation procedure for *Streptococcus thermophilus*: the use of transfection to define a regeneration medium. *Biochimie*, **70**, 567–577.

Mercenier, A., Robert, C., Romero, D.A., Slos, P. and Lemoine, Y. (1987) In J.J. Ferretti & R.C. Curtiss (eds.), *Streptococcal Genetics*, Washington, D.C., ASM Publications, pp. 234–239.

Mercenier, A., Slos, P., Faelen, M. and Lecocq, J.-P. (1988) Plasmid transduction in *Streptococcus thermophilus*. *Molec. Gen. Genet.*, **212**, 386–389.

Meyer, J. and Jordi, R. (1987) Purification and characterization of X-prolyl-dipeptidyl-aminopeptidase from *Lactobacillus lactis* and from *Streptococcus thermophilus*. *J. Dairy Sci.*, **70**, 738–745.

Mills, O.E. and Thomas, T.D. (1981). Nitrogen sources for growth of lactic streptococci in milk. *N.Z.J. Dairy Sci. Technol.*, **15**, 43–55.

Mollet, B. and Delley, M. (1990) Spontaneous deletion formation within the β-galactosidase gene of *Lactobacillus bulgaricus*. *J. Bacteriol.*, **172**, 5670–5676.

Mollet, B. and Delley, M. (1991) A β-galactosidase deletion mutant of *Lactobacillus bulgaricus* reverts to generate an active enzyme by internal DNA sequence duplication. *Molec. Gen. Genet.*, **227**(1), 17–21.

Mollet, B., Knol, J., Poolman, B., Marciset, O. and Delley, M. (1993) Directed genomic integration, gene replacement and integrative gene expression in *Streptococcus thermophilus*. *J. Bacteriol.*, **175**, 4315–4324.

Moran Jr., C.P., Lang, N., LeGrice, S.F.J., Lee, G., Stephens, M., Soneshein, A.L., Pero, J. and Losick, R. (1982) Nucleotide sequences that signal the initiation of transcription and translation in *Bacillus subtilis*. *Mol. Gen. Genet.*, **186**, 339–346.

Morelli, L., Cocconcelli, P.S., Bottazzi, V., Damiani, G., Ferretti, L. and Sgaramella, V. (1987). *Lactobacillus* protoplast transformation. *Plasmid*, **17**, 73–79.

Morelli, L., Vescovo, M. and Bottazzi, V. (1983) Identification of chloramphenicol resistance plasmids in *Lactobacillus reuteri* and *Lactobacillus acidophilus*. *Int. J. Microbiol.*, **1**, 1–5.

Morse, M.L., Hill, K.L., Egan, J.B. and Hengstenberg, W. (1968) Metabolism of lactose by *Staphylococcus aureus* and its genetic basis. *J. Bacteriol.*, **95**, 2270–2274.

Mortvedt, C.I. and Nes, I.F. (1990) Plasmid-associated bacteriocin production by a *Lactobacillus sake* strain. *J. of Gen. Microbiol.*, **136**(8), 1601–1608.

Muriana, P. and Klaenhammer, T.R. (1987) Conjugal transfer of plasmid-encoded determinants for bacteriocin production and immunity in *Lactobacillus acidophilus* 88. *Appl. Environ. Microbiol.*, **53.**, 553–560.

Muriana, P.M. and Klaenhammer, T.R. (1991) Cloning, phenotypic expression, and DNA sequence of the gene for Lactocin F, an antimicrobial peptide produced by *Lactobacillus* spp. *J. Bacteriol.*, **173**(5), 1779–1788.

Naes, H. and Nissen-Meyer, J. (1992) Purification and amino-terminal amino acid sequence determination of the cell-wall-bound proteinase from *Lactobacillus paracasei* spp. *paracasei*. *J. Gen. Microbiol.*, **138**, 313–318.

Nakamura, L.K. (1981) *Lactobacillus amylovorus*, a new starch-hydrolyzing species from cattle waste-corn fermentations. *Int. J. Syst. Bacteriol.*, **31**(1), 56–63.

Nakamura, L.K. and Crowell, C.D. (1979) *Lactobacillus amylophilus*, a new starch-hydrolyzing species from swine waste-corn fermentation. *Dev. Indust. Microbiol.*, **20**, 531–540.

Natori, Y., Kano, Y. and Imamoto, F. (1990) Nucleotide sequences and genomic constitution of five tryptophan genes of *Lactobacillus casei*. *J. Biochem. (Tokyo)*, **107**(2), 248–255.

Nel, L., Wingfield, D., Van der Meer, L.J. and Van Vuuren, H.J.J. (1987) Isolation and characterization of *Leuconostoc oenos* bacteriophages from wine and sugarcane. *FEMS Micro. Lett.*, **44**, 63–67.

Neve, H., Krush, U. and Teuber, M. (1990) Virulent and temperate bacteriophages of thermophilic lactic acid streptococci. *FEMS Microbiol.*, **87**, P58.

O'Sullivan, T. and Daly, C. (1982) Plasmid DNA in Leuconostoc species. *Irish J. Food Sci. Technol.*, **6**, 206.

Oppenheim, D.S. and Yanofsky, C. (1980) Transitional coupling during expression of the tryptophan operon of *Escherichia coli*. *Genetics.*, **5**, 785–795.

Orberg, P.K. and Sandine, W.E. (1984). Common occurrence of plasmid DNA and vancomycin resistance in *Leuconostoc* spp. *Appl. Environ. Microbiol.*, **48**, 1129–1133.

Oskouian, B. and Stewart, G.C. (1990) Expression and catabolite repression of the lactose operon of *Staphylococcus aureus.*, *J. Bacteriol.*, **172**, 3804–3812.

Park, S.F. and Stewart, G.S.A.B. (1990) High efficiency transformation of *Listeria monocytogenes* by electroporation of penicillin-treated cells. *Gene*, **94**, 129–132.

Platt, T. (1986) Transcription termination and the regulation of gene expression. *Ann. Rev. Biochem.*, **55**, 339–372.

Poolman, B., Royer, T.J., Mainzer, S.E. and Schmidt, B.F. (1989) Lactose transport system of *Streptococcus thermophilus*: A hybrid protein with homology to the melibiose carrier and enzyme III of phosphoenolpyruvate-dependent phosphotransferase systems. *J. Bacteriol.*, **171**(1), 244–253.

Poolman, B., Royer, T.J., Mainzer, S.E. and Schmidt, B.F. (1990) Carbohydrate utilization in *Streptococcus thermophilus*: characterization of the genes for aldolase–1-epimerase (mutarotase) and UDP-glucose 4-epimerase. *J. Bacteriol.*, **172**, 4037–4047.

Porter, E.V. and Chassy, B.M. (1988) Nucleotide sequence of the β-D-phospho-galactoside galactohydrolase gene of *Lactobacillus casei*: comparison to analogous *pbg* genes of other Gram-positive organisms. *Gene*, **62**, 263–276.

Posno, M., Heuvelmans, P.T.H.M., Van Giezen, M.J.F., Lokman, B.C., Leer, R.J. and Pouwels, P.H. (1991) Complementation of the inability of *Lactobacillus* strains to utilize D-xylose with D-xylose catabolism-encoding genes of *Lactobacillus pentosus*. *Appl. Environ. Microbiol.*, **57**, 2764–2766.

Posno, M., Leer, R.J., VanLuijk, N., VanGeizen, M.J.F. and Heuvelmans, P.T.H.M. (1991b). Incompatibility of *Lactobacillus* vectors with replicons derived from small cryptic *Lactobacillus* plasmids and segregational instability of the introduced vectors. *Appl. Environ. Microbiol.*, **57**(6), 1822–1828.

Pouwels, P.H., Leer, R.J. and Posno, M. (1992) Genetic modification of *Lactobacillus*: A new approach toward strain improvement. In *Actes du Colloque Lactic 91*, pp. 133–148.

Pouwels, P.H., Vanluijk, N., Leer, R.J. and Posno, M. (1993) Control of replication of the *Lactobacillus pentosus* plasmid p353–2: Evidence for a mechanism involving transcriptions attenuation of the gene coding for the replication protein. *Molec. Gen. Genet.*, in press.

Powell, I.B., Achen, M.G., Hillier, A.J. and Davidson, B.E. (1988) A simple and rapid method for genetic transformation of lactic streptococci by electroporation. *Appl. Environ. Microbiol.*, **54**, 655–660.

Pucci, M.J., Monteschio, M.E. and Kemker, C.L. (1988) Intergeneric and intrageneric conjugal transfer of plasmid-encoded antibiotic resistance determinants in *Leuconostoc* spp. *Appl. Environ. Microbiol.*, **54**, 281–287.

Raya, R.R., Fremaux, C., de Antoni, G.L. and Klaenhammer, T.R. (1992) Site-specific integration of the temperate bacteriophage φADH into the *Lactobacillus gasseri* chromosome and molecular characterization of the phage *att*P and bacterial *att*B attachment sites. *J. Bacteriol.*, **174**, 5584–5592.

Raya, R.R. and Klaenhammer, T.R. (1992) High-frequency plasmid transduction by *Lactobacillus gasseri* bacteriophage PHI-ADH. *Appl. Environ. Microbiol.*, **58**, 187–193.

Raya, R.R., Kleeman, E.G., Luchansky, J.B. and Klaenhammer, T.R. (1989) Characterization of the temperate bacteriphage φADH and plasmid transduction in *Lactobacillus acidophilus* ADH. *Appl. Environ. Microbiol.*, **55**(9), 2206–2213.

Redondo-Lopez, V. (1990) Emerging role of lactobacilli in the control and maintenance of vaginal bacterial microflora. *Rev. Infec. Dis.*, **12**, 856–872.

Reid, G., Bruce, A., McGroarty, J.A., Cheng, K.-J. and Costerton, J.W. (1990). Is there a role for lactobacilli in prevention of urogenital and intestinal infections? *Clin. Microbiol. Rev.*, **3**, 335–344.

Reizer, J., Reizer, A. and Saier, M.H. (1990) The cellobiose permease of *Escherichia coli* consists of three proteins and is homologous to the lactose permease of *Staphylococcus aureus*. *Res. in Microbiol.*, **141**, 1061–1067.

Rinckel, L.A. and Savage, D.C. (1990) Characterization of plasmids and plasmid-borne macrolide resistance from *Lactobacilus* sp. strain 100–33. *Plasmid*, **23**, 119–125.

Rixon, J.E. Hazlewood, G.P. and Gilbert, H.J. (1990) Integration of an unstable plasmid into the chromosome of *Lactobacillus plantarum*. *FEMS Microbiol. Lett.*, **71**(1–2), 105–110.

Romero, D.A., Slos, P., Robert, C., Castellino, I. and Mercenier, A. (1987). Conjugative mobilization as an alternative vector delivery system for lactic streptococci. *Appl. Environ. Microbiol.*, **53**, 2405–2413.

Roussel, Y., Colmin, C., Simonet, J.M. and Decaris, B. (1993) Strain characterization, genome size and plasmid content in the *Lactobacillus acidophilus* group (Hansen and Marquot). *J. Appl. Bacteriol.*, **74**, 549–556.

Rygus, T. and Hillen, W. (1992) Catabolite repression of the *xyl* operon of *Bacillus megaterium*. *J. Bacteriol.*, **174**, 3049–3055.

Sanders, M.E. (1988) Phage resistance in lactic acid bacteria. *Biochimie*, **70**, 411–422.

Sarra, *et al.* (1989) Antagonism and adhesion among isogenic strains of *Lactobacillus reuteri* in the caecum of gnotobiotic mice. *Microbiologica*, **12**, 69–74.

Scheirlinck, T., Mahillon, J., Joos, H., Dhaese, P. and Michiels, F. (1989) Integration and expression of a-amylase and endoglucanase genes in the *Lactobacillus plantarum* chromosome. *Appl. and Environ. Microbiol.*, **55**(9), 2130–2137.

Scheler, A., Rygus, T., Allmansberger, R. and Hillen, W. (1991) Molecular cloning, structure, promoters, and regulatory elements for transcription of the *Bacillus licheniformis* encoded regulon for xylose utilization. *Arch. Microbiol.*, **155**, 526–534.

Schillinger, U. and Lücke, F.K. (1989) Antimicrobial activity of *Lactobacillus sake* isolated from meat. *Appl. Environ. Microbiol.*, **55**, 1901–1906.

Schleifer, K.H. (1987) Recent changes in the taxonomy of lactic acid bacteria. *FEMS Microbiol. Rev.*, **46**, 201–203.

Schleifer, K.H., Ludwig, W., Amman, R., Heitel, C., Ehrmann, M., Köhler, W. and Krause, A. (1992) Phylogenetic relationships of lactic acid bacteria and their identification with nucleic acid probes. In *Lactic 91*, Caen: Centre de Publications de l'Universite de Caen, pp. 23–32.

Schleifer, K.L., Ehrmann, M., Krush, U. and Neve, H. (1991) Revival of the Species of *Streptococcus thermophilus* (ex Orla-Jensen, 1919). *Syst. Appl. Microbiol.*, **14**, 386–388.

Schmidt, B.F., Adams, R.M., Requadt, C., Power, S. and Mainzer, S.E. (1989) Expression and nucleotide sequence of the *Lactobacillus bulgaricus* β-galactosidase gene cloned in *Escherichia coli*. *J. Bacteriol.*, **171**(2), 625–635.

Schroeder, C.J., Robert, C., Lenzen, G., McKay, L.L. and Mercenier, A. (1991) Analysis of the *lacZ* sequences from 2 *Streptococcus thermophilus* strains – comparison with the *Escherichia coli* and *Lactobacillus bulgaricus* β-galactosidase sequences. *J. Gen. Microbiol.*, **137**, 369–380.

Sechaud, L. (1990) *Caractérisation de 35 bactériophages de Lactobacillus helveticus*. Doctoral Dissertation, INRA, Jouy en Joses, France.

Sechaud, L., Cluzel, J.-P., Rousseau, M., Baumgartner, A. and Accolas, J.-P. (1988) Bacteriophages of lactobacilli. *Biochimie*, **70**, 401–410.

Shahbal, S., Hemme, D. and Desmazeaud, J. (1991) High cell wall-associated proteinase activity of some *Streptococcus thermophilus* strains (H-strains) correlated with a high acidification rate in milk. *Le Lait*, **71**, 351–357.

Shahbal, S., Hemme, D. and Renault, P. (1993) Characterization of a cell envelope-associated proteinase activity of *Streptococcus thermophilus* H-strains. *Appl. Environ. Microbiol.*, **59**, 177–182.

Shankar, P.A. and Davies, F.L. (1977) Amino acid peptide utilization by *Streptococcus thermophilus* in relation to yogurt manufacture. *J. Appl. Bacteriol.*, **43**, 8–13.

Shay, B.J., Egan, A., Wright, M. and Rogers, P. (1988) Cysteine metabolism in an isolate of *Lactobacillus sake*: plasmid composition and cysteine transport. *FEMS Microbiol. Lett.*, **56**, 183–188.

Sherman, J.M. (1937) The streptococci. *Bacteriol. Rev.*, **1**, 3–97.

Shimizu-Kadota, M. (1987) Properties of lactose plasmid pLY101 in *Lactobacillus casei*. *Appl. Environ. Microbiol.*, **53**, 2987–2991.

Shimizu-Kadota, M., Flickinger, J.L. and Chassy, B.M. (1988) Evidence that *Lactobacillus casei* insertion element ISLI has a narrow host range. *J. Bacteriol.*, **170**(10), 4976–4978.

Shimizu-Kadota, M., Kiwaki, M., Hirokawa, H. and Tsuchida, N. (1985) ISL1: a new transposable element in *Lactobacillus casei*. *Mol. Gen. Genet.*, **200**, 193–198.

Shimizu-Kadota, M. and Kudo, S. (1984) Liposome-mediated transfection of *Lactobacillus case* spheroplasts. *Agric. Biol. Chem.*, **48**, 1105–1107.

Shimizu-Kadota, M., Shibahara-Sone, H. and Ishiwa, H. (1991) Shuttle plasmid vectors for *Lactobacillus casei* and *Escherichia coli* with a minus origin. *Appl. Environ. Microbiol.*, **57**, 3292–3300.

Shine, J. and Dalgarno, L. (1974) The 3'-terminal sequence of *Escherichia coli* 16S RNA: complementarity to non-sense triplets and ribosome binding sites. *Proc. Natl. Acad. Sci USA*, **71**, 5463–5467.

Shrago, A.W., Chassy, B.M. and Dobrogosz, W.J. (1986) Conjugal plasmid transfer (pAMβ1) in *Lactobacillus plantarum*. *Appl. Environ. Microbiol.*, **52**, 574–576.

Shrago, A.W. and Dobrogosz, W.J. (1988) Conjugal transfer of group B streptococcal plasmids and comobilization of *Escherichia coli-Streptococcus* shuttle plasmids to *Lactobacillus plantarum*. *Appl. Environ. Microbiol.*, **54**, 824–826.

Sibakov, M., Koivula, T., Von Wright, A. and Palva, I. (1991) Secretion of TEM β-lactamase with signal sequences isolated form the chromosome of *Lactococcus lactis* spp. *lactis*. *Appl. and Environ. Microbiol.*, **57**(2), 341–348.

Simon, D. and Chopin, A. (1988) Construction of a vector plasmid family and its use for molecular cloning in *Streptococcus lactis*. *Biochimie*. **70**, 559–566.

Skaugen, M. (1989) The complete nucleotide sequence of a small cryptic plasmid from *Lactobacillus plantarum*. *Plasmid.*, **22**, 175–179.

Slos, P., Bourquin, J.C., Lemoine, Y. and Mercenier, A. (1991) Isolation and characterization of chromosomal promoters of *Streptococcus salivarius* spp. *thermophilus*. *Appl. Environ. Microbiol.*, **57**(5), 1333–1339.

Smiley, M.B. and Fryder, V. (1978) Plasmids, lactic acid production, and N-acetyl-D-glucosamine fermentation in *Lactobacillus helveticus* Subsp. *jugurti*. *Appl. Environ. Microbiol.*, **35**, 777–781.

Smith, M.D. and Clewell, D.B. (1984) Return of *Streptococcus faecalis* DNA cloned in *Escherichia coli* to its original host via transformation of *Streptococcus sanguis* followed by conjugative mobilization. *J. Bacteriol.*, **160**, 1109–1114.

Solaiman, D.K.Y. and Somkuti, G.A. (1990) Isolation and characterization of a type II restriction endonuclease from *Streptococcus thermophilus*. *FEMS Micro. Lett.*, **67**, 261–266.

Solaiman, D.K.Y. and Somkuti, G.A. (1991) A type-II restriction endonuclease of *Streptococcus thermophilus*. *FEMS Microbiol. Lett.*, **80**(1), 75–80.

Solaiman, D.K.Y. and Somkuti, G.A. (1992). DNA structures contributing to the instability of recombinant plasmids in *Streptococcus thermophilus*. *Biotechnol. Lett.*, **14**, 753–758.

Somkuti, G.A., Solaiman, D.K.Y. and Steinberg, D.H. (1992) Expression of *Streptomyces* sp. cholesterol oxidase in *Lactobacillus casei*. *Appl. Microbiol. Biotechnol.*, **37**, 330–334.

Somkuti, G.A. and Steinberg, D.H. (1988) Genetic transformation of *Streptococcus thermophilus* by electroporation. *Biochimie*, **70**, 503–517.

Somkuti, G.A. and Steinberg, D.H. (1991) DNA-DNA hybridization analysis of *Streptococcus thermophilus* plasmids. *FEMS Microbiol. Lett.*, **78**, 271–276.

Sozzi, T., Maret, R. and Poulin, J.M. (1976). Mise en évidence de bactériophage dans le vin. *Experientia*, **32**, 568–569.

Stackebrandt, E., Fowler, V.J. and Woese, C.R. (1983) A phylogenetic analysis of lactobacilli, *Pediococcus pentosaceus* and *Leuconostoc mesenteroides*. *Syst. Appl. Microbiol.*, **4**, 326–337.

Stackenbrandt, E. and Teuber, M. (1988) Molecular taxonomy and phylogenetic position of lactic acid bacteria. *Biochimie*, **70**, 317–324.

Swenson, J.M., Facklam, R.R. and Thornsberry, C. (1990) Antimicrobial susceptibility of vancomycin-resistant *Leuconostoc*, *Pediococcus* and *Lactobacillus* species. *Antimicrobial Agents and Chemotherapy*, **34**(4), 543–549.

Takiguchi, R., Hashiba, H., Aoyama, K. and Ishii, S. (1989) Complete nucleotide sequence and characterization of a cryptic plasmid from *Lactobacillus helveticus* spp. *jugurti*. *Appl. Environ. Microbiol.*, **55**(6), 1653–1655.

Tannock, G.W. (1987) Conjugal transfer of plasmid pAMβ1 in *Lactobacillus reuteri* and between lactobacilli and *Enterococcus faecalis*. *Appl. Environ. Microbiol*, **53**, 2693–2695.

Thomas, T.D. and Pritchard, G.G. (1987). Proteolytic enzymes of dairy starter cultures. *FEMS Microbiol Rev.*, **46**, 245–268.

Thompson, J. (1987) Regulation of sugar transport and metabolism in lactic acid bacteria. *FEMS Microbiol. Rev.*, **46**, 221–231.

Thompson, J.K. and Collins, M.A. (1989) Evidence for the conjugal transfer of a plasmid pVA797: :pSA3 co-integrate into strains of *Lactobacillus helveticus*. *Lett. Appl. Microbiol.*, **9**(2), 61–64.

Thompson, K. and Collins, M. (1991) Molecular cloning in *Lactobacillus helveticus* by plasmid pSA3::pVA797 co-integrate formation and conjugal transfer. *Appl. Microbiol. and Biotech.*, **35**(3), 334–338.

Tohyama, J.B., Sakurai, T. and Arai, H. (1971) Transduction by temperate phage PLS–1 in *Lactobacillus salivarius*. *Jpn. J. Bacteriol.*, **26**, 482–489.

Toy, J. and Bognar, A.L. (1990) Cloning and expression of the gene encoding *Lactobacillus casei* Folylpoly-γ-glutamate synthetase in *Escherichia coli* and determination of its primary structure. *J. Biol. Chem.*, **265**(5), 2492–2499.

Trevors, J.T., Chassy, B.M., Dower, W.J. and Blaschek, H.P. (1992) Electrotransformation of bacteria by plasmid DNA. In D.C. Chang, B.M. Chassy, J.A. Saunders, & A.E. Sowers (eds), *Guide to Electroporation and Electrofusion* San Diego, Academic Press, pp. 265–290.

Tsai, H.J. and Sandine, W.E. (1987) Conjugal transfer of lactose-fermenting ability from *Streptococcus lactis* C2 to *Leuconostoc cremoris* CAF7 yields leuconostoc that ferment lactose and produce diacetyl. *J. Indust. Microbiol.*, **2**, 25–33.

Tsai, H.J. and Sandine, W.E. (1987b) Conjugal transfer of nisin plasmid genes from *Streptococcus lactis* 7962 to *Leuconostoc dextranicum* 181. *Appl. Environ. Microbiol.*, **53**, 352–357.

Van der Lelie, D., Van der Vossen, J.M.B.M. and Venema, G. (1988) Effect of plasmid incompatibility on DNA transfer to *Streptococcus cremoris*. *Appl. Environ. Microbiol*, **54.**, 865–871.

Van Rooijen, R.J. and de Vos, W.M. (1991) Molecular cloning, transcriptional analysis, and nucleotide sequence of *lac*R, a gene encoding the repressor of the lactose phosphotransferase system of *Lactococcus lactis*. *J. of Biol. Chem.*, **265**(30), 18499–18503.

Van Rooijen, R.J., Van Schalkwijk, S. and de Vos, W.M. (1991) Molecular cloning, characterization, and nucleotide sequence of the tagatose–6-phosphate pathway gene cluster of the lactose operon of *Lactococcus lactis*. *J. Biol. Chem.*, **266**(11), 7176–7181.

Vanderslice, P., Copeland, W. and Robertus, J. (1986) Cloning and nucleotide sequence of wild type and a mutant histidine decarboxylase from *Lactobacillus* 30a. *J. Biol. Chem.*, **261**, 15186–15191.

Vescovo, M., Morelli, L. and Bottazzi, V. (1982) Drug resistance plasmids in *Lactobacillus acidophilus* and *Lactobacillus reuteri*. *Appl. Environ. Microbiol.*, **43**, 50–56.

Vescovo, M., Morelli, L., Bottazzi, V. and Gasson, M.J. (1983) Conjugal transfer of broad-host-range plasmid pAMβ1 into enteric species of lactic acid bacteria. *Appl. Environ. Microbiol.*, **46**, 753–755.

Vidgrén, G., Palva, I., Pakkanen, R., Lounatmaa, K. and Palva, A. (1992) S-layer of *Lactobacillus brevis*: PCR cloning and determination of the nucleotide sequence. *J. Bacteriol.*, **174**, 7419–7427.

Vogel, R.F., Gaier, W. and Hammes, W.P. (1990) Expression of the lipase gene from *Staphylococcus hyicus* in *Lactobacilus curvatus* Lc2-c *FEMS Microbiol. Lett.*, **69**, 289–292.

Vujcic, M. and Topisirovic, L. (1993) Molecular analysis of the Rolling-Circle replicating plasmid pA1 of *Lactobacillus plantarum* A112. *Appl. Environ. Microbiol.*, **59**, 274–280.

Weickert, M.J. and Chambliss, G.H. (1990) Site-directed mutagenesis of a catabolite repression operator sequence in *Bacillus subtilis*. *Proc. Natl. Acad. Sci. USA*, **87**, 6238–6242.

Weisburg, W.G., Tully, J.G., Rose, D.L., Petzel, J.P., Oyaizu, H., Yang, D., Mandelco, L., Sechrest, J., Lawrence, T.G., Van Etten, J., Maniloff, J. and Woese, C.R. (1989) A phylogenetic classification of the mycoplasmas: Basis for their classification. *J. Bacteriol.*, **171**, 6455–6467.

Index